D1391427

Enterpri

BIOMETRICS
Personal Identification in Networked Society

THE KLUWER INTERNATIONAL SERIES IN ENGINEERING AND COMPUTER SCIENCE

BIOMETRICS
Personal Identification
in Networked Society

edited by

Anil K. Jain
Michigan State University
E.Lansing, Michigan

and

Ruud Bolle and Sharath Pankanti
IBM, T.J. Watson Research Center
Yorktown Heights, New York

Kluwer Academic Publishers
Boston/Dordrecht/London

Distributors for North, Central and South America:
Kluwer Academic Publishers
101 Philip Drive
Assinippi Park
Norwell, Massachusetts 02061 USA
Tel: 781-871-6600
Fax: 781-871-6528
E-mail: kluwer@wkap.com

Distributors for all other countries:
Kluwer Academic Publishers Group
Distribution Centre
Post Office Box 322
3300 AH Dordrecht, THE NETHERLANDS
Tel: 31 78 6576 000
        Fax: 31 78 6576 254
        E-mail: orderdept@wkap.nl

Electronic Services: http://www.wkap.nl

Library of Congress Cataloging-in-Publication Data

Biometrics : personal identification in networked society / edited by
    Anil K. Jain and Ruud Bolle and Sharath Pankanti.
        p.   cm. -- (The Kluwer international series in engineering and
    computer science ; SECS 479)
    Includes bibliographical references and index.
    ISBN 0-7923-8345-1
    1. Pattern recognition systems. 2. Biometry. 3. Identification-
-Automation. 4. Computer vision. I. Jain, Anil K., 1948-
II. Bolle, Ruud. III. Pankanti, Sharath. IV. Series.
TK7882.P3B36 1998
006.4--dc21
                                                                98-31272
                                                                CIP

Contents

Foreword

The need to authenticate ourselves to machines is ever increasing in today's networked society and is necessary to close the air gap between man and machine to secure our transactions and networks. Only biometrics (automatically recognizing a person using distinguishing traits) can recognize you as you.

This first book on biometrics advances the science of biometrics by laying a foundation and theoretical framework in contributed chapters by leading experts from industry and academia. Biometric technology has advanced tremendously over the last few years and has moved from research labs and Hollywood to real-world applications. Like any technology with commercial applications, it has been difficult, until now, to assess the state of the art in biometrics in the open literature.

As Mark Twain said, "First get your facts; then you can distort them at your leisure." We are moving away from the Twain era of evaluation to the science of evaluation with a chapter on scientifically based performance evaluation. Understanding the performance of biometric systems in various real-world situations is key to their application.

Most of today's major biometric technologies are represented here by chapters on face, fingerprint, and speaker recognition, among others. In keeping with the networked society theme of this book, included are chapters on systems, networking, related technologies, and privacy issues. After the benefits of biometrics exceed their cost and their performance is understood, social, legal, and ethical issues are crucial to society's accepting the ubiquitous application of biometrics. This book encompasses all these aspects of biometrics, which are introduced by the editors.

Joseph P. Campbell, Jr.
Chair, Biometric Consortium, 1994-1998
http://www.biometric.org/

Preface

Determining the identity of a person is becoming critical in our vastly interconnected information society. As increasing number of biometrics-based identification systems are being deployed for many civilian and forensic applications, biometrics and its applications have evoked considerable interest. The current state of affairs is that the technical and technological literature about the overall state-of-the-art in biometrics is dispersed across a wide spectrum of books, journals, and conference proceedings. As biometrics emerges as a multi-billion dollar industry, there is a growing need for a comprehensive, consolidated, fair, and accessible overview of the biometrics technology and its implications to society from well-reputed information sources. This edited book is an attempt to disseminate the technological aspects and implications of biometrics. In particular, this book addresses the following needs.

- Survey the biometrics methods in commercial use and in research stage.
- Assess the capabilities and limitations of different biometrics.
- Understand the general principles of design of biometric systems and the underlying trade-offs.
- Understand the issues underlying the design of biometric systems.
- Identify issues in the realistic evaluation of biometrics-based systems.
- To recognize personal privacy and security implications of biometrics-based identification technology.
- To nurture synergies of biometric technology with the other existing and emerging technologies.

The book is organized as follows: Chapter 1 is a brief overview of the biometric technology and the research issues underlying the biometrics-based identification applications. A number of biometrics-based technologies are commercially available today and many more are being developed in the educational and commercial research laboratories world wide. Currently, there are mainly eight different biometrics including face, fingerprint, hand geometry, iris, retinal pattern, signature, voice-print, and thermograms have actually been deployed for identification. In each of the next eight chapters (Chapters 2-9), the leading experts and pioneers of biometric technology describe a particular biometric, its characteristics, the specific problems underlying the design of an identification/authentication system based on that biometric, performance evaluation of the existing systems and open issues which need to be addressed. The next five chapters (Chapters 10-14) describe biometrics which are not yet commercially available but which are under active research for on-line identification: keystroke dynamics, dait, odor, ear, and DNA.

A number of emerging civilian applications involve a very large number of identities (e.g., several million) and at the same time have demanding performance requirements (e.g., scalability, speed, accuracy). Chapter 15 addresses the research issues underlying design of a large identification and authentication system. To accomplish and engineer the design of highly reliable and accurate biometrics-based identification systems, it may often be necessary to effectively integrate discriminatory information contained in several different biometrics. These integration issues are dealt with in Chapter 16. A large cross-section of the population interested in biometrics is overwhelmed by the quickly growing pace of the technology, by the hype rampant in the media, and by the unsubstantiated claims/counter-claims heard from unreliable sources. Issues underlying performance metrics and fair evaluations are described in Chapter 17. Often, the biometric technologies need to be embedded in other technologies (e.g., smart card). The related integration issues are discussed in Chapter 18 in the context of smartcards.

More than anybody, the technical community needs to be aware of the threats posed by foolproof identification schemes to our rights of freedom and privacy. Biometrics could potentially unleash methods of covert and unwanted identification that might endanger each individual's privacy. How could we prevent the abuse of acquired biometric measurement from its unintended use? Is there a need to legislate the legitimate use of biometrics for identification applications? If so, what is the most effective method to harness the capabilities of legitimate use of biometrics without compromising the rights of individuals? Could a biometric measurement shed information about an individual which she would like to keep to herself? Implications of biometrics on lives in our society cannot be fully comprehended without learning the capabilities and limitations of each biometric. Privacy and security issues are discussed in Chapter 19.

A good starting point for additional biometric related resources for the interested readers is the biometric consortium's homepage: http://www.biometrics.org/. Its list server provides a forum for discussing the contemporary biometric related topics.

The book is primarily intended for the technical community interested in biometrics: scientists, engineers, technologists, biometrics application developers, and system integrators. However, a serious attempt has been made to maintain as much a non-technical description of each topic as is possible. Where this was not possible, a technical discussion is followed up with a non-technical summary. Additonally, readers may further refer to http://members.aol.com/afb31/af00001.htm for a glossary of all biometric terms and jargon. As a result, most of the material covered in this book should be comprehensible to anyone with a moderate scientific background.

A number of people helped make this edited book a reality. Lin Hong and Salil Prabhakar helped us through every phase of editing this book. Rick Kjeldsen's expertise in MS Word troubleshooting was invaluable. We take this opportunity to thank IBM Research management for providing the infrastructure support without which this book would not have been possible. In particular, we are grateful to Sharon Nunes for her encouragment and enthusiastic support of this project. Many thanks to Euklyn Elvy, Maurice Klapwald, Kathleen Pathe, Jim Leonard, and Jennifer Zago for their support. We are grateful to the leading biometric experts who agreed to write chapters for this book and who extended their full cooperation for expediting a timely publication of this book. We also thank Alex Greene at Kluwer Academic Publishers for his help in resolving copyright related issues.

IBM T. J. Watson. Research Center
Hawthorne
August 1998

Anil K. Jain
Ruud Bolle
Sharath Pankanti

1 INTRODUCTION TO BIOMETRICS

Anil Jain
Michigan State University
East Lansing, MI
jain@cse.msu.edu

Ruud Bolle and Sharath Pankanti
IBM T. J. Watson Research Center
Yorktown Heights, NY
{bolle,sharat}@us.ibm.com

Abstract *Biometrics deals with identification of individuals based on their biological or behavioral characteristics. Biometrics has lately been receiving attention in popular media. it is widely believed that biometrics will become a significant component of the identification technology as (i) the prices of biometrics sensors continue to fall, (ii) the underlying technology becomes more mature, and (iii) the public becomes aware of the strengths and limitations of biometrics. This chapter provides an overview of the biometrics technology and its applications and introduces the research issues underlying the biometrics.*

Keywords: *Biometrics, identification, verification, access control, authentication, security, research issues, evaluation, privacy.*

1. Introduction

Associating an identity with an individual is called personal identification. The problem of resolving the identity of a person can be categorized into two fundamentally distinct types of problems with different inherent complexities: (i) verification and (ii) recognition (more popularly known as identification[1]). Verification (authentication) refers to the problem of confirming or denying a person's

---

[1] The term identification is used in this book either to refer to the general problem of identifying individuals (identification and authentication) or to refer to the specific problem of identifying an individual from a database which involves one to many search. We rely on the context to disambiguate the reference.

claimed identity (Am I who I claim I am?). Identification (Who am I?) refers to the problem of establishing a subject's identity - either from a set of already known identities (closed identification problem) or otherwise (open identification problem). The term *positive personal identification* typically refers (in both verification as well as identification context) to identification of a person with high certainty.

Human race has come a long way since its inception in small tribal primitive societies where every person in the community knew every other person. In today's complex, geographically mobile, increasingly electronically inter-connected information society, accurate identification is becoming very important and the problem of identifying a person is becoming ever increasingly difficult. A number of situations require an identification of a person in our society: have I seen this applicant before? Is this person an employee of this company? Is this individual a citizen of this country? Many situations will even warrant identification of a person at the far end of a communication channel.

2. Opportunities

Accurate identification of a person could deter crime and fraud, streamline business processes, and save critical resources. Here are a few mind boggling numbers: about $1 billion dollars in welfare benefits in the United States are annually claimed by "double dipping" welfare recipients with fraudulent multiple identities [10]. MasterCard estimates the credit card fraud at $450 million per annum which includes charges made on lost and stolen credit cards: unobtrusive positive personal identification of the legitimate ownership of a credit card at the point of sale would greatly reduce the credit card fraud; about 1 billion dollars worth of cellular telephone calls are made by the cellular bandwidth thieves - many of which are made from stolen pins and/or cellular telephones. Again, an identification of the legitimate ownership of the cellular telephones would prevent cellular telephone thieves from stealing the bandwidth. A reliable method of authenticating legitimate owner of an ATM card would greatly reduce ATM related fraud worth approximately $3 billion annually [11]. A positive method of identifying the rightful check payee would also reduce billions of dollars misappropriated through fraudulent encashment of checks each year. A method of positive authentication of each system login would eliminate illegal break-ins into traditionally secure (even federal government) computers. The United States Immigration and Naturalization service stipulates that it could each day detect/deter about 3,000 illegal immigrants crossing the Mexican border without delaying the legitimate people entering the United States if it had a quick way of establishing positive personal identification.

3. Identification Methods

The problem of authentication and identification is very challenging. In a broad sense, establishing an identity (either in a verification context or an identification context) is a very difficult problem; Gertrude Stein's [12] quote *"rose is a rose is a rose is a rose"* summarizes the essence of the difficulty of a positive identification problem: an identity of a person is so much woven into the fabric of everything that a person represents and believes that the answers to the identity of a person transcend the scope of an engineering system and the solutions could (perhaps) only be sought in a philosophical realm. For example, can a brain-dead person be identified as her fully sane counterpart for authenticating an electronic fund transfer? Engineering approach to the (abstract) problem of authentication of a person's identity is to reduce it to the problem of authentication of a concrete entity related to the person (Figure 1.1). Typically, these entities include (i) a person's possession (*"something that you possess"*), e.g., permit physical access to a building to all persons whose identity could be authenticated by possession of a key; (ii) person's knowledge of a piece of information (*"something that you know"*), e.g., permit login access to a system to a person who knows the user-id and a password associated with it. Some systems, e.g., ATMs, use a combination of "something that you have" (ATM card) and "something that you know" (PIN) to establish an identity. The problem with the traditional approaches of identification using possession as a means of identity is that the possessions could be lost, stolen, forgotten, or misplaced. Further, once in control of the identifying possession, by definition, any other "unauthorized" person could abuse the privileges of the authorized user. The problem with using knowledge as an identity authentication mechanism is that it is difficult to remember the passwords/PINs; easily recallable passwords/PINs (e.g., pet's name, spouse's birthday)

Figure 1.1 Prevalent methods of identification based on possession and knowledge: Keys, employee badge, driver license, ATM card, and credit card.

could be easily guessed by the adversaries. It has been estimated that about 25% of the people using ATM cards write their ATM PINs on the ATM card [13], thereby defeating possession/knowledge combination as a means of identification. As a result, these techniques cannot distinguish between an authorized person and an impostor who acquires the knowledge/possession, enabling the access privileges of the authorized person.

Yet another approach to positive identification has been to reduce the problem of identification to the problem of identifying physical characteristics of the person. The characteristics could be either a person's physiological traits, e.g., fingerprints, hand geometry, etc. or her behavioral characteristics, e.g., voice and signature. This method of identification of a person based on his/her physiological/behavioral characteristics is called biometrics[2]. The primary advantage of such an identification method over the methods of identification utilizing "something that you possess" or "something that you know" approach is that a biometrics cannot be misplaced or forgotten; it represents a tangible component of "something that you are". While biometric techniques are not an identification panacea, they, especially, when combined with the other methods of identification, are beginning to provide very powerful tools for problems requiring positive identification.

4. Biometrics

What biological measurements qualify to be a biometric? Any human physiological or behavioral characteristic could be a biometrics provided it has the following desirable properties [15]: (*i*) *universality*, which means that every person should have the characteristic, (*ii*) *uniqueness*, which indicates that no two persons should be the same in terms of the characteristic, (*iii*) *permanence*, which means that the characteristic should be invariant with time, and (*iv*) *collectability*, which indicates that the characteristic can be measured quantitatively. In practice, there are some other important requirements [15,16]: (*i*) *performance*, which refers to the achievable identification accuracy, the resource requirements to achieve an acceptable identification accuracy, and the working or environmental factors that affect the identification accuracy, (*ii*) *acceptability*, which indicates to what extent people are willing to accept the biometric system, and (*iii*) *circumvention*, which refers to how easy it is to fool the system by fraudulent techniques.

5. Biometrics Technology: Overview

No single biometrics is expected to effectively satisfy the needs of all identification (authentication) applications. A number of biometrics have been proposed, researched, and evaluated for identification (authentication) applications. Each biometrics has its strengths and limitations; and accordingly, each biometric appeals to a particular identification (authentication) application. A summary of the existing and burgeoning biometric technologies is described in this section.

- Voice
  Voice is a characteristic of an individual [17]. However, it is not expected to be sufficiently unique to permit identification of an individual from a large database of identities (Figure 1.2). Moreover, a voice signal available for authentication is

---

[2] Note the distinction between the terms biometrics and biometry: biometry encompasses a much broader field involving application of statistics to biology and medicine [14].

typically degraded in quality by the microphone, communication channel, and digitizer characteristics. Before extracting features, the amplitude of the input signal may be normalized and decomposed into several band-pass frequency channels. The features extracted from each band may be either time-domain or frequency domain features. One of the most commonly used features is cepstral feature - which is a logarithm of the Fourier Transform of the voice signal in each band. The matching strategy may typically employ approaches based on hidden Markov model, vector quantization, or dynamic time warping [17]. Text-dependent speaker verification authenticates the identity of a subject based on a fixed predetermined phrase. Text-independent speaker verification is more difficult and verifies a speaker identity independent of the phrase. Language-independent speaker verification verifies a speaker identity irrespective of the language of the uttered phrase and is even more challenging.

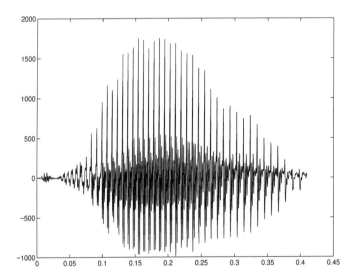

Figure 1.2 Voice signal representing an utterance of the word "seven". X and Y axes represent time and signal amplitude, respectively.

Voice capture is unobtrusive and voice print is an acceptable biometric in almost all societies. Some applications entail authentication of identity over telephone. In such situations, voice may be the only feasible biometric. Voice is a behavioral biometrics and is affected by a person's health (e.g., cold), stress, emotions, etc. To extract features which remain invariant in such cases is very difficult. Besides, some people seem to be extraordinarily skilled in mimicking others. A reproduction of an earlier recorded voice can be used to circumvent a voice authentication system in the remote unattended applications. One of the methods of combating this problem is to prompt the subject (whose identity is to be authenticated) to utter a different phrase each time.

- Infrared Facial and Hand Vein Thermograms

Figure 1.3 Identification based on facial thermograms [1]. The image is obtained by sensing the infrared radiations from the face of a person. The graylevel at each pixel is characteristic of the magnitude of the radiation.

Human body radiates heat and the pattern of heat radiation is a characteristic of each individual body [18]. An infrared sensor could acquire an image indicating the heat emanating from different parts of the body (Figure 1.3). These images are called thermograms. The method of acquisition of the thermal image unobtrusively is akin to the capture of a regular (visible spectrum) photograph of the person. Any part of the body could be used for identification. The absolute values of the heat radiation are dependent upon many extraneous factors and are not completely invariant to the identity of an individual; the raw measurements of heat radiation need to be normalized, e.g., with respect to heat radiating from a landmark feature of the body. The technology could be used for covert identification solutions and could distinguish between identical twins. It is also claimed to provide enabling technology for identifying people under the influence of drugs: the radiation patterns contain signature of each narcotic drug [19]. A thermogram-based system may have to address sensing challenges in uncontrolled environments, where heat emanating surfaces in the vicinity of the body, e.g., room heaters and vehicle exhaust pipes, may drastically affect the image acquisition phase. Infrared facial thermograms seem to be acceptable since their acquisition is a non-contact and non-invasive sensing technique.

Identification systems using facial thermograms are commercially available [1]. A related technology using near infrared imaging [2] is used to scan the back of a clenched fist to determine hand vein structure (Figure 1.4). Infrared sensors are prohibitively expensive which is a factor inhibiting wide spread use of thermograms.

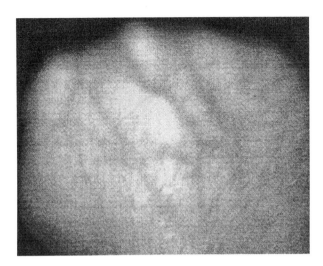

Figure 1.4 Identification based on hand veins [2]. An infrared image of the back of a clenched human fist. The structure of the vasculature could be used for identification.

- Fingerprints

   Fingerprints are graphical flow-like ridges present on human fingers. Their formations depend on the initial conditions of the embryonic development and they are believed to be unique to each person (and each finger). Fingerprints are one of the most mature biometric technologies used in forensic divisions worldwide for criminal investigations and therefore, have a stigma of criminality associated with them. Typically, a fingerprint image is captured in one of two ways: (i) scanning an inked impression of a finger or (ii) using a live-scan fingerprint scanner (Figure 1.5).

   Major representations of the finger are based on the entire image, finger ridges, or salient features derived from the ridges (minutiae). Four basic approaches to identification based on fingerprint are prevalent: (i) the invariant properties of the gray scale profiles of the fingerprint image or a part thereof; (ii) global ridge patterns, also known as fingerprint classes; (iii) the ridge patterns of the fingerprints; (iv) fingerprint minutiae – the features resulting mainly from ridge endings and bifurcations.

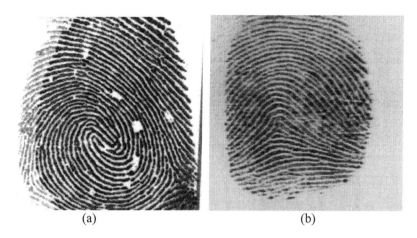

(a)   (b)

Figure 1.5 A fingerprint image could be captured from the inked impression of a finger or directly imaging a finger using frustrated total internal reflection technology. The former is called an inked fingerprint (a) and the latter is called a live-scan fingerprint (b).

- Face

  Face is one of the most acceptable biometrics because it is one of the most common method of identification which humans use in their visual interactions (Figure 1.6). In addition, the method of acquiring face images is non-intrusive. Two primary approaches to the identification based on face recognition are the following: (i) Transform approach [20, 21]: the universe of face image domain is represented using a set of orthonormal basis vectors. Currently, the most popular basis vectors are eigenfaces: each eigenface is derived from the covariance analysis of the face image population; two faces are considered to be identical if they are sufficiently "close" in the eigenface feature space. A number of variants of such an approach exist. (ii) Attribute-based approach [22]: facial attributes like nose, eyes, etc. are extracted from the face image and the invariance of geometric properties among the face landmark features is used for recognizing features.

  Facial disguise is of concern in unattended authentication applications. It is very challenging to develop face recognition techniques which can tolerate the effects of aging, facial expressions, slight variations in the imaging environment and variations in the pose of face with respect to camera (2D and 3D rotations) [23].

- Iris

  Visual texture of the human iris is determined by the chaotic morphogenetic processes during embryonic development and is posited to be unique for each person and each eye [24]. An iris image is typically captured using a non-contact imaging process (Figure 1.7). The image is obtained using an ordinary CCD camera with a resolution of 512 dpi. Capturing an iris image involves cooperation from the user, both to register the image of iris in the central imaging area and to ensure that the iris is at a predetermined distance from the focal plane of the camera. A position-invariant constant length byte vector feature is derived from an annular part of the iris image based on its texture. The identification error rate using iris technology is believed to be extremely small and the constant length position invariant code permits an extremely fast method of iris recognition.

Figure 1.6 Identification based on face is one of the most acceptable methods of biometric-based identification.

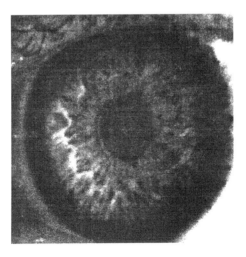

Figure 1.7 Identification Based on Iris. The visual texture of iris could be used for positive person identification.

- Ear

  It is known that the shape of the ear and the structure of the cartilegenous tissue of the pinna are distinctive[3]. The features of an ear are not expected to be unique to each individual. The ear recognition approaches are based on matching vectors of distances of salient points on the pinna from a landmark location (Figure 1.8) on the ear [3]. No commercial systems are available yet and authentication of individual identity based on ear recognition is still a research topic.

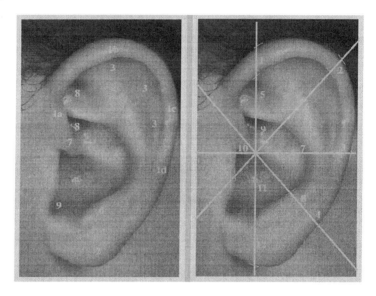

Figure 1.8 An image of an ear and the features used for ear-based identification [3]. Feature vector consists of the distances of various salient locations on the pinna from a landmark location.

- Gait

  Gait is the peculiar way one walks and is a complex spatio-temporal behavioral biometrics. Gait is not supposed to be unique to each individual, but is sufficiently characteristic to allow identity authentication. Gait is a behavioral biometric and may not stay invariant especially over a large period of time, due to large fluctuations of body weight, major shift in the body weight (e.g., waddling gait during pregnancy [25], major injuries involving joints or brain (e.g., cerebellar lesions in Parkinson disease [25]), or due to inebriety (e.g., drunken gait [25]).

  Humans are quite adept at recognizing a person at a distance from his gait. Although, the characteristic gait of a human walk has been well researched in biomechanics community to detect abnormalities in lower extremity joints, the

---

[3] Department of Immigration and Naturalization in the United States specifically requests photographs of individuals with clearly visible right ear.

use of gait for identification purposes is very recent. Typically, gait features are derived from an analysis of a video-sequence footage (Figure 1.9) of a walking person [26] and consist of characterization of several different movements of each articulate joint. Currently, there do not exist any commercial systems for performing gait-based authentication. The method of input acquisition for gait is not different from that of acquiring facial pictures, and hence gait may be an acceptable biometric. Since gait determination involves processing of video, it is compute and input intensive.

Figure 1.9 Authentication based on gait typically uses a sequence of images of a walking person. One of the frames in the image sequence is illustrated here.

- Keystroke Dynamics
  It is hypothesized that each person types on a keyboard in a characteristic way. This behavioral biometrics is not expected to be unique to each individual but it offers sufficient discriminatory information to permit identity authentication [27]. Keystroke dynamics is a behavioral biometric; for some individuals, one may expect to observe a large variations from typical typing patterns. The keystrokes of a person using a system could be monitored unobtrusively as that person is keying in other information. Keystroke dynamic features are based on time durations between the keystrokes. Some variants of identity authentication use features based on inter-key delays as well as dwell times - how long a person holds down a key. Typical matching approaches use a neural network architecture to associate identity with the keystroke dynamics features. Some commercial systems are already appearing in the market.
- DNA
  DNA (DeoxyriboNucleic Acid) is the one-dimensional ultimate unique code for one's individuality - except for the fact that identical twins have the identical DNA pattern. It is, however, currently used mostly in the context of forensic applications for identification [4]. Three issues limit the utility of this biometrics for other applications: (i) contamination and sensitivity: it is easy to steal a piece

of DNA from an unsuspecting subject to be subsequently abused for an ulterior purpose; (ii) automatic real-time identification issues: the present technology for genetic matching is not geared for online unobtrusive identifications. Most of the human DNA is identical for the entire human species and only some relatively small number of specific locations (polymorphic loci) on DNA exhibit individual variation. These variations are manifested either in the number of repetitions of a block of base sequence (length polymorphism) or in the minor non-functional perturbations of the base sequence (sequence polymorphism) [70]. The processes involved in DNA based personal identification determine whether two DNA samples originate from the same/different individual(s) based on the distinctive signature at one or more polymorphic loci. A major component of these processes now exist in the form of cumbersome chemical methods (wet processes) requiring an expert's skills. There does not seem to be any effort directed at a complete automation of all the processes.(iii) privacy issues: information about susceptibilities of a person to certain diseases could be gained from the DNA pattern and there is a concern that the unintended abuse of genetic code information may result in discrimination in e.g., hiring practices.

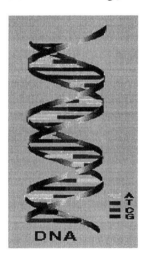

Figure 1.10 DNA is double helix structure made of four bases: Adenine (A), Thymine (T), Cytosine (C), and Guanine (G) [4]. The sequence of bases is unique to each individual (with the exception of identical twins) and could be used for positive person identification.

• Signature and Acoustic Emissions

The way a person signs her name is known to be a characteristic of that individual (Figure 1.11). Although signatures require contact and effort with the writing instrument, they seem to be acceptable in many government, legal, and commercial transactions[4] as a method of personal authentication. Signatures are a behavioral biometric, evolve over a period of time and are influenced by physical and emotional conditions of the signatories. Signatures of some people vary a lot: even the successive impressions of their signature are significantly different. Further, the professional forgers can reproduce signatures to fool the unskilled eye. Although, the human experts can discriminate genuine signatures from the forged ones, modeling the invariance in the signatures and automating signature recognition process are challenging. There are two approaches to signature verification: static and dynamic. In static signature verification, only geometric (shape) features of the signature are used for authenticating an identity [28]. Typically, the signature impressions are normalized to a known size and decomposed into simple components (strokes). The shapes and relationships of strokes are used as features. In dynamic signature verification, not only the shape features are used for authenticating the signature but the dynamic features like acceleration, velocity, and trajectory profiles of the signature are also employed. The signature impressions are processed as in a static signature verification system. Invariants of the dynamic features augment the static features, making forgery difficult since the forger has to not only know the impression of the signature but also the way the impression was made.

A related technology is authentication of an identity based on the characteristics of the acoustic emissions emitted during a signature scribble. These acoustic emissions are claimed to be a characteristic of each individual [29].

Figure 1.11 Identification based on signature. Signatures have long been accepted as a legitimate means of identification.

- Odor
  It is known that each object exudes an odor that is characteristic of its chemical composition and could be used for distinguishing various objects. Among other things, the automatic odor detection technology [30] is presently being investigated for detecting land mines [31]. A whiff of air surrounding an object

---

[4] In some developing countries with low literacy rates, "thumbprint" is accepted as a legal signature.

is blown over an array of chemical sensors, each sensitive to a certain group of (aromatic) compounds. The feature vector consists of the signature comprising of the normalized measurements from each sensor. After each act of sensing, the sensors need to be initialized by a flux of clean air.

Body odor serves several functions including communication, attracting mates, assertion of territorial rights, and protection from a predator. A component of the odor emitted by a human (or any animal) body is distinctive to a particular individual. It is not clear if the invariance in a body odor could be detected despite deodorant smells, and varying chemical composition of the surrounding environment. Currently, no commercial odor-based identity authentication systems exist.

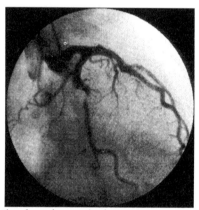

Figure 1.12 Identification based on retinal scan is perceived to be the most secure method of authenticating an identity.

- Retinal Scan
  The retinal vasculature is rich in structure and is supposed to be a characteristic of each individual and each eye (Figure 1.12). It is claimed to be the most secure biometrics since it is not easy to change or replicate the retinal vasculature. Retinal scans, glamorized in movies and military installations, are mostly responsible for the "high-tech-expensive" impression of the biometric technology[5]. The image capture requires a person to peep into an eye-piece and focus on a specific spot in the visual field so that a predetermined part of the retinal vasculature could be imaged. The image acquisition involves cooperation of the subject, entails contact with the eyepiece, and requires a conscious effort on the part of the user. All these factors adversely affect the public acceptability of retinal biometric. A number of retinal scan-based identity authentication installations are in operation which boast *zero* false positives in all the installations to-date[6]. Retinal vasculature can reveal some medical conditions, e.g., hypertension, which is another factor standing in the way of public acceptance of retinal scan based-biometrics.

[5] Although, iris scanning appears to be more expensive than retinal scanning.
[6] These systems were operating at an unknown high false negative rates [32].

Figure 1.13 Authentication based on hand geometry. Although two-dimensional profile of a hand is illustrated here, in commercial hand geometry-based authentication systems, three-dimensional profile of the hand is sensed.

- Hand and Finger Geometry
  In recent years, hand geometry (Figure 1.13) has become a very popular access control biometrics which has captured almost half of the physical access control market [33]. Some features related to a human hand, e.g., length of fingers, are relatively invariant and peculiar (although, not unique) to each individual. The image acquisition system requires cooperation of the subject and captures frontal and side view images of the palm flatly placed on a panel with outstretched fingers. The registration of the palm is accomplished by requiring the subject's fingers to be aligned with a system of pegs on the panel which is not convenient for subjects with limited flexibility of palm, e.g., those suffering from arthritis. The representational requirements of the hand are very small (9 bytes) which is an attractive feature for bandwidth and memory limited systems. The hand geometry is not unique and cannot be scaled up for systems requiring identification of an individual from a large population of identities. In spite of this, hand geometry has gained acceptability in a number of the installations in last few years for identity authentication applications.

  Finger geometry [34] is a variant of hand geometry and is a relatively new technology which relies only on geometrical invariants of fingers (index and middle). A finger geometry acquisition device closely resembles that for hand geometry but is more compact. It is claimed to be more accurate than hand geometry. However, the technology for finger geometry based authentication is not as mature as that for hand geometry.

Biometrics	Universality	Uniqueness	Permanence	Collectability	Performance	Acceptability	Circumvention
Face	High	Low	Medium	High	Low	High	Low
Fingerprint	Medium	High	High	Medium	High	Medium	High
Hand Geometry	Medium	Medium	Medium	High	Medium	Medium	Medium
Keystrokes	Low	Low	Low	Medium	Low	Medium	Medium
Hand Vein	Medium	Medium	Medium	Medium	Medium	Medium	High
Iris	High	High	High	Medium	High	Low	High
Retinal Scan	High	High	Medium	Low	High	Low	High
Signature	Low	Low	Low	High	Low	High	Low
Voice Print	Medium	Low	Low	Medium	Low	High	Low
F. Thermograms	High	High	Low	High	Medium	high	High
Odor	High	High	High	Low	Low	Medium	Low
DNA	High	High	High	Low	High	Low	Low
Gait	Medium	Low	Low	High	Low	High	Medium
Ear	Medium	medium	High	medium	Medium	High	Medium

Table 1.1 Comparison of biometrics technologies. The data are based on perception of three biometrics experts.

6. Biometrics Technologies: A Comparison

Each biometric technology has its strengths and limitations. No single biometrics is expected to effectively meet the needs of all the applications. A brief comparison of 14 different biometric techniques that are either widely used or under investigation, including face, fingerprint, hand geometry, keystroke dynamics, hand vein, iris, retinal pattern, signature, voice-print, facial thermograms, odor, DNA, gait, and ear [15, 35, 24, 16, 36, 1, 2, 27, 31, 4, 3] is provided in Table 1.1. Although each of these biometric techniques, to a certain extent, possesses the above mentioned desirable properties and has been used in practical systems [15, 35, 24, 16] or has the potential to become a valid biometric technique [16], not many of them are acceptable (in court of law) as indisputable evidence of identity.

Which biometrics should be used for a given application? The match between a biometrics and an application is determined depending upon the requirements of the given application, the characteristics of the application, and properties of the biometrics. In the context of biometrics-based identification (authentication) systems, an application is characterized by the following properties: (i) does the application need identification or authentication? The applications requiring an identification of a subject from a large database of identities need scalable and relatively more unique biometrics. (ii) Is it attended (semi-automatic) or unattended (completely automatic)? An application may or may not afford a human operator at or near the biometric acquisition stage. In the applications deployed at remote locations with unfriendly or unsafe climate, for instance, the use of biometrics requiring an operator assistance for the capture of physiological or behavioral measurement may not be feasible. (iii) Are the users habituated (or willing to be habituated) to the given biometrics? Performance of a biometrics-based system improves steadily as the subjects instinctively learn to give "good" biometric measurements. This is more true for some biometrics than others; e.g., it is more difficult to give bad retinal image than a fingerprint image. Some applications may tolerate the less effective learning phase of the application deployment for a longer time than others. (iv) Is the application covert or overt? Not all biometrics can be captured without the knowledge of the subject to be identified. Even the biometrics which could be captured without the knowledge of a subject may not be used in some countries due to privacy legislations. (v) Are the subjects cooperative or non-cooperative? Typically, applications involving non-cooperative subjects warrant the use of physiological biometrics which cannot be easily changed. For instance, it is easy to change one's voice compared to changing one's retinal vasculature. (vi) What are the storage requirement constraints? Different applications impose varying limits on the size of the internal representation for the chosen biometrics. (vii) How stringent are the performance requirement constraints? For example, applications demanding higher accuracies need more unique biometrics. (viii) What types of biometrics are acceptable to the users? Different biometrics are acceptable in applications deployed in different demographics depending on the cultural, ethical, social, religious, and hygienic standards of that society. The acceptability of a biometrics in an application is often a compromise between the sensitivity of a community to various perceptions/taboos and the value/convenience offered by a biometrics-based identification.

7. Automatic Identification

The concept of individuality of personal traits has a long history, and identification of a person based on his physical characteristics is not new. Humans (and other animals) recognize each other based on their physical characteristics. As we pick up the telephone, we expect our friends to recognize us based on our voice. We know from a number of archeological artifacts (Figure 1.14) that our ancestors recognized the individuality of fingerprint impressions [5] on their picture drawings. Prehistoric Chinese have been known to recognize that fingerprints can help establish the identity of an individual uniquely [5]. In 1882 Alphonse Bertillon, chief of criminal identification of Paris police department developed a very detailed method of identification based on a number of bodily measurements, physical description, and photographs [37]. The Bertillon System of Anthropometric Identification system gained wide acceptance before getting superseded by fingerprint based identification systems. Some of the physical characteristics, e.g., DNA, fingerprints, and signatures, have gained a legal status and these characteristics could be used as evidence in the court of law to establish a proof of identity. Having gained the legitimacy, elaborate systems of rules have been developed for (i) matching these biometrics to decide whether a pair of biometric measurements, e.g., two fingerprints, belong to the same person or not; (ii) searching a given biometric measurement in a database consisting of a number of other measurements of the same biometrics. These rules are derived from manual systems of matching and indexing because of historical reasons, and require trained experts for operation of manual/semi-automatic identification systems. For example, the traditional fingerprint identification systems used in the forensic applications require well-trained experts in acquisition of fingerprints, classifying/indexing the fingerprint, and fingerprint matching.

On the other hand, use of biometrics in *fully* automated applications is a relatively new and emerging phenomenon. There is a growing interest in biometrics from a wide cross-section of society: engineers, technologists, scientists, and, government and corporate executives. The excitement of the emergence of biometrics-based technology is evident by publication of dozens of biometrics articles in the popular press [38], organization of exclusive technical conferences devoted to biometrics [39, 40, 41, 42, 43], organization of biometrics related workshops [44, 45], increasing focus on biometrics in security and financial trade-shows [46, 47], institution of biometric consortia [48, 49], special issues of reputed technical journals [50], publication of a few periodicals devoted to biometrics [51, 52], and even establishment of an exclusive biometric shop [38] in the last couple of years.

The perception that biometric technologies are hi-tech, high-cost systems and can only be afforded in forensics and high-security military installations is rapidly changing. Spiraling increase in the availability of inexpensive computing resources, advances in image understanding, better matching strategies provided by progress in pattern recognition and computer vision field, cheaper sensing technologies, and increasing demand for the identification needs, are forcing biometric technology into new applications/markets requiring positive personal identification.

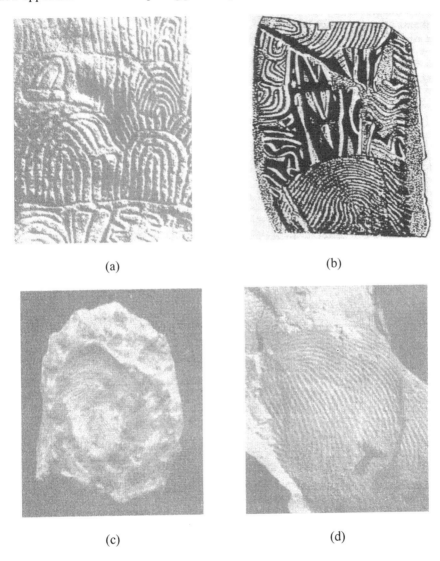

(a)

(b)

(c)

(d)

Figure 1.14 Archeological artifacts depicting fingerprint impressions: (a) Neolithic Carvings at Gavrinis Island [5]; (b) Standing Stone at Goat Island (2,000 B.C.) [72]; (c) A Chinese clay seal (300 B.C.) [71]; and (d) An impression on a Palestinian lamp (400 A. D.) [5]. The Chinese clay seal and Palestinian lamp impressions indicate the identity of their respective owners. Reprinted with permissions of A. Moenssens and J.Berry.

The biometrics-based identification (authentication) technology is being either adopted or contemplated in a very broad range of civilian applications: (*i*) *banking security* such as electronic fund transfers, ATM security, internet commerce, check cashing, and credit card transactions, (*ii*) *physical access control* such as airport access control, (*iii*) *information system security* like access to databases via login privileges, (*iv*) *government benefits distribution* such as welfare disbursement programs [53], (*v*) *customs and immigration* such as *INS* Passenger Accelerated Service System (INSPASS) which permits faster immigration procedures based on hand geometry [54], (*vi*) *national ID systems* which provide a unique ID to the citizens and integrate different government services [55], (*vii*) *voter and driver registration* providing registration facilities for voters and drivers. (*viii*) *customer loyalty/preference* schemes providing incentives to repeat/preferred customers of a business establishment, and (*ix*) *Telecommunications* such as cellular bandwidth access control.

8. Research Issues

The general problem of personal identification raises a number of important research issues: what identification technologies are the most effective to achieve accurate and reliable identification of individuals? In this section, we summarize the challenges in biometrics research [56]. Some of these problems are well-known open problems in the allied areas (e.g., pattern recognition and computer vision), while the others need a systematic cross-disciplinary effort.

We believe that biometrics technology alone may not be sufficient to resolve these issues effectively; the solutions to the outstanding open problems may lie in innovative engineering designs exploiting constraints otherwise unavailable to the applications and in harnessing the biometric technology in combination with other allied technologies.

Design

It is not clear whether the use of the features and philosophies underlying the identification systems heavily tuned for human use (e.g., faces and fingerprints) is as effective for fully automatic processes (Figure 1.15). Nor do we know whether identification technologies inspired and used by humans are indeed as amenable and effective for completely automatic identification systems. In fact, it is not even clear if the solutions solely relying on biometrics-based identifications are the most desirable engineering solutions in many real-world applications. Both, a different set of functional requirements demanded by the emerging market applications and the retrospective wisdom of futility of myopic dependence on human intuition for engineering designs suggest that full automation of the biometrics-based identification systems warrant a careful examination of all the underlying components of the positive identifications of the emerging applications.

A biometric-based identification (authentication) system operates in two distinct modes (Figure 1.16): enrollment and identification (authentication). During enrollment, biometric measurements are captured from a given subject, relevant information from the raw measurement is gleaned by the feature extractor, and (feature, person) information is stored in a database. Additionally, some form of ID for the subject may be generated for the subject (along with the visual/machine representation of the biometrics). In identification mode, the system senses the biometric measurements from the subject, extracts features from the raw measurements, and searches the database using the features thus extracted. The system may either be able to determine the identity of the subject or decide the person is not represented in the database. In authentication mode of operation, the subject presents his system assigned ID and the biometric measurements, the system extracts (input) features from the measurements, and attempts to match the input features to the (template) features corresponding to subject's ID in the system database. The system may, then, either determine that the subject is who he claims to be or may reject the claim. In some situations, a single system operates as both an identification and an authentication system with a common database of (identity, feature) associations.

Figure 1.15 Human vision is fooled by many subtle perceptual tricks and it is hoped that machine vision may be better equipped in correctly recognizing the deceit in such situations. Although, a typical human subject may wrongly believe the faces shown in this picture to belong to Al Gore and President Bill Clinton, on closer inspection, one could recognize that both the faces in the picture identically show Bill Clinton's facial features and the crown of hair on one of the faces has been digitally manipulated to appear similar to that of Al Gore [6].

Design of a biometrics-based identification system could essentially be reduced to the design of a pattern recognition system. The conventional pattern recognition system designers have adopted a sequential phase-by-phase modular architecture (Figure 1.17). Although, it is generally known in the research community that more integrated, parallel, active system architectures involving feedback/feed-forward control have a number of advantages, these concepts have not yet been fully exploited in commercial biometrics-based systems.

Given the speed, accuracy, and cost performance specifications of an end-to-end identification system, the following design issues need to addressed: (i) how to acquire the input data/measurements (biometrics)? (ii) what internal representation (features) of the input data is invariant and amenable for an automatic feature

extraction process?   (iii) given the input data, how to extract the internal
representation from it?   (iv) given two input samples in the selected internal
representation, how to define a matching metric that translates the intuition of
``similarity" among the patterns?   (v) how to implement the matching metric?
Additionally, for reasons of efficiency, the designer may also need to address the
issues involving (vi) organization of a number of (representations) input samples into
a database and (vii) effective methods of searching a given input sample
representation in the database.

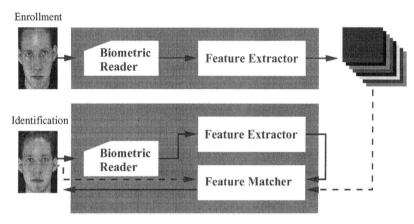

Figure 1.16 Architecture of a typical biometric system.

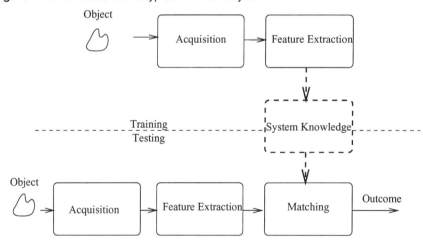

Figure 1.17 Architecture of a typical pattern recognition system.

Researchers in pattern recognition have realized that effectively resolving these
issues is very difficult and there is a need to constrain the environment to engineer
feasible solutions. We will describe some of the research problems in the design of
biometrics-based identification systems.

1. **Acquisition.** Acquiring relevant data for the biometrics is one of the critical processes which has not received adequate attention. The amount of care taken in acquiring the data (often) determines the performance of the entire system. Two of the associated tasks are: (a) quality assessment; automatically assessing the suitability of the input data for automatic processing and (b) segmentation; separation of the input data into foreground (object of interest) and background (irrelevant information).

    A number of opportunities exist for incorporating (i) the context of the data capture which may further help improve the performance of the system and (ii) avoiding undesirable measurements (and subsequent recapture of desirable measurements). With inexpensive desktop computing and large input bandwidth, typically the context of the data capture could be made richer to improve the performance. For instance, a fingerprint is traditionally captured from its 2D projection on a flat surface. Why not capture a 3D image? Why not take a color image? Why not use active sensing? Such enhancements may often improve the performance of the biometric systems.

    Although a number of existing identification systems routinely assign a quality index to the input measurement indicating its desirability for matching (Figure 1.18), the approach to such a quality assessment metric is subjective, debatable, and typically inconsistent. A lot of research effort needs to be focussed in this area to systematize both (i) the rigorous and realistic models of the input measurements and (ii) metrics for assessment of quality of a measurement. When the choice of rejecting a poor quality input measurement is not available (e.g., in legacy databases), the system may optionally attempt at gleaning useful signal from the noisy input measurements. Such operation is referred to as signal/image enhancement (Figure 1.19) and is computationally intensive. How to enhance the input measurements without introducing any artifacts is an active research topic.

    Similarly, the conventional foreground/background separation (Figure 1.20) typically relies on an *ad hoc* processing of input measurements and enhancing the information bandwidth of input channel (e.g., using more sensory channels) often provides very effective avenues for segmentation. Further, robust and realistic models of the object of interest often facilitate cleaner and better design of segmentation algorithms.

2. **Representation**

Which machine-readable representations completely capture the invariant and discriminatory information in the input measurements? This representation issue constitutes the essence of system design and has far reaching implications on the design of the rest of the system. The unprocessed measurement values are typically not invariant over the time of capture and there is a need to determine salient features of the input measurement which both discriminate between the identities as well as remain invariant for a given individual. Thus, the problem of representation is to determine a measurement (feature) space which is invariant (less variant) for the input signals belonging to the same identity and which differ maximally for those belonging to different identities (high *interclass* variation and low *intraclass* variation). To systematically determine the discriminatory power of an information source and arrive at an effective feature space is a challenging problem.

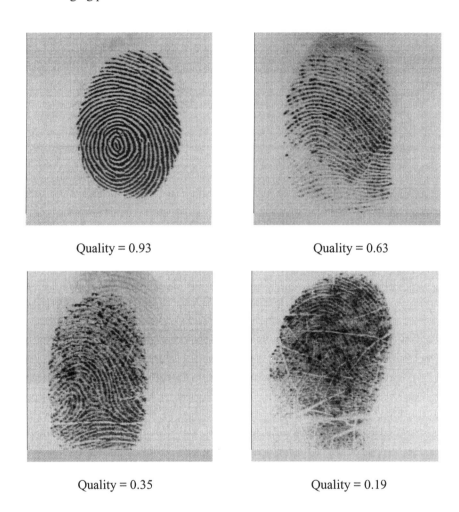

Quality = 0.93   Quality = 0.63

Quality = 0.35   Quality = 0.19

Figure 1.18 Fingerprint Quality: Automatically and consistently determining suitability of a given input measurement for automatic identification is a challenging problem. A fingerprint quality assessment algorithm quantifies suitability of fingerprint images for automatic fingerprint identification system by assigning a quality index in the range of [0,1]; numbers closer to zero indicate poor quality images.

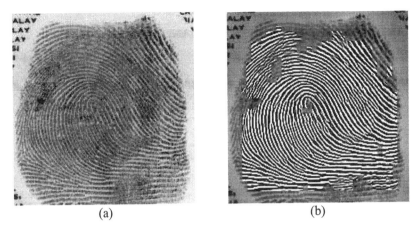

(a)   (b)

Figure 1.19 Enhancement: Automatically enhancing fingerprint images without introducing artifacts is a challenging problem: (a) a poor quality fingerprint image; (b) result of image enhancement of fingerprint image shown in (a) [8].

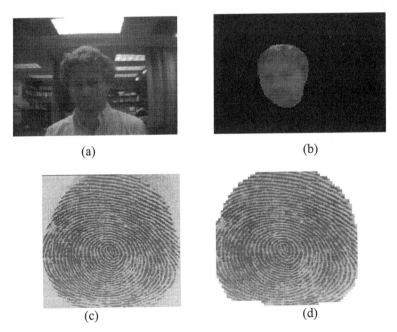

Figure 1.20 Segmentation: Determining the region containing the object of interest from a given image is a challenging problem. (a) image showing a face; (b) a face detection algorithm identifying the region of interest in (a) for a face recognition system; (c) a fingerprint image; and (d) foreground/background separation of the fingerprint image shown in (c) [7].

A related issue about representation is the *saliency* of a measurement signal and its representation. More distinctive biometric signals offer more reliable identity authentication. Less complex measurement signals inherently offer a less reliable identification. This phenomenon has a direct impact in many biometrics-based identification, e.g., signature, where less distinctive signatures could be easily forged. A systematic method of quantifying distinctiveness of a specific signal associated with an identity and its representation is needed for effective identification systems.

Additionally, in some applications, storage space is at a premium, e.g., in a smart card application, typically, about 2K bytes of storage is available. In such situations, the representation also needs to be parsimonious. The issues of most salient features of an information source also need to be investigated.

Representation issues cannot be completely resolved independent of a specific biometric domain and involve complex trade-offs. Take, for instance, the fingerprint domain. Representations based on the entire gray scale profile of a fingerprint image are prevalent among the verification systems using optical matching [57, 58]. However, the utility of the systems using such representation schemes may be limited due to factors like brightness variations, image quality variations, scars, and large global distortions present in the fingerprint image because these systems are essentially resorting to template matching strategies for verification. Further, in many verification applications terser representations are desirable which preclude representations that involve the entire gray scale profile of fingerprint images. Some system designers attempt to circumvent this problem by restricting that the representation be derived from a *small* (but consistent) part of the finger [57]. However, if this same representation is also being used for identification applications, then the resulting systems might stand at a risk of restricting the number of unique identities that could be handled, simply because of the fact that the number of distinguishable templates is limited. On the other hand, an image-based representation makes fewer assumptions about the application domain (fingerprints) and, therefore, has the potential to be robust to wider varieties of fingerprint images. For instance, it is extremely difficult to extract a landmark-based representation from a (degenerate) finger devoid of any ridge structure.

3. Feature Extraction

Given raw input measurements, automatically extracting the given representation is an extremely difficult problem, especially where input measurements are noisy (see Figure1.21).

A given arbitrarily complex representation scheme should be amenable to automation without any human intervention. For instance, the manual system of fingerprint identification uses as much as a dozen features [59]. However, it is not feasible to incorporate these features into a fully automatic fingerprint system

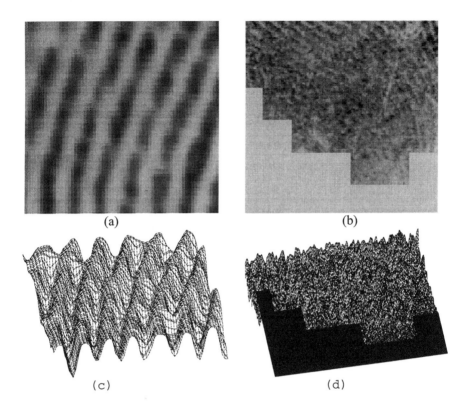

Figure 1.21 Automatically gleaning finger features from the fingerprint images is extremely difficult, especially, when the fingerprint is of poor quality (a) a portion of good quality fingerprint image; (b) a portion of poor quality fingerprint image; (c) 3-dimensional visualization of (a); and (d) 3-dimensional visualization of (b).

because it not easy to reliably detect these features using state-of-the-art image processing techniques. Determining features that are amenable to automation has not received much attention in computer vision and pattern recognition research and is especially important in biometrics which are entrenched in the design philosophies of an associated mature manual system of identification.

Traditionally, the feature extraction system follows a staged sequential architecture which precludes effective integration of extracted information available from the measurements. Increased availability of inexpensive computing and sensing resources makes it possible to use better architectures/methods for information processing to detect the features reliably.

Once the features are determined, it is also a common practice to design feature extraction process in a somewhat ad hoc manner. The efficacy of such methods is limited especially when input measurements are noisy. Rigorous models of feature representations are helpful in a reliable extraction of the features from the input measurements, especially, in noisy situations.

Determining terse and effective models for the features is a challenging research problem.

4. Matching

The crux of a matcher is a similarity function which quantifies the intuition of similarity between two representations of the biometric measurements. Determining an appropriate similarity metric is a very difficult problem since it should be able to discriminate between the representations of two different identities despite noise, structural and statistical variations in the input signals, aging, and artifacts of the feature extraction module. In many biometrics, say signature verification, it is difficult to even define the ground truth [28]: do the given two signatures belong to the same person or different persons?

A representation scheme and a similarity metric determine the accuracy performance of the system for a given population of identities; hence the selection of appropriate similarity scheme and representation is critical.

Given a complex operating environment, it is critical to identify a set of valid assumptions upon which the matcher design could be based. Often, there is a choice between whether it is more effective to exert more constraints by incorporating better engineering design or to build a more sophisticated similarity function for the given representation. For instance, in a fingerprint matcher, one could constrain the elastic distortion altogether and design the matcher based on a rigid transformation assumption or allow arbitrary distortions and accommodate the variations in the input signals using a clever matcher. Where to strike the compromise between the complexity of the matcher and controlling the environment is an open problem.

Consider design of a matcher in the domain of fingerprint-based identification systems (see Figure 1.22). Typically, the fingerprint imaging system presents a number of peculiar and challenging situations some of which are unique to fingerprint image capture scenario: (I) Inconsistent contact: the act of sensing distorts the finger. The three-dimensional shape of the finger gets mapped onto the two-dimensional surface of the glass platen. Typically, this (non-homogeneous) mapping function is determined by the pressure and contact of the finger on the glass platen (see Figure 1.23). (ii) Non-uniform contact: the ridge structure of a finger would be completely captured if ridges of the part of the finger being imaged are in complete optical contact with the glass platen. However, the dryness of the skin, skin disease, sweat, dirt, humidity in the air all confound the situation resulting in a non-ideal contact situation: some parts of the ridges may not come in complete contact with the platen and regions representing some valleys may come in contact with the glass platen. This results in "noisy" low contrast images, leading to either spurious minutiae or missing minutiae. (iii) Irreproducible contact: vigorous manual work, accidents etc. inflict injuries to the finger, thereby, changing the ridge structure of the finger either permanently or semi-permanently. This may introduce additional spurious minutiae. (iv) Feature extraction artifacts: the feature extraction algorithm is imperfect and introduces measurement errors. Various image processing operations might introduce inconsistent biases to perturb the location and orientation estimates of the reported minutiae from their gray scale counterparts. (vi) The act of sensing itself adds noise to the image. For example, residues are leftover on the glass platen from the previous fingerprint capture. A typical imaging system distorts the image of the object being sensed due to imperfect imaging conditions. In the frustrated total internal reflection (FTIR) sensing scheme, for example, there is a geometric distortion because the image plane is not parallel to the glass platen.

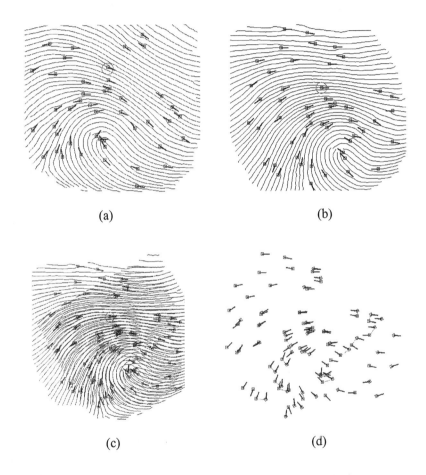

(a)   (b)

(c)   (d)

Figure 1.22 Fingerprint Matcher: Results of applying the matching algorithm [7] to an input and a template minutiae set; (a) input minutiae set; (b) template minutiae set; (c) input and template fingerprint are aligned based on the minutiae marked with green circles; and (d) matching result where template minutiae and their correspondences are connected by green lines. The matching score for the fingerprints was 37. The score range was 0--100; scores closer to 100 indicate better match.

Figure 1.23 Impressions of a finger captured by exerting different (magnitude/direction) forces on the finger during fingerprint acquisition results in a significant non-homogenous distortion of the ridge structures; consequently, the fingerprints are difficult to match.

In light of the operational environments mentioned above, the design of the similarity functions and matching algorithms needs to establish and characterize a realistic model of the variations among the representations of mated pairs. Is the distortion, for instance, significant for the given imaging? Is it easier to prevent distortion or is it more effective to take into account all the distortions possible and formulate a clever similarity function?

5.   Search, Organization, and Scalability

Systems dealing with a large number of identities should be able to effectively operate as the number of users in the system increases to its operational capacity and should only gracefully degrade as the system accommodates more users than envisaged at the time of its design. As civilian applications (e.g., driver and voter registration, National ID systems and IDs for health, medical, banking, cellular, transportation, and e-commerce applications) enrolling a very large number of identities (e.g., tens of millions) are being designed and integrated, we are increasingly looking toward biometrics to solve authentication and identification problems.

In identity authentication systems, biometrics are cost effective and are easier to maintain because these systems do not have to critically depend on issuing/reissuing other identity (magnetic stripe/smart/2D bar code) cards. Tasks like maintaining the database of identities, selection of a record etc. may require more resources, but the technical complexity of matching a biometric representation offered by the user to that stored in the system does not increase as the number of identities handled by the system increases arbitrarily.

On the other hand, identification of an individual among a large number of identities becomes increasingly complex as the number of identities stored in the system increases. Many applications like National ID systems, passport and visa issuance further require a constant throughput and a very small turnaround time. A designer of such systems needs to adopt radically different strategies and mode of operation than those adopted by traditional forensic identification systems. This has a profound influence on every aspect of the system, including the choice of biometrics, features, metric of similarity, matching criteria, operating point, etc. None of these design issues have been rigorously studied, neither in biometrics nor even in pattern recognition research.

All these criteria point to using those biometrics which remain invariant over a long period of time. Designing constant length, one-dimensional, indexable features will become increasingly important for identification applications involving a large number of identities.

Evaluation

An end-user is interested in determining the performance of the biometric system for *his specific application*: does the system make an accurate identification? Is the system sufficiently fast? How much would be the cost of the system? Among these issues, characterizing the accuracy performance is the most difficult; we will only address accuracy performance issues here.

No metric is sufficiently adequate to give a reliable and convincing indication of the identification accuracy of a biometric system. A decision made by a biometric system is either a *genuine individual* type of decision or an *impostor* type of decision, which can be represented by two statistical distributions called genuine distribution and impostor distribution, respectively (see Figure 1.24). For each type of decision, there are two possible decision outcomes, true or false. Therefore, there are a total of four possible outcomes: (*i*) a genuine individual is accepted, (*ii*) a genuine individual is rejected, (*iii*) an impostor is rejected, and (*iv*) an impostor is accepted. Outcomes (*i*) and (*iii*) are correct whereas (*ii*) and (*iv*) are incorrect. In principle, we can use the false (impostor) acceptance rate (FAR), the false (genuine individual) reject rate (FRR) and the equal error rate (EER)[7] to indicate the identification accuracy of a biometric system [24, 60]. Unfortunately, the performance of a system in the context of a given population (database) is a random variable and, strictly speaking, it cannot be computed or measured but can only be estimated from empirical data and the estimates of the performance are very data dependent. Therefore, they are meaningful only for a specific database in a specific test environment. For example, the performance of a particular biometric system claimed by its manufacturer had a FRR of 0.3% and a FAR of 0.1%. An independent test by the Sandia National Lab. found that the same system had a FRR of 25% with an unknown FAR [61]! In order to provide a more reliable assessment of a biometric system, some more descriptive performance measures are necessary. Receiver operating curve (ROC) and *d'* are the two other commonly used measures. A receiver operating curve provides an empirical assessment of the system performance at different operating points which is more informative than FAR and FRR. The statistic *d'* gives an indication of the separation between the genuine distribution and impostor distribution [60]. The

---

[7] Equal error rate is defined as the value of FAR/FRR at an operating point on ROC where FAR and FRR are equal.

existing performance metrics are empirical and provide us means of estimating performance with respect to a specific database. For such empirical performance metric to be able to precisely generalize to the entire population of interest, the test data should (*i*) be large enough to represent the population and (*ii*) contain enough samples from each category of the population [60].

To obtain fair and honest test results, enough samples should be available, the samples should be representative of the population, and adequately represent all the categories (impostors and genuine). In reality, especially for emerging applications, we do not have access to a sufficient number of test samples nor are the samples representative of the actual population of interest. In such situations, there is a need to obtain predictive models of performance in terms of controllable and measurable parameters of the available data. Such predictive models of performance may be useful both for bootstrapping a small number of available samples as well as obtaining realistic estimates of the performance of a given biometric technology to a given application.

Irrespective of the choice of a performance metric, error bounds that indicate the confidence of the estimates are valuable for understanding the significance of the test results. Estimating confidence measures without using unrealistic naive models of the hypothesized population distributions is challenging.

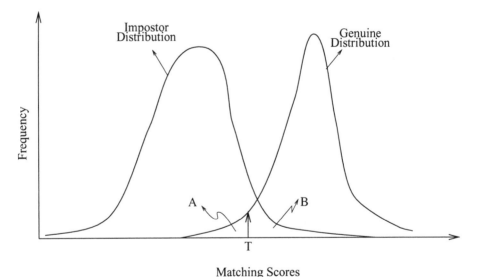

Figure 1.24 The distributions of matching scores obtained by matching input measurements from the same identities (genuine distribution) and those obtained from different identities (impostor distribution). Given a matching score threshold (T), the areas A and B represent false rejection rate (FRR) and false acceptance rate (FAR), respectively.

Integration

The accuracy of an identification system obviously will improve as we effectively utilize and integrate an increasing number of information sources related to an individual to confirm her identity. It also becomes increasingly difficult to abuse the system privileges. However, the challenge of integration is to ascertain that the system performance degrades gracefully as some of the information sources become unavailable or unreliable. As better performance is demanded of identification systems and as a variety of different sensors become affordable, integration of different biometrics will become an important issue [62, 63, 64, 65, 66]. Integration of different technologies is also becoming critical for imparting capabilities to the identification system. For instance, biometric sensor integrated smart cards could provide facilities for identity authentication without divulging any information about biometric measurements.

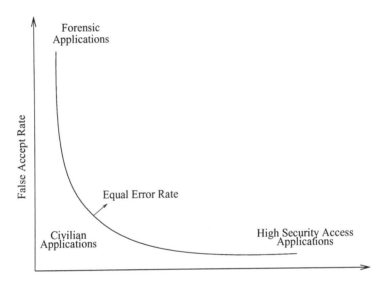

Figure 1.25 Receiver operating characteristics (ROC) of a system illustrates false reject rate (FRR) and false acceptance rate (FAR) of a matcher at all operating points. Each point on an ROC defines FRR and FAR for a given matcher operating at a particular matching score threshold. High security access applications are concerned about break-ins and hence operate the matcher at a point on ROC with small FAR. Forensic applications desire to catch a criminal even at the expense of examining a large number of false accepts and hence operate their matcher at a high FAR. Civilian applications attempt to operate their matchers at the operating points with both, low FRR and low FAR.

Deciding efficient architectures for integration is an open research problem and it is perhaps the single most factor in determining the behavior of an integrated system. Decision level, feature level, and measurement level integration architectures have been studied in the literature [67, 68]. Determination of which integration strategies are appropriate for a given identification application needs more focussed research.

Circumvention

Some problems plague all identification technologies (based on possession, knowledge, or biometrics) alike. Fraud in an identification system is possible in different forms. Some forms of fraud are characterized as loopholes in the system: possibilities of illegitimate access to a system not envisaged by its designers. Other forms involve transcending the means and mechanisms of identification used by the system (super-system) and hence, in principle, cannot be completely eliminated using any strategies embedded inside the system (intra-system). The latter type of fraud could be categorized as follows:

- Collusion: In any application, some operators of the system will have a super-operator status which allows them to bypass the identification component of the processing and to overrule the decision made by the system. This facility is incorporated in the system work-flow to permit handling of exceptional situations, e.g., processing of individuals with no fingers in a fingerprint-based identification system. This could potentially lead to an abuse of the system by way of collusion between the super-operators and the users.
- Coercion: The genuine users could be potentially coerced to identify themselves to the system. The identification means could be forcibly extracted from a genuine user to gain access to the system with concomitant privileges. For instance, an ATM user could be forced to give away her ATM card and PIN at a gun point. It is desirable to reliably detect instances of coercion without endangering the lives of genuine users and take an appropriate action.
- Denial: It is possible that a genuine user may identify himself to the system using the legitimate means of the identification to gain access to the privileges and is subsequently denied such an access.
- Covert Acquisition: It is possible that the means of identification could be compromised without the knowledge of a legitimate user and be subsequently abused. For instance, a significant amount of fraud in telecommunication theft is ascribed to video-snooping: video-recording the scenes of users punching their pins at, say, a public telephone.

As mentioned earlier, many of these problems may not be fully eliminated. Currently, attempts to reduce fraud in an identification system are process-based and ad hoc. There is a need to focus research effort on systematic and technology-intensive approaches to combat fraud in the system. This is especially true in terms of biometrics-based identification systems where the captured biometric measurements and context may have sufficient information to deter and detect some forms of fraud. In particular, multi-biometrics may show promise in approaching solutions to many of the above mentioned problems.

Some other problems related to identification are more specific to biometrics-based systems. For instance, skilled humans have an uncanny ability to disguise their identity and are able to assume (forge/mimic) a different (specific) identity (Figure 1.26). The "chameleon identities" pose an additional problem to the reliability of the identification systems based on some biometrics and warrant more research.

9. Privacy

- Privacy: Any biometrics-based technology is traditionally perceived as dehumanizing and as a threat to an individual's privacy rights (see Figure 1.27). As identification technologies become more and more foolproof, the process of getting identified itself leaves trails of undeniable private information. e.g., where is an individual? What is the individual buying?, etc. In case of biometrics-based identification, this problem is even more serious because the biometric features may additionally inform others about the medical history or susceptibilities of an individual, e.g., retinal vasculature may divulge information about diabetes or hypertension [69]. Consequently, there is a legitimate concern about privacy issues associated with the biometrics-based identification.

Figure 1.26 Multiple Personalities: All the people in this image are the same person (The New York Times Magazine, September 1, 1996/section 6, pages 48-49, reproduced with permission of Robert Trachtenberg).

- Proscription: This issue is somewhat related to the previous issue. When a biometric measurement is offered to a given system, the information contained in it should not be used for any other purpose than its intended use. In any (networked) information processing system, it is difficult to ensure that the biometric measurements captured will only be used for its intended purpose.

Wide-spread use of biometrics-based identification systems should not only address the above mentioned issues from technical standpoint but also from the public perception point of view. This is especially true for assuring the users that their biometric information will remain private and will only be used for the expressed purpose for which it was collected.

Figure 1.27 "The true terror is in the card", an illustration from Robert E. Smith's article [9] in the New York Times Magazine summarizes the essence of public perception about biometric technology: it is dehumanizing and is a threat to privacy rights of an individual. The original picture by Jana Sterback, "Generic Man", 1989. Copyright © 1996 by The New York Times. Reprinted by permission.

10. Novel Applications

As biometric technology matures, there will be an increasing interaction among the (biometric) market, (biometric) technology, and the (identification) applications. The emerging interaction is expected to be influenced by the added value of the technology, the sensitivities of the population, and the credibility of the service provider. It is too early to predict where, how, and which biometric technology would evolve and be mated with which applications.

Applications like automating identification for more convenient travel, for transactions via e-commerce, etc. seem to be ready for commercialization, but perhaps, biometric technology could open up a whole new genre of futuristic hi-tech applications that were not foreseen before.

Take for instance, the application of content-based search of digital libraries and in particular, video. One of the tasks in content-based search involves ascription of sound bytes to individuals (identities) depicted in the corresponding video segment. The association of the sound to identity is essentially a closed identification problem. What makes this problem interesting is the opportunity to exploit the context and clues offered from the vision-based processing (e.g., number of people in the video and lip movements) of the video. Voice-based clues could generate plausible hypotheses about identities of visual entities. From the visual input, the hypothesis could either be accepted or rejected depending on the coherency of the sound and vision based results.

Or imagine, in a hi-tech mall, the features extracted from DNA of millions of cells shredded by the body of a passing individual (and a potential customer) would be instantly matched to determine the exact identity or a possible category of population. That individual would then be treated exclusively depending upon his spending pattern. Perhaps, there would be data mining based on the biometric characteristics!

Interesting scenarios might materialize as a number of civilian applications of identification are integrated based on a single or multiple biometric technologies. This will certainly have a profound influence on the way we conduct our business.

11. Summary

Biometrics is a science of automatically identifying individuals based on their unique physiological or behavioral characteristics. A number of civilian and commercial applications of biometrics-based identification are emerging. At the same time, a number of legitimate concerns are being raised against the use of biometrics for various applications; three of them appear to be the most significant: cost, privacy, and performance.

As more and more legislations are brought into effect, both, protecting the privacy rights of the individuals as well as endorsing the use of biometrics for legitimate uses and as the prices of the biometric sensors continues to fall, the added value of the biometrics-based systems will continue to attract more applications. It is expected that in the next five years, the rising number of applications may increase the demand for the biometric sensors to drive a volume-based pricing.

For the wide-spread use of the biometrics to materialize, it is necessary to undertake systematic studies of the fundamental research issues underlying the design and evaluation of identification systems. Further, it is critical to engineer the match between the application needs and the available technologies.

Acknowledgments

We would like to thank Biometric Consortium and discussions on its list server for providing a rich and up to date source of information. We are grateful to Tulasi Perali for pointing out specific medical pathologies. Jon Connell provided the face foreground/background separation images shown in Figures 1.20 (a) and (b). Iris (Figure1.7) and retina (Figure 1.12) images were scanned from the product literatures from IriScan and EyeDentify, respectively. We thank Lin Hong for reviewing this chapter and lending a number of images from his PhD thesis. We are grateful to

Profs. Tomaso Poggio and Pawan Sinha for letting us use Fig. 1.15 and to Drs. Andre Moenssens and John Berry for granting us permission to use illustrations in Fig. 1.14. Thanks to Norm Haas for pointing out illustrations in the New York Times Magazine.

References

[1] "Technology recognition systems homepage," *http://www.betac.com/~imagemap/subs?-155,150*, 1997.

[2] J. Rice, "Veincheck homepage," *http://innotts.co.uk/~joerice/*, 1997.

[3] M. Burge and W. Burger, "Ear biometrics for machine vision," in *21st Workshop of Austrian Association for Pattern Recognition, http://www.cast.uni-linz.ac.at/st/vision/-Papers/oagm-97/*, (Hallstatt), ÖAGM, Verlag R Oldenbourg, May 1997.

[4] Federal Bureau of Investigation Educational Internet Publication, "DNA testing," *http://www.fbi.gov/kids/dna/dna.htm*, 1997.

[5] A. Moenssens, *Fingerprint Techniques*. Chilton Book Company, London, 1971.

[6] P. Sinha and T. Poggio, "I think I know that face...," *Nature*, Vol. 384, No. 6608, p. 406, 1996.

[7] A. Jain, L. Hong, S. Pankanti, and R. Bolle, "On-line identity authentication system using fingerprints," *Proceedings of IEEE (Special Issue on Automated Biometrics)*, Vol. 85, pp. 1365-1388, September 1997.

[8] L. Hong, Y. Wan, and A. Jain, "Fingerprint image enhancement: Algorithm and performance evaluation," *IEEE Trans. Pattern Analysis and Machine Intelligence*, 1998 (To appear).

[9] R. E. Smith, "The true terror is in the card," in *The New York Times Magazine, September 8, 1996*, pp. 58.

[10] J. D. Woodward, "Biometrics: Privacy's foe or privacy's friend?," *Proceedings of the IEEE (Special Issue on Automated Biometrics)*, Vol. 85, pp. 1480-1492, September 1997.

[11] L. Lange and G. Leopold, "Digital identification: It's now at our fingertips," *EEtimes at http://techweb.cmp.com/eet/823/, March 24*, Vol. 946, 1997.

[12] G. Stein, "Sacred emily," in *The Oxford Book of American Light Verse* (W. Harmon, ed.), pp. 286-294, Oxford University Press, 1979.

[13] J. R. Parks, "Personal identification - biometrics," *in Information Security* (D. T. Lindsay and W. L. Price, eds.), pp. 181-191, North Holland: Elsevier Science, 1991.

[14] R. G. D. Steel and J. H. Torrie, *Principles and Procedures of Statistics: A Biometrical Approach* (McGraw-Hill Series in Probability and Statistics), New York: McGraw-Hill, third ed., 1996.

[15] R. Clarke, "Human identification in information systems: Management challenges and public policy issues," *Information Technology & People*, Vol. 7, No. 4, pp. 6-37, 1994.

[16] E. Newham, *The Biometric Report*. http://www.sjb.com/: SJB Services, New York, 1995.

[17] S. Furui, "Recent advances in speaker recognition," in *Lecture Notes in Computer Science 1206, Proceedings of Audio- and Video Biometric Person Authentication AVBPA'97, First International Conference, Crans-Montana, Switzerland, March 12-14*, pp. 237-252, Springer-Verlag, Berlin, 1997.

[18] F. J. Prokoski, R. B. Riedel, and J. S. Coffin, "Identification of individuals by means of facial thermography," in *Proceedings of The IEEE 1992 International Carnahan Conference on Security Technology: Crime Countermeasures, Atlanta, GA, USA 14-16 Oct.*, pp. 120-125, IEEE, 1992.

[19] F. J. Prokoski, "NC-TEST: noncontact thermal emissions screening technique for drug and alcohol detection," in *Proc. SPIE: Human Detection and Positive Identification: Methods and Technologies* (L. A. Alyea and D. E. Hoglund, eds.), Vol. 2932, pp. 136-148.

[20] M. Turk and A. Pentland, "Eigenfaces for recognition", *Journal of Cognitive Neuroscience*, Vol. 3, No. 1, pp. 71-86, 1991.

[21] D. Swets and J. J. Weng, "Using discriminant eigenfeatures for image retrieval," *IEEE Trans. Pattern Analysis and Machine Intelligence*, Vol. 18, pp. 831-836, August 1996.

[22] J. J. Atick, P. A. Griffin, and A. N. Redlich, "Statistical approach to shape from shading: Reconstruction of 3-dimensional face surfaces from single 2-dimensional images," *Neural Computation*, Vol. 8, pp. 1321-1340, August 1996.

[23] P. J. Phillips, P. J. Rauss, and S. Z. Der, "The FERET (Face Recognition Technology) evaluation methodology," in *Proceedings of the IEEE Conference on Computer Vision and Pattern Recognition 97, June 17-19*, (San Juan, Puerto Rico), 1997.

[24] J. G. Daugman, "High confidence visual recognition of persons by a test of statistical independence," *IEEE Trans. Pattern Analysis and Machine Intelligence*, Vol. 15, No. 11, pp. 1148-1161, 1993.

[25] T. Julian, L. A. Vontver, and D. Dumesic, *Review of Obstetrics & Gynecology*. Stanford, CT: Appleton and Lange, 1995.

[26] D. Cunado, M. S. Nixon, and J. N. Carter, "Using gait as a biometric, via phase-weighted magnitude spectra," in *Lecture Notes in Computer Science 1206, Proceedings of Audio- and Video- Biometric Person Authentication AVBPA'97, First International Conference, Crans-Montana, Switzerland, March 12-14* (J. Bigun, G. Chollet, and G. Borgefors, eds.), pp. 95-102, Springer-Verlag, Berlin, 1997.

[27] S. A. Bleha and M. Obaidat, "Computer users verification using the perceptron algorithm," *IEEE Trans. Systems Man Cybernetics*, Vol. 23, pp. 900-902, 1993.

[28] V. Nalwa, "Automatic on-line signature verification," *Proceedings of the IEEE*, Vol. 85, pp. 213-239, February 1997.

[29] British Technology Group, "Automatic signature verification using acoustic emissions," *http://www.btgusa.com/security/prod1.html*, 1997.

[30] T. A. Dickinson, J. White, J. S. Kauer, and D. R. Walt, "A chemical-detecting system based on a cross-reactive optical sensor array," *Nature*, Vol. 382, pp. 697-700, 1996.

[31] M. Howard, "Artificial nose may be able to sniff out land mines," *http://www.tufts.edu/communications/tech.html*, 1997.

[32] RAYCO Security, "Eyedentify retina biometric reader," http://www.raycosecurity.com/-hirsch/EyeDentify.html, 1997.

[33] "Biometric Technology Today," *http://www.sjb.co.uk/*, November 1996.

[34] Biomet Partners Inc., "Positive verification of a person's identity: Digi-2 3-dimensional finger geometry," *http://www.webconsult.ch/biomet.htm*, 1997.

[35] S. C. Davies, "Touching big brother: How biometric technology will fuse flesh and machine," *Information Technology & People*, Vol. 7, No. 4, pp. 60-69, 1994.

[36] B. Miller, "Vital signs of identity," *IEEE Spectrum*, Vol. 31, No. 2, pp. 22-30, 1994.

[37] H. T. F. Rhodes, *Alphonse Bertillon: Father of Scientific Detection*. Abelard-Schuman, New York, 1956.

[38] America Online, "Biometrics in news," *http://members.aol.com/biometric/news.html*, 1997.

[39] J. Bigun, C. Chollet, and C. Borgefors, (editors), *Lecture Notes in Computer Science 1206, Proceedings of Audio- and Video- Biometric Person Authentication AVBPA'97, First International Conference, Crans-Montana, Switzerland, March 12-14*. Springer-Verlag, Berlin, 1997.

[40] "Asian Conference on Computer Vision: Special Session on Biometrics, January 8-11, Hong Kong," *http://www.vic.ust.hk/accv98*, 1998.

[41] *Proceedings of Biometric Consortium Eighth Meeting, San Jose, California*. June 1996.

[42] *Proceedings of Biometric Consortium Ninth Meeting, Crystal City, Virginia*. April 1997.

[43] "31st Asilomar Conference on Signals, Systems, and Computers, Pacific Grove, CA, November 2-5," *http://dubhe.cc.nps.navy.mil/asilomar/*, 1997.

[44] "Workshop on Automatic Identification and Technology, November 6-7, Stony Brook, NY," *http://www.cs.sunysb.edu/simtheo/waiat*, 1998.

[45] "Face Recognition: From Theory to Applications, NATO Advanced Study Institute (ASI) Program, June 23 - July 4, Stirling, Scotland, UK," *http://chagall.gmu.edu/faces97-/natoasi/*, 1997.

[46] *CardTech/SecurTech: Technology and Applications*. May 1997.

[47] "Biometricon'97, Arlington, VA, March 13-14," *http://www.ncsa.com/cbdc/cbdc-c.html*, 1997.

[48] "The Biometric Consortium," http://www.biometrics.org/.

[49] "Commercial Biometrics Developer Consortium (CBDC)," *http//www.ncsa.com/cbdc/*, 1997.

[50] *Proceedings of the IEEE (Special Issue on Automated Biometrics)*, Vol. 85, September 1997.

[51] "Biometric Technology Today," http://www.sjb.co.uk/.

[52] "Automatic I. D. News homepage," *http://www.autoidnews.com/*, 1997.

[53] D. Mintie, "Welfare ID at the point of transaction using fingerprint and 2D bar codes," in *Proc. CardTech/SecurTech, Volume II: Applications*, (Atlanta, Georgia), pp. 469-476, May 1996.

[54] "INS Passenger Accelerated Service System (INSPASS)," *http://www.biometrics.org:8080/~BC/REPORTS/INSPASS.html*, 1996.

[55] S. Hunt, "National ID programs around the world," in *Proc. CardTech/SecurTech, Vol. II: Applications*, (Atlanta, Georgia), pp. 509-520, May 1996.

[56] R. Bolle, N. Ratha, and S. Pankanti, "Research issues in biometrics," in *Proceedings of Asian Conference on Computer Vision: Special Session on Biometrics, January, 8-11, Hong Kong*, (http://www.vic.ust.hk/accv98), 1998.

[57] Mytec Technologies, "Access control applications using optical computing," *http//www.mytec.com/*, 1997.

[58] R. Bahuguna, "Fingerprint verification using hologram matched filterings," in *Proceedings Biometric Consortium Eighth Meeting*, (San Jose, California), June 1996.

[59] Federal Bureau of Investigation, *The Science of Fingerprints: Classification and Uses*. Washington, D.C.: U.S. Government Printing Office, 1984.

[60] J. G. Daugman and G. O. Williams, "A proposed standard for biometric decidability," in *Proc. CardTech/SecureTech Conference*, (Atlanta, CA), pp. 223-234, 1996.

[61] J. P. Campbell, L. A. Alyea, and J. S. Dunn, "Biometric security: Government applications and operations," *http://www.biometrics.org/*, 1996.

[62] J. Kittler, Y. P. Li, J. Matas, and M. U. Ramos Sànchez, "Combining evidence in multimodal personal identity recognition systems," in *Lecture Notes in Computer Science 1206, Proceedings of Audio- and Video- Biometric Person Authentication AVBPA'97, First International Conference, Crans-Montana, Switzerland, March 12-14*, pp. 327-334, Springer-Verlag, Berlin, 1997.

[63] R. Brunelli and D. Falavigna, "Person identification using multiple cues," *IEEE Trans. Pattern Analysis and Machine Intell*igence, Vol. 10, pp. 955-966, October 1995.

[64] U. Dieckmann, P. Lankensteiner, R. Schamburger, B. Froba, and S. Meller, "SESAM: A biometric person identification system using sensor fusion," in *Lecture Notes in Computer Science 1206, Proceedings of Audio- and Video- Biometric Person Authentication AVBPA'97, First International Conference, Crans-Montana, Switzerland, March 12-14*, pp. 301-310, Springer-Verlag, Berlin, 1997.

[65] E. S. Bigun, J. Bigun, B. Duc, and S. Fischer, "Expert conciliation for multi modal person authentication system by Bayesian statistics," in *Lecture Notes in Computer Science 1206, Proceedings of Audio- and Video- Biometric Person Authentication AVBPA'97, First International Conference, Crans-Montana, Switzerland, March 12-14*, pp. 291-300, Springer-Verlag, Berlin, 1997.

[66] L. Hong and A. Jain, "Integrating faces and fingerprints for personal identification," in *Proceedings of Asian Conference on Computer Vision: Special Session on Biometrics, January, 8-11, Hong Kong*, (http://www.vic.ust.hk/accv98), 1998.

[67] R. R. Tenney and N. R. Sandell, "Structures for distributed decision making," *IEEE Trans. on Systems, Man, and Cybernetics*, Vol. 11, pp. 517-526, August 1981.

[68] R. R. Tenney and N. R. Sandell, "Strategies for distributed decision making," *IEEE Trans. on Systems, Man, and Cybernetics*, Vol. 11, pp. 527-538, August 1981.

[69] S. J. McPhee, M. A. Papadakis, L. M. Tierney, and R. Gonzales, *Current Medical Diagnosis and Treatment.* Stamford, CT: Appleton and Lange, 1997.

[70] K. Inman, and N. Rudin, *An Introduction to Forensic DNA Analysis.* CRC Press, Boca Raton, Florida, 1997.

[71] H. Cummins and C. Midlo, *Finger Prints, Palms and Soles*, Dover, New York, 1961.

[72] J. Berry, "The history and development of fingerprinting," H. C. Lee and R. E. Gaensslen, Eds., *Advances in Fingerprint Technology*, CRC Press, London, 1994.

2 FINGERPRINT VERIFICATION

Lawrence O'Gorman
Veridicom Inc.
Chatham, NJ
log@veridicom.com

Abstract *The use of fingerprints for identification has been employed in law enforcement for about a century. A much broader application of fingerprints is for personal authentication, for instance to access a computer, a network, a bank-machine, a car, or a home. The topic of this chapter is fingerprint verification, where "verification" implies a user matching a fingerprint against a single fingerprint associated with the identity that the user claims. The following topics are covered: history, image processing methods, enrollment and verification procedures, system security considerations, recognition rate statistics, fingerprint capture devices, combination with other biometrics, and the future of fingerprint verification.*

Keywords: *fingerprint verification, fingerprint matching, biometric, image enhancement, image filtering, feature detection, minutiae, security, fingerprint sensor.*

1. Introduction[1]

The use of fingerprints as a biometric is both the oldest mode of computer-aided, personal identification and the most prevalent in use today. However, this widespread use of fingerprints has been and still is largely for law enforcement applications. There is expectation that a recent combination of factors will favor the use of fingerprints for the much larger market of personal authentication. These factors include: small and inexpensive fingerprint capture devices, fast computing hardware, recognition rate and speed to meet the needs of many applications, the explosive growth of network and Internet transactions, and the heightened awareness of the need for ease-of-use as an essential component of reliable security.

This chapter contains an overview of fingerprint verification methods and related issues. We first describe fingerprint history and terminology. Digital image processing

---

[1] Portions of this chapter have previously appeared in, L. O'Gorman, "Overview of fingerprint verification technologies," *Elsevier Information Security Technical Report,* Vol. 3, No. 1, 1998.

methods are described that take the captured fingerprint from a raw image to match result. Systems issues are discussed including procedures for enrollment, verification, spoof detection, and system security. Recognition statistics are discussed for the purpose of comparing and evaluating different systems. We describe different fingerprint capture device technologies. We consider fingerprints in combination with other biometrics in a multi-modal system and finally look to the future of fingerprint verification.

It is necessary to state at the onset that there are many different approaches used for fingerprint verification. Some of these are published in the scientific literature, some published only as patents, and many are kept as trade secrets. We attempt to cover what is publicly known and used in the field, and cite both the scientific and patent literature. Furthermore, while we attempt to be objective, some material is arguable and can be regarded that way.

2. History

There is archaeological evidence that fingerprints as a form of identification have been used at least since 7000 to 6000 BC by the ancient Assyrians and Chinese. Clay pottery from these times sometimes contain fingerprint impressions placed to mark the potter. Chinese documents bore a clay seal marked by the thumbprint of the originator. Bricks used in houses in the ancient city of Jericho were sometimes imprinted by pairs of thumbprints of the bricklayer. However, though fingerprint individuality was recognized, there is no evidence this was used on a universal basis in any of these societies.

In the mid-1800's scientific studies were begun that would established two critical characteristics of fingerprints that are true still to this day: no two fingerprints from different fingers have been found to have the same ridge pattern, and fingerprint ridge patterns are unchanging throughout life. These studies led to the use of fingerprints for criminal identification, first in Argentina in 1896, then at Scotland Yard in 1901, and to other countries in the early 1900's.

Computer processing of fingerprints began in the early 1960s with the introduction of computer hardware that could reasonably process these images. Since then, automated fingerprint identification systems (AFIS) have been deployed widely among law enforcement agencies throughout the world.

In the 1980s, innovations in two technology areas, personal computers and optical scanners, enabled the tools to make fingerprint capture practical in non-criminal applications such as for ID-card programs. Now, in the late 1990s, the introduction of inexpensive fingerprint capture devices and the development of fast, reliable matching algorithms has set the stage for the expansion of fingerprint matching to personal use.

Why include a history of fingerprints in this chapter? This history of use is one that other types of biometric do not come close to. Thus there is the experience of a century of forensic use and hundreds of millions of fingerprint matches by which we can say with some authority that fingerprints are unique and their use in matching is extremely reliable. For further historical information, see [2].

3. Matching: Verification and Identification

 Matching can be separated into two categories: verification and identification. *Verification* is the topic of this chapter. It is the comparison of a *claimant* fingerprint against an *enrollee* fingerprint, where the intention is that the claimant fingerprint matches the enrollee fingerprint. To prepare for verification, a person initially enrolls his or her fingerprint into the verification system. A representation of that fingerprint is stored in some compressed format along with the person's name or other identity. Subsequently, each access is authenticated by the person identifying him or herself, then applying the fingerprint to the system such that the identity can be verified. Verification is also termed, *one-to-one matching*.

Identification is the traditional domain of criminal fingerprint matching. A fingerprint of unknown ownership is matched against a database of known fingerprints to associate a crime with an identity. Identification is also termed, *one-to-many matching*.

There is an informal third type of matching that is termed *one-to-few matching*. This is for the practical application where a fingerprint system is used by "a few" users, such as by family members to enter their house. A number that constitutes "few" is usually accepted to be somewhere between 5 and 20.

4. Feature Types

The lines that flow in various patterns across fingerprints are called *ridges* and the spaces between ridges are *valleys*. It is these ridges that are compared between one fingerprint and another when matching. Fingerprints are commonly matched by one (or both) of two approaches. We describe the fingerprint features as associated with these approaches.

The more microscopic of the approaches is called *minutia matching*. The two minutia types that are shown in Figure 2.1 are a ridge *ending* and *bifurcation*. An ending is a feature where a ridge terminates. A bifurcation is a feature where a ridge splits from a single path to two paths at a Y-junction. For matching purposes, a minutia is attributed with features. These are type, location *(x, y)*, and direction (and some approaches use additional features).

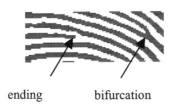

ending   bifurcation

Figure 2.1 Fingerprint minutiae: ending and bifurcation.

The more macroscopic approach to matching is called *global pattern matching* or simply *pattern matching*. In this approach, the flow of ridges is compared at all locations between a pair of fingerprint images. The ridge flow constitutes a global pattern of the fingerprint. Three fingerprint patterns are shown in Figure 2.2. (Different classification schemes can use up to ten or so pattern classes, but these three are the basic patterns.)

Two other features are sometimes used for matching: *core* and *delta*. (Figure 2.2.) The core can be thought of as the center of the fingerprint pattern. The delta is a singular point from which three patterns deviate. The core and delta locations can be used as landmark locations by which to orient two fingerprints for subsequent matching – though these features are not present on all fingerprints.

There may be other features of the fingerprint that are used in matching. For instance, pores can be resolved by some fingerprint sensors and there is a body of work (mainly research at this time) to use the position of the pores for matching in the same manner that the minutiae are used. Size of the fingerprint, and average ridge and valley widths can be used for matching, however these are changeable over time. The

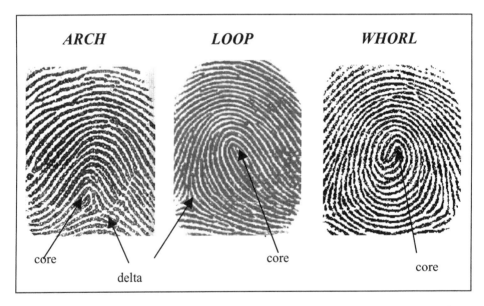

Figure 2.2 Fingerprint patterns: arch, loop, and whorl. Fingerprint landmarks are also shown: core and delta. (No delta locations fall within the captured area of the whorl here.)

positions of scars and creases can also be used, but are usually not used because they can be temporary or artificially introduced.

5. Image Processing and Verification

Following image capture to obtain the fingerprint image, image processing is performed. The ultimate objective of image processing is to achieve the best image by which to produce the correct match result. The image processing steps are the following: image noise reduction and enhancement, feature detection, and matching.

This section is organized to describe first the sequence of processing and verification via a "common" minutia-based approach. This is described without variants and optional methods (of which there are many) for the sake of reading flow and simplicity. It is important to note that, though many researchers and product developers follow this approach, all do not, and even the choice of what constitutes "common" may be contentious. In the final subsections of this section, variations of this approach, both minutia-based and non-minutia-based, are described.

Image Specifications

Depending upon the fingerprint capture device, the image can have a range of specifications. Commonly, the pixels are 8-bit values, and this yields an intensity range from 0 to 255. The image resolution is the number of pixels per unit length, and this ranges from 250 dots per inch (100 dots per centimeter) to 625 dots per inch (250 dots per centimeter), with 500 dots per inch (200 dots per centimeter) being a common standard. The image area is from 0.5 inches square (1.27 centimeter) to 1.25 inches (3.175 centimeter), with 1 inch (2.54 centimeter) being the standard. We discuss more on image capture devices in Section 8.

Image Enhancement

A fingerprint image is one of the noisiest of image types. This is due predominantly to the fact that fingers are our direct form of contact for most of the manual tasks we perform: finger tips become dirty, cut, scarred, creased, dry, wet, worn, etc. The image enhancement step is designed to reduce this noise and to enhance the definition of ridges against valleys. Two image processing operations designed for these purposes are the adaptive, matched filter and adaptive thresholding. The stages of image enhancement, feature detection, and matching are illustrated in Figure 2.3.

There is a useful side to fingerprint characteristics as well. That is the "redundancy" of parallel ridges. Even though there may be discontinuities in particular ridges, one can always look at a small, local area of ridges and determine their flow. We can use this "redundancy of information" to design an adaptive, matched filter. This filter is applied to every pixel in the image (spatial convolution is the technical term for this operation). Based on the local orientation of the ridges around each pixel, the matched filter is applied to enhance ridges oriented in the same direction as those in the same locality, and decrease anything oriented differently. The latter includes noise that may be joining adjacent ridges, thus flowing perpendicular to the local flow. These incorrect "bridges" can be eliminated by use of the matched filter. Figure 2.3(b) shows an orientation map where line sectors represent the orientation of ridges in each locality. Thus, the filter is adaptive because it orients itself to local ridge flow. It is matched because it should enhance – or match – the ridges and not the noise.

After the image is enhanced and noise reduced, we are ready to extract the ridges. Though the ridges have gradations of intensity in the original grayscale image, their true information is simply binary: ridges against background. Simplifying the image to this binary representation facilitates subsequent processing. The binarization operation takes as input a grayscale image and returns a binary image as output. The image is reduced in intensity levels from the original 256 (8-bit pixels) to 2 (1-bit pixels).

The difficulty in performing binarization is that all the fingerprint images do not have the same contrast characteristics, so a single intensity threshold cannot be chosen. Furthermore, contrast may vary within a single image, for instance if the finger is pressed more firmly at the center. Therefore, a common image processing tool is used, called locally adaptive thresholding. This operation determines thresholds adaptively to the local image intensities. The binarization result is shown in Figure 2.3(c).

The final image processing operation usually performed prior to minutia detection is thinning. Thinning reduces the widths of the ridges down to a single pixel. See Figure 2.3(d). It will be seen in the next section how these single-pixel width ridges facilitate the job of detecting endings and bifurcations. A good thinning method will reduce the ridges to single-pixel width while retaining connectivity and minimizing the number of artifacts introduced due to this processing. These artifacts are comprised primarily of spurs, which are erroneous bifurcations with one very short branch. These artifacts are removed by recognizing the differences between legitimate and erroneous minutiae in the feature extraction stage described below.

Image enhancement is a relatively time-consuming process. A 500x500-pixel fingerprint image has 250,000 pixels; several multiplications and other operations are applied at each pixel. Both matched filtering and thinning contribute largely to this time expenditure. Consequently, many fingerprint systems are designed to conserve operations at this stage to reach a match result more quickly. This is not a good tradeoff. The results of all subsequent operations depend on the quality of the image as captured by the sensor and as processed at this stage. Economizing for the sake of speedup will result in degraded match results, which in turn will result in repeated attempts to verify or false rejections. Therefore, it is our contention that a system offering reasonable speed with a correct answer is much better than a faster system that yields poorer match results.

Feature Extraction

The fingerprint minutiae are found at the feature extraction stage. Operating upon the thinned image, the minutiae are straightforward to detect. Endings are found at termination points of thin lines. Bifurcations are found at the junctions of three lines. See Figure 2.3(e).

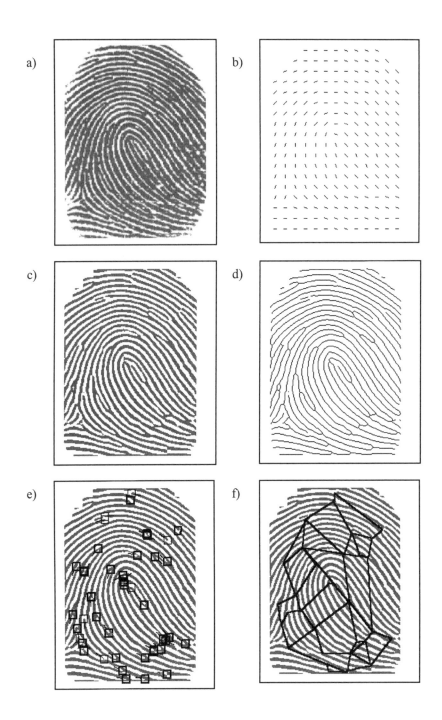

Figure 2.3 Sequence of fingerprint processing steps: a) original, b) orientation, c) binarized, d) thinned, e) minutiae, f) minutia graph.

There will always be extraneous minutiae found due to a noisy original image or due to artifacts introduced during matched filtering and thinning. These extraneous features are reduced by using empirically determined thresholds. For instance, a bifurcation having a branch that is much shorter than an empirically determined threshold length is eliminated because it is likely to be a spur. Two endings on a very short isolated line are eliminated because this line is likely due to noise. Two endings that are closely opposing are eliminated because these are likely to be on the same ridge that has been broken due to a scar or noise or a dry finger condition that results in discontinuous ridges. Endings at the boundary of the fingerprint are eliminated because they are not true endings but rather the extent of the fingerprint in contact with the capture device.

Feature attributes are determined for each valid minutia found. These consist of: ridge ending or bifurcation type, the *(x,y)* location, and the direction of the ending or bifurcation. Although minutia type is usually determined and stored, many fingerprint matching systems do not use this information because discrimination of one from the other is often difficult.

The result of the feature extraction stage is what is called a *minutia template*. This is a list of minutiae with accompanying attribute values. An approximate range on the number of minutiae found at this stage is from 10 to 100. If each minutia is stored with type (1 bit), location (9 bits each for *x* and *y*), and direction (8 bits), then each will require 27 bits – say 4 bytes – and the template will require up to 400 bytes. It is not uncommon to see template lengths of 1024 bytes.

Verification

At the verification stage, the template from the claimant fingerprint is compared against that of the enrollee fingerprint. This is done usually by comparing neighborhoods of nearby minutiae for similarity. A single neighborhood may consist of three or more nearby minutiae. Each of these is located at a certain distance and relative orientation from each other. Furthermore, each minutia has its own attributes of type (if it is used) and minutia direction, which are also compared. If comparison indicates only small differences between the neighborhood in the enrollee fingerprint and that in the claimant fingerprint, then these neighborhoods are said to match. This is done exhaustively for all combinations of neighborhoods and if enough similarities are found, then the fingerprints are said to match. Template matching can be visualized as graph matching, that is comparing the shapes of graphs joining fingerprint minutiae. This is illustrated in Figure 2.3(f).

Note that the word, "similar" is used in the paragraph above instead of "same". Neighborhoods will rarely match exactly because of two factors. One is the noisy nature of a fingerprint image. The other is that the skin is an elastic surface, so distances and minutia directions will vary.

One result of the verification stage is a match score, usually a number between 0 and 1 (or 10 or 100). Higher values in the range indicate higher confidence in a match. This match score is then subject to a user-chosen threshold value. If the score is greater than the threshold, the match result is said to be true (or 1) indicating a correct verification, otherwise the match is rejected and the match result is false (or 0). This threshold can be chosen to be higher to achieve greater confidence in a match result, but the price to pay for this is a greater number of false rejections. Conversely, the threshold can be chosen lower to reduce the number of false rejections, but the price to pay in this case is a greater number of false acceptances. The trade-off between false acceptance and false rejection rates is further discussed in Section 7.

The user has control of only one parameter, the threshold, for most commercial verification products. This customization procedure is called *back-end adjustment*, because a match score is calculated first and a threshold can be chosen after to determine the match result. There are systems that, in addition to offering back-end adjustment, offer *front-end adjustment* as well. This enables the user to adjust some of the parameter values before the match score is calculated, then to adjust the threshold after. Systems with front-end adjustment offer more versatility in obtaining the best results for different conditions, but are more complex for the user to adjust. This is why, for most systems, the vendor sets the optimum front-end parameter values and the user has control only of the matching threshold value via back-end adjustment.

Identification and One-to-Few Matching

Although the emphasis in this chapter is verification, we briefly mention identification and one-to-few matching methods. For identification, the objective is to determine a match between a test fingerprint and one of a database of fingerprints whose size may be as high as 10,000 to tens of millions. One cannot simply apply the verification techniques just described to all potential matches because of the prohibitive computation time required. Therefore, identification is usually accomplished as a two-step process. Fingerprints in the database are first categorized by pattern type, or *binned*. The same is done for the test fingerprint. Pattern comparison is done between test fingerprint and database fingerprints. This is a fast process that can be used to eliminate the bulk of non-matches. For those fingerprints that closely match in pattern, the more time-consuming process of minutia-based verification is performed.

One-to-few matching is usually accomplished simply by performing multiple verifications of a single claimant fingerprint against the 5 to 20 potential matches. Thus the execution time is linear in the number of potential matches. This time requirement becomes prohibitive if "few" becomes too large, then an approach akin to identification must be used.

Variations on the Common Approach: Other Methods

Since one of the most vexing challenges of fingerprint processing is obtaining a clean image upon which to perform matching, there are various methods proposed to perform image enhancement. Most of these involve filtering that is adaptively matched to the local ridge orientations [23, 19, 25, 22, 24, 37, 26, 27, 14]. The orientation map is first determined by dividing the image into windows (smaller regions) and calculating the local ridge orientations within these. The orientation can be determined in each window by spatial domain processing or by frequency domain processing after transformation by a 2-dimensional fast Fourier transform.

After image enhancement and binarization of the fingerprint image, thinning is usually performed on the ridges. However, a different approach eliminates the binarization and thinning stages (both computationally expensive and noise producing) [20]. This approach involves tracing ridges not from the binary or thinned image, but from the original grayscale image. The result of grayscale ridge-following is the endpoint and bifurcation minutiae similar to the common approach.

Instead of using only a single size window to determine the orientation map, multiple window sizes can be used via a multi-resolution approach [24, 15]. Local orientation values are determined first throughout the image at a chosen, initial resolution level – that is a chosen window size of pixels within which the orientation is calculated. A measure of consistency of the orientation in each window is calculated. If the consistency is less than a threshold, the window is divided into four smaller sub-windows and the same process is repeated until consistency is above threshold for each window or sub-window. This multi-resolution process is performed to avoid smoothing over small areas of local orientation, as will be the case especially at the fingerprint core.

Because of the difficulty of aligning minutiae of two fingerprints, neighborhood matching was one of the earliest methods of facilitating a match [28, 1, 42]. Groups of neighboring minutiae are identified in one fingerprint, usually two to four minutiae to a neighborhood, and each of these is compared against prospective neighborhoods of another fingerprint. There are two levels to matching. One is matching the configurations of minutiae within a neighborhood against another neighborhood. The other is matching the global configurations formed by the separate neighborhoods between enroll and verify fingerprints.

Because it is time-consuming to compare all neighborhood combinations between enroll and verify fingerprints, methods have been proposed to align the fingerprints to reduce the number of comparisons. A common method, and also a traditional method used for visual matching, is to locate a core and delta and align the fingerprints based on these landmarks [29]. The core and delta are usually found on the basis of their position with respect to the ridge flow, therefore the orientation map is determined and used for this [41]. An elegant method to locate singular points in a flow field is the Poincaré index [17, 36, 16]. For each point in the orientation map, the orientation angles are summed for a closed curve in a counter-clockwise direction around that point. For non-singular points, the sum is equal to 0 degrees; for the core, the sum is equal to 180 degrees; for a delta, the sum is equal to -180 degrees.

Other methods have been proposed to reduce the computational load of minutia matching. One approach is to sort the list of minutiae in some order conducive to efficient comparisons prior to matching. (This is especially appropriate for one-to-many matching, since sorting is done once per fingerprint, but matching many times.) A linearly sorted list of minutiae can be compiled by scanning the fingerprint from a selected center point outward by a predetermined scanning trajectory such as a spiral [39]. In this way, one-dimensional vectors of minutiae, including their characteristics, can be compared between enroll and verify fingerprints. Another method to linearize the minutia comparison is the "hyperladder" matcher [11]. This hyperladder is constructed sequentially by comparing minutia pairs in enroll and verify fingerprints and adding more rungs as consecutive neighboring minutiae match. In another approach, an attributed graph can be constructed where branches constitute nearest-neighbor minutiae and these emanate like "stars" on the graph [10]. These stars are compared between fingerprint pairs, the number of matching branches constituting the degree of confidence in the match.

Because there is so little discriminating information at a single minutia (even the type is unreliable), a different approach is to describe minutiae by more features [47, 40]. For instance, a minutia can be described by the length and curvature of the ridge it is on and of similar features on neighboring ridges.

Variations on the Common Approach: Correlation Matching

This discussion of matching has been minutia-focused to this point, to the exclusion of the global pattern matching approach mentioned in Section 4. Instead of using minutiae, some systems perform matches on the basis of the overall ridge pattern of the fingerprint. This is called *global matching, correlation,* or simply *image multiplication* or *image subtraction.*

It is visibly apparent that a pair of fingerprints of different pattern types, for instance whorl and arch, does not match. Global matching schemes go beyond the simple (and few) pattern categories to differentiate one whorl from a different whorl, for instance. Simplistically, this can be thought of as a process of aligning two fingerprints and subtracting them to see if the ridges correspond. There are four potential problems (corresponding to three degrees of freedom and another factor).

1. The fingerprints will likely have different locations in their respective images (translational freedom). We can establish a landmark such as a core or delta by which to register the pair, however if these are missing or not found reliably, subsequent matching steps will fail.

2. The fingerprints may have different orientations (rotational freedom). If a proper landmark has been found in (1), the fingerprint can be rotated around this, but this is error-prone, computationally expensive, or both. It is error-prone because the proper center of rotation depends on a single, reliably determined landmark. It is computationally expensive because performing correlation for many orientations involves repeatedly processing the full image.

3. Because of skin elasticity (non-linear warping), even if matching fingerprints are registered in location and orientation, all sub-regions may not align.

4. Finally, there is the inevitable problem of noise. Two images of matching fingerprints will have different image quality, ridges will be thicker or thinner, discontinuities in ridges will be different depending on finger dryness, the portion of the fingerprint captured in each image will be different, etc.

The descriptions below are more sophisticated modifications and extensions to the basic correlation approach to deal with the problems listed.

Strictly speaking, correlation between two images involves translating one image over another and performing multiplication of each corresponding pixel value at each translation increment [38]. When the images correspond at each pixel, the sum of these multiplications is higher than if they do not correspond. Therefore, a matching pair will have a higher correlation result than a non-matching pair. A threshold is chosen to determine whether a match is accepted, and this can be varied to adjust the false acceptance rate versus false rejection rate tradeoff similarly to the case for minutia matching.

Correlation matching can be performed in the spatial frequency domain instead of in the spatial domain as just described [12]. The first step is to perform a 2-dimensional fast Fourier transform (FFT) on both the enrollee and claimant images. This operation transforms the images to the spatial frequency domain. The two transformed images are multiplied pixel-by-pixel, and the sum of these multiplications is equivalent to the spatial domain correlation result. An advantage of performing frequency domain transformation is that the fingerprints become translation-independent; that is, they do not have to be aligned translationally because the origin of both transformed images is the zero-frequency location, (0,0). There is a trade-off to this advantage however, that is the cost of performing the 2-dimensional FFT.

Frequency domain correlation matching can be performed optically instead of digitally [43, 44, 21]. This is done using lenses and a laser light source. Consider that a glass prism separates projected light into a color spectrum, that is it performs frequency transformation. In a similar manner, the enrollee and claimant images are projected via laser light through a lens to produce their Fourier transform. Their superposition leads to a correlation peak whose magnitude is high for a matching pair and lower otherwise. An advantage of optical signal processing is that operations occur at the speed of light, much more quickly than for a digital processor. However, the optical processor is not as versatile – as programmable – as a digital computer, and because of this few or no optical computers are used in commercial personal verification systems today.

One modification of spatial correlation is to perform the operation not upon image pixels but on grids of pixels or on local features determined within these grids [8, 6]. The enrollee and claimant fingerprint images are first aligned, then (conceptually) segmented by a grid. Ridge attributes are determined in each grid square: average pixel intensity, ridge orientation, periodicity, or number of ridges per grid. Corresponding grid squares are compared for similar attributes. If enough of these are similar, then this yields a high match score and the fingerprints are said to match.

The relative advantages and disadvantages between minutia matching and correlation matching differ between systems and algorithmic approaches. In general, minutia matching is considered by most to have a higher recognition accuracy. Correlation can be performed on some systems more quickly than minutia matching, especially on systems with vector-processing or FFT hardware. Correlation matching is less tolerant to elastic, rotational, and translational variances of the fingerprint and of extra noise in the image.

6. Systems Issues

The effectiveness of a complete fingerprint verification system depends on more than the verification algorithms just described. There are other, higher level considerations, which we will call systems issues. These include enrollment and verification procedures, speed and ergonomics, user-feedback, anti-spoofing, and security considerations.

It is essential to the goal of high recognition rate that the enrollment procedure results in the capture of the highest quality fingerprint image(s) obtainable because enrollment occurs once while verification occurs many times. Therefore, a well-designed verification system will require the user to go through more time and effort for enrollment than for verification. A fingerprint may be captured multiple times and the best taken or some combination of each taken as the enrolled fingerprint.

There are options in the design of the verification procedure as well. The fingerprint can be captured once or a few times until a positive match is made. A procedure such as this will decrease false rejections, but increase false acceptances. Verification can be performed on not just one, but two or more fingers. This will enhance the recognition rate, however it will also cause the user to expend more time.

System ergonomics are important. For instance, there are limits to the amount of time that a person is willing to wait in personal authentication applications. That amount of time varies with the particular application and depends on what the person is also doing during processing, for instance swiping a bankcard or entering an identification number. Between 0.5 and 1 second are usually regarded as an acceptable range for processing time. Other user ergonomics considerations include: the number of repeated attempts in case of false rejections, the procedures for enrollment and verification, the design of the capture device, and the recognition setting that determines the trade-off between false acceptance and false rejection.

Quality feedback is useful when an image is captured to indicate to the user how to place the finger for the best possible image quality. The type of feedback includes: "finger is placed too high", "finger is not pressed hard enough", etc.

Anti-spoofing deterrents must be built into a fingerprint system to prevent use of an artificial fingerprint, a dead finger, or latent fingerprint. A latent fingerprint sometimes remains on a sensor surface due to skin oil residue from the previously applied fingerprint. Countermeasures are built into some sensors, such as the ability to distinguish true skin temperature, resistance, or capacitance.

Since the fingerprint system is only as secure as its weakest link, a complete, secure system must be designed. For instance, minutia templates must be secured by some means such as encryption to prevent impostors from inserting their templates into the database in place of properly enrolled users. The end result of fingerprint verification is a "yes" or "no" that is used to gain access. If it is simple just to circumvent the fingerprint system to send a "yes", then the system provides little security. A solution to this problem is to ensure that the host receiving the recognition decision knows that this is from the trusted client, such as by digitally signing the information passed to the host. (For further information on encryption, see reference [33].)

7. Recognition Rate

Terminology and Measurement

The ultimate measure of utility of a fingerprint system for a particular application is recognition rate. This can be described by two values. The *false acceptance rate (FAR)* is the ratio of the number of instances of pairs of different fingerprints found to (erroneously) match to the total number of match attempts. The *false rejection rate (FRR)* is the ratio of the number of instances of pairs of the same fingerprint are found not to match to the total number of match attempts. FAR and FRR trade off against one another. That is, a system can usually be adjusted to vary these two results for the particular application, however decreasing one increases the other and vice versa. FAR is also called, *false match rate* or *Type II error*, and FRR is also called *false non-match rate* or *Type I error*. These are expressed as values in [0, 1] interval or as percentage values.

The ROC-curve plots FAR versus FRR for a system. (ROC stands for Receiver Operating Curve for historical reasons. Yes, "ROC-curve" is redundant, but this is the common usage.) ROC-curves are shown in Figure 2.4. The FAR is usually plotted on the horizontal axis as the independent variable. The FRR is plotted on the vertical axis as the dependent variable. Because of the range of FAR values, this axis is often on a logarithmic scale. Figure 2.4 contains two solid curves and three dotted curves. The solid curves do not represent any particular data; they are included for illustrative purposes to show better and worse curve placements. The typical ROC-curve has a shape whose "elbow" points toward (0,0) and whose asymptotes are the positive *x*- and *y*-axes. The sharper the elbow and (equivalently) the closer is the ROC-curve to the *x*- and *y*-axes, the lower is the recognition error and the more desirable is the result.

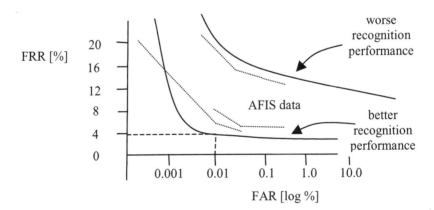

Figure 2.4 ROC-curves. The 2 solid curves are of hypothetical data illustrating desirable and less desirable recognition performance. The 3 dotted curves are of real data measuring the performance of 3 commercial AFIS [46].

The procedure for using the ROC-curve is as follows. Choose an acceptable level of FAR. On Figure 2.4, a dashed line is shown at 0.01% FAR. The FRR corresponding to this choice is the attainable FRR, in this example about 4%. Alternatively, the FRR can be specified and the FAR found on the curve.

There is no single set of FAR and FRR specifications useful for all different applications. If the fingerprint system is specified for very high security situations such as for military installations, then the FAR will be chosen to be very low (e.g., <0.001%). However, this results in higher FRR, sometimes in the range from 5% to 20%. Typical customer applications such as for automatic teller machines cannot afford to alienate users with such a high FRR. Therefore, the choice in these applications is low FRR (e.g., <0.5%), at the sacrifice of higher FAR. (An FRR specification that is sometimes quoted for automatic teller machines is less than 1 per 100,000 false rejection.)

Third-Party Benchmarking

In Figure 2.4, we include three ROC-curves of AFIS data from third-party benchmarking [46]. The database was compiled from employees of the Philippine Social Security System, mostly white-collar workers. The database consists of 600 people, 8 fingers per person, and two sets per person, where enrollment and verification sets were captured with an intervening interval of 2 to 8 weeks. From this database, 3278 matching fingerprint pairs and 4129 non-matching pairs were tested. These images were captured with an Identicator DF-90 optical scanner at 500dpi, 512x512 pixels, 1x1" image size.

We include these AFIS data from a respected third-party tester for the reader to compare against other data whose validity may be suspect. There is much misleading information in the commercial biometric industry regarding recognition rates. In general, these AFIS can be expected to yield better recognition results than most verification systems (though AFIS generally have higher cost, they are slower for 1-to-1 matching, and require more memory). Note that this AFIS test is only for single image comparisons. A verification system can take advantage of the real-time nature of its application to perform multiple verification attempts·so as to improve the recognition rate.

Specifying and Evaluating Recognition Rate Statistics

For statistical results to be properly evaluated, they must be accompanied by the following information: sample size, description of population, and testing description. The *sample size* should contain the following information: the number of subjects (people), the number of fingers, the number of images per finger, and the total number of fingerprint images. In addition, the image number should be broken out into number of match and non-match images. For example, a test might consist of 100 subjects, 2 fingers per subject and 4 images per finger. The total number of images is 100x2x4=800. If each finger (200) is compared against each of the other images from the same finger (4 choose 2 = 6 pairs per unique finger), there are 200x6=1200 matching pairs. If one image from each finger (200) is compared against all images from different subjects (99x4x2=792), there are 200x792=158,400 non-match pairs.

The *description of population* states the type of subjects included in the sample. Of particular importance in judging fingerprint statistics is the type of work engaged in by the subjects. A study involving masons will have different statistical results than that involving white-collar workers whose hands are subject to less abuse. The age statistics should be described, at least stating a relative breakdown on the number of children, adults, and elderly people included in the sample. The proportion of males and females should also be stated.

Finally, the *test design* should be described. Of particular interest is who performed the tests. The strong preference is that a reputable third-party conducts and reports the test. Was the capture procedure supervised or not? Were the subjects given training or visual feedback to place the finger correctly on the fingerprint capture device? Was the sample manually filtered in any way to remove "goats" (people whose fingerprints are very difficult to capture and match with reliable quality)? Was the procedure adjusted using a practice sample of fingerprints, then tested separately on different images to yield the published results? What was the range of rotational and translational variance allowed, or were the fingerprints manually centered in the image? What were the make and specifications of the capture device? Where and when were the tests conducted (e.g., Florida humid summer or Minnesota dry winter)? What components of the system were involved in the test: just matching algorithms, just sensor, full system? Most test results do not list all these conditions, but the most possible information enables more valid evaluation.

It is important to emphasize that results cannot be compared if determined under different test conditions. It is a misrepresentation of test data to state that a matcher achieved certain results for test design A, so it can be compared against the results from test design B. Valid comparisons between results can be done only for the same database under the same conditions.

8. Image Capture Devices

We organize image capture devices into three categories: optical, solid-state, and other. There is yet another category, fingerprint acquisition via inking, which is the traditional mode of criminal fingerprint capture. It is evident that this is inappropriate for fingerprint verification due to the inconvenience involved with ink, the need for subsequent digitization, and perhaps the stigma of this type of capture. The type of image acquisition for fingerprint verification is also called "live-scan fingerprint capture".

Optical fingerprint capture devices have the longest history and use of these categories, dating back to the 1970s. These operate on the principal of frustrated total internal reflection (FTIR). A laser light illuminates a fingerprint placed on a glass surface (platen). The reflectance of this light is captured by a CCD array (solid-state camera). The amount of reflected light is dependent upon the depth of ridges and valleys on the glass and the finger oils between the skin and glass. The light that passes through the glass into valleys is not reflected to the CCD array, whereas light that is incident upon ridges on the surface of the glass (more precisely, the finger oils on the ridges that constitute the ridge-to-glass seal) is reflected.

Innovations in optical devices have been made recently, primarily in an effort to reduce the size of these devices. Whereas an optical sensor was housed in a box about 6x3x6 inches as recently as the mid-1990s, smaller devices have recently appeared that are in the order of 3x1x1 inches. Different optical technologies than FTIR have also been developed. For instance, fiber optics has been proposed to capture the fingerprint [7]. A bundle of optical fibers is aimed perpendicularly to the fingerprint surface. These illuminate the fingerprint and detect reflection from it to construct the image. Another proposal is a surface containing an array of microprisms mounted upon an elastic surface [4]. When a fingerprint is applied to the surface, the different ridge and valley pressures alter the planar surfaces of the microprisms. This image is captured optically via the reflected light (or absence of it) from the microprisms.

Solid-state sensors have appeared on the marketplace recently, though they have been proposed in the patent literature for almost two decades. These are microchips containing a surface that images the fingerprint via one of several technologies. Capacitive sensors have been designed to capture the fingerprint via electrical measurements [45, 18, 48, 13]. Capacitive devices incorporate a sensing surface composed of an array of about 100,000 conductive plates over which is a dielectric surface. When the user places a finger on this surface, the skin constitutes the other side of an array of capacitors. The measure of voltage at a capacitor drops off with the distance between plates, in this case the distance to a ridge (closer) or a valley (further). Pressure-sensitive surfaces have been proposed where the top layer is of an elastic, piezoelectric material to conform to the topographic relief of the fingerprint and convert this to an electronic signal [30, 31, 9, 34]. Temperature sensitive sensors have been designed to respond to the temperature differential between the ridges touching the surface of the device and the valleys more distant from them [9].

Ultrasonic scanning falls into the final category of fingerprint capture technologies [32]. An ultrasonic beam is scanned across the fingerprint surface much like laser light for optical scanners. In this case, it is the echo signal that is captured at the receiver, which measures range, thus ridge depth. Ultrasonic imaging is less affected by dirt and skin oil accumulation than is the case for optical scanning, thus the image can be a truer representation of the actual ridge topography.

Two of the three most important factors that will decide when fingerprint verification will be commercially successful in the large-volume personal verification market are low cost and compact size. (The other factor is recognition rate, discussed in Section 7.) Capture device prices have fallen over an order of magnitude between the early to late 1990s (from approximately $1500 (US) to $100), and manufacturers promise close to another order of magnitude decrease in the next few years. As far as size, we have mentioned the reduction of optical sensor size from 6x3x6 inches to 3x1x1. Solid-state sensor systems are this size or smaller, and as further integration places more circuitry on the chip (such as digitizer circuitry to convert the fingerprint measurements to digital intensities), these systems are becoming even smaller. Solid-state sensors are approaching the lower limit of size needed to capture the surface area of the finger, about 1x1 inch with a fraction of an inch depth.

A functionality that has not been available before solid-state sensors is locally adjustable, software-controlled, automatic gain control (AGC). For most optical devices, gain can be adjusted only manually to change the image quality. Some solid-state sensors, however, offer the capability to automatically adjust the sensitivity of a pixel or row or local area to provide added control of image quality. AGC can be combined with feedback to produce high quality images over different conditions. For instance, a low-contrast image (e.g., dry finger) can be sensed and the sensitivity increased to produce an image of higher contrast on a second capture. With the capability to perform local adjustment, a low-contrast region in the fingerprint image can be detected (e.g., where the finger is pressed with little pressure) and sensitivity increased for those pixel sensors on a second capture.

Optical scanners also have advantages. One advantage of larger models is in image capture size. It is costly to manufacture a large, solid-state sensor, so most current solid-state products have sub-1 inch square image area, whereas optical scanners can be 1 inch or above. However, this advantage is not true for some of the smaller optical scanners. The small optical scanners also have smaller image capture areas because a larger area would require a longer focal length, thus larger package size. Optical scanners are subject to linear distortion at the image edges when larger image capture area is combined with smaller package size.

9. Multi-Modal Biometrics

Multi-modal biometrics refers to the combination of two or more biometric modalities into a single system. The most compelling reason to combine different modalities is to improve recognition rate. This can be done when features of different biometrics are statistically independent. For the different modalities listed in Table 2.1, it is likely that each is largely independent from the other (though we know of no research study to date that confirms this).

There are other reasons to combine biometrics. One is that different modalities are more appropriate in different situations. For a home banking application for instance, a customer might enroll both with fingerprint and voice. Then, the fingerprint can be used from a home or laptop sensor; while voice and a PIN (personal identification number) can be used over the phone. Another reason is simply customer preference. For instance, an automatic teller machine could offer eye and fingerprint and face biometrics, or a combination of two of these for the customer to choose.

Although fingerprints can be combined with other modalities, there are reasons to suggest that this would not be the first biometric to require complementing. One reason is that, along with eye systems, fingerprint systems already have very high recognition rates. This contrasts with less reliable modalities where combining one with another or with a PIN is more advantageous. Another reason is that a single person has up to ten statistically independent samples in ten fingers, compared to two for eye and hand, and one for face, voice, and signature.

Table 2.1 shows selected features of each modality and can be used to determine complementary modalities for multi-modal systems. A few notes on this table:

☐   Biometric technologies are changing rapidly; for the most up-to-date information, check company literature and industry reports such as at reference [3] and review issues such as [35, 5].

☐   The row for the eye biometric describes features applying to either iris or retinal scanning technologies.

☐   In the matching column, whereas all technologies are appropriate for 1-to-1 matching, only fingerprint and eye technologies are proven to have acceptable recognition rates to be practical for 1-to-many matching. This is an indication that these two modalities provide the highest recognition rates for verification as well.

☐   Variation of the salient features used for recognition is very different for different modalities. Fingerprint and eye features remain consistent for a lifetime, whereas the others change with growth. On a day-to-day basis, there is far less variation for all modalities, though voice can change with illness and signature with demeanor.

☐   As far as sensor cost, eye systems are currently more costly than the others; voice systems can be zero cost to the user if a telephone is used.

☐   Fingerprint and voice systems have the smallest comparative sizes with eye systems currently the largest.

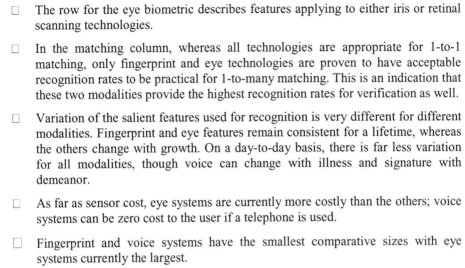

Biometric	Matching 1-to-1, 1-to-many	Variation: Lifetime, Day-to-Day	Maximum Independent Samples per Person	Sensor Cost [$US]	Sensor Size
fingerprint	yes, yes	none, little	10	$10-10^2$	very small
eye	yes, yes	none, very little	2	10^2-10^3	medium
hand	yes, no	much, very little	2	10^2	medium
face	yes, no	much, medium	1	10^2	small
voice	yes, no	much, medium	1	$0-10^2$	very small
signature	yes, no	much, medium	1	10^2	medium

Table 2.1 Features of different biometric modalities.

10. Future

Where is biometric technology going? System price will continue to decrease along with size, while recognition rates will improve (at a slower rate than price and size changes). Recognition rate will be a deciding factor in acceptance for demanding applications such as automatic teller machines (requiring a very low rate of false rejections), and military (requiring a very low rate of false acceptances). For especially demanding applications, multi-modal systems will evolve to combine biometrics to provide an optimum level of security and convenience to users. Alternatively, multiple verifications, such as by using multiple fingers, will be used to enhance recognition reliability. If costs plummet as the industry projects, personal use of biometric systems will grow to replace the current reliance on passwords, PINs, and door keys that are used for computers, home security systems, restricted entry, ATMs, credit cards, Internet access, corporate networks, confidential databases, etc. The biometrics promise is to make access much simpler while at the same time providing a higher level of security.

References

[1] K. Asai, H. Izumisawa, H. Owada, S. Kinoshita, and S. Matsuno, "Method and Device for Matching Fingerprints with Precise Minutia Pairs Selected from Coarse Pairs," *US Patent 4646352*, 1987.

[2] J. Berry, "The history and development of fingerprinting," *in Advances in Fingerprint Technology*, (H. C. Lee and R. E. Gaensslen, ed.s), CRC Press, Florida, 1994, pp. 1-38, 1994.

[3] Biometric Consortium Web page: www.biometrics.org.

[4] W. S. Chen and C. L. Kuo, "Apparatus for Imaging Fingerprint or Topgraphic Relief Pattern on the Surface of an Object," *US Patent 5448649*, 1995.

[5] C. Ciechanowicz, *Special issue on biometric technologies*, Elsevier Information Security Technical Report, Vol. 3, No. 1, 1998.

[6] L. Coetzee and E. C. Botha, "Fingerprint Recognition in Low Quality Images," *Pattern Recognition*, Vol. 26, No. 10, pp. 1441-1460, 1993.

[7] R. F. Dowling Jr. and K. L. Knowlton, "Fingerprint Acquisition System With a Fiber Optic Block," *US Patent 4785171*, 1988.

[8] E. C. Driscoll Jr., C. O. Martin, K. Ruby, J. J. Russell, and J. G. Watson, "Method and Apparatus for Verifying Identity Using Image Correlation," *US Patent 5067162*, 1991.

[9] D. G. Edwards, "Fingerprint Sensor," *US Patent 4429413*, 1984.

[10] M. A. Eshera and R. E. Sanders, "Fingerprint Matching System," *US Patent 5613014*, 1997.

[11] S. Ferris, R. L. Powers, and T. Lindh, "Hyperladder Fingerprint Matcher," *US Patent 5631972*, 1997.

[12] R. C. Gonzalez and Richard E. Woods, *Digital Image Processing*, Addison-Wesley, Massachusetts, 1992.

[13] D. Inglis, L. Manchanda, R. Comizzoli, A. Dickinson, E. Martin, S. Mendis, P. Silverman, G. Weber, B. Ackland, and L. O'Gorman, ``A robust, 1.8V, 250 microWatt, direct contact 500dpi fingerprint sensor," *IEEE Solid State Circuits Conference*, San Francisco, 1998.

[14] A. K. Jain, L. Hong, and R. Bolle, "On-line fingerprint verification," *IEEE Trans. Pattern Analysis and Machine Intelligence*, Vol. 19, No. 4, pp. 302-313, 1997.

[15] A. K. Jain, L. Hong, S. Pankanti, and R. Bolle, "An Identity-Authentication System Using Fingerprints," *Proceedings of the IEEE*, Vol. 85, No. 9, pp. 1365-1388, 1997.

[16] K. Karu and A. K. Jain, "Fingerprint Classification," *Pattern Recognition*, Vol. 29, No. 3, pp. 389-404, 1996.

[17] M. Kawagoe and A. Tojo, "Fingerprint Pattern Classification," *Pattern Recognition*, Vol. 17, pp. 295-303, 1984.

[18] A. G. Knapp, "Fingerprint Sensing Device and Recognition System Having Predetermined Electrode Activation," *US Patent 5325442*, 1994.

[19] H. E. Knutsson, R. Wilson, and G. H. Granlund, "Anisotropic Nonstationary Image Estimation and its Applications: Part I – Restoration of Noisy Images," *IEEE Trans. Communications*, Vol. 31, pp. 388-397, 1983.

[20] D. Maio and D. Maltoni, "Direct Gray-Scale Minutiae Detection in Fingerprints," *IEEE Trans. Pattern Analysis and Machine Intelligence*, Vol. 19, No. 1, pp. 27-40, 1997.

[21] R. A. Marsh and George S. Petty, "Optical Fingerprint Correlator," *US Patent 5050220*, 1991.

[22] B. M. Mehtre, N. N. Murthy, and S. Kapoor, "Segmentation of Fingerprint Images Using the Directional Image," *Pattern Recognition*, Vol. 20, No. 4, pp. 429-435, 1987.

[23] O. Nakamura, K. Goto, and T. Minami, "Fingerprint Classification by Directional Distribution Patterns," *Systems, Computers, and Controls*, Vol. 13, pp. 81-89, 1982.

[24] L. O'Gorman and J. V. Nickerson, "An approach to fingerprint filter design", *Pattern Recognition*, Vol. 22, No. 1, pp. 29-38, 1989.

[25] E. Peli, "Adaptive Enhancement Based on a Visual Model," *Optical Engineering*, Vol. 26, No. 7, pp. 655-660, 1987.

[26] N. K. Ratha, S. Chen, and A. K. Jain, "Adaptive Flow Orientation-Based Feature Extraction in Fingerprint Images," *Pattern Recognition*, Vol. 28, No. 11, pp. 1657-1672, 1995.

[27] N. K. Ratha, K. Karu, S. Chen, and A. K. Jain, "A Real-Time Matching System for Large Fingerprint Databases", *IEEE Trans. Pattern Analysis and Machine Intelligence*, Vol. 18, No. 8, pp. 799-813, 1996.

[28] J. P. Riganati and V. A. Vitols, "Minutiae Pattern Matcher," *US Patent 4135147*, 1979.

[29] J. P. Riganati and V. A. Vitols, "Automatic Pattern Processing System," *US Patent 4151512*, 1979.

[30] H. Ruell, "Input Sensor Unit for a Fingerprint Identification System," *US Patent 4340300*, 1982.

[31] H. Ruell, "Fingerprint Sensor," *US Patent 4394773*, 1983.

[32] J. K. Schneider and W. E. Glenn, "Surface Feature Mapping Using High Resolution C-span Ultrasonography," *US Patent 5587533*, 1996.

[33] B. Schneier, *Applied Cryptography*, John Wiley and Sons, Inc., New York, 1996.

[34] T. Scheiter, M. Biebl, and H. Klose, "Sensor for Sensing Fingerprints and Method for Producing the Sensor," *US Patent 5373181*, 1994.

[35] W. Shen and R. Khanna (eds.), "Special issue on automated biometrics," *Proceedings of the IEEE*, Vol. 85, No. 9, Sept., pp. 1343-1492, 1997.

[36] B. G. Sherlock and D. M. Munro, "A Model for Interpreting Fingerprint Topology," *Pattern Recognition*, Vol. 26, No. 7, pp. 1047-1055, 1993.

[37] B. G. Sherlock, D. M. Munro, and K. Millard, "Algorithm for Enhancing Fingerprint Images," *Electronics Letters*, Vol. 28, No. 18, pp. 1720-1721, 1992.

[38] A. Sibbald, "Method and Apparatus for Fingerprint Characterization and Recognition Using Auto-correlation Pattern," *US Patent 5633947*, 1997.

[39] M. K. Sparrow, "Fingerprint Recognition and Retrieval System," *US Patent 4747147*, 1988.

[40] M. K. Sparrow, "Vector Based Topological Fingerprint Matching," *US Patent 5631971*, 1997.

[41] V. S. Srinivasan and N. N. Murthy, "Detection of Singular Points in Fingerprint Images," *Pattern Recognition*, Vol. 25, No. 2, pp. 139-153, 1992.

[42] K. E. Taylor and J. B. Glickman, "Apparatus and Method for Matching Image Characteristics Such as Fingerprint Minutiae," *US Patent 4896363*, 1990.

[43] C. E. Thomas, "Method and Apparatus for personal Identification," *US Patent 3704949*, 1972.

[44] G. J. Tomko, "Method and Apparatus for Fingerprint Verification," *US Patent 4876725*, 1989.

[45] C. Tsikos, "Capacitive Fingerprint Sensor," *US Patent 4353056*, 1982.

[46] J. L. Wayman, "Biometric Identification Standards Research," *Report to U.S. Federal Highway Administration (FHWA)*, San Jose State University, December 1997.

[47] M. Yamada, A. Kodata, and H. Tominaga, "A Method of Describing Fingerprint Structure and Identification Algorithm Using Geometric Characteristics," *Systems and Computers in Japan*, Vol. 25, No. 5, pp. 100-112, 1994.

[48] N. D. Young, G. Harkin, R. M. Bunn, D. J. McCulloch, R. W. Wilks, and A. K. Knapp, "Novel fingerprint scanning arrays using polysilicon TFT's on glass and polymer substrates," *IEEE Electronic Device Letters*, Vol. 18, No. 1, pp. 19-20, 1997.

3 FACE RECOGNITION

John J. Weng
Michigan State University
East Lansing, MI
weng@cse.msu.edu

Daniel L. Swets
Augustana College
Sioux Falls, SD
swets@inst.augie.edu

Abstract *Identifying an individual from his or her face is one of the most nonintrusive modalities in biometrics. However, it is also one of the most challenging ones. This chapter discusses why it is challenging and the factors that a practitioner can take advantage of in developing a practical face recognition system. Some of the well known approaches are discussed along with some algorithmic considerations. A face recognition algorithm is presented as an example with some experimental data. Some possible future research directions are outlined at the end of the chapter.*

Keywords: *Face recognition, face detection, appearance-based methods, principal component analysis, linear discriminant analysis, recursive partition trees, incremental learning.*

1. Introduction

Face recognition from images is a sub-area of the general object recognition problem. It is of particular interest in a wide variety of applications. Applications in law enforcement for mugshot identification, verification for personal identification such as driver's licenses and credit cards, gateways to limited access areas, surveillance of crowd behavior are all potential applications of a successful face recognition system.

The environment surrounding a face recognition application can cover a wide spectrum – from a well controlled environment to an uncontrolled one. In a controlled environment, frontal and profile photographs of human faces are taken, complete with a uniform background and identical poses among the participants.

These face images are commonly called mug shots. Each mug shot can be manually or automatically cropped to extract a normalized subpart called a canonical face image, as shown in Fig. 3.1. In a canonical face image, the size and position of the face are normalized approximately to the predefined values and the background region is minimized. Face recognition techniques for canonical images have been successfully developed by many face recognition systems.

Figure 3.1 A few examples of canonical frontal face images.

General face recognition, a task which is done by humans in daily activities, comes from a virtually uncontrolled environment. Systems to automatically recognize faces from uncontrolled environment must first detect faces in sensed images. A scene may or may not contain a set of faces; if it does, their locations and sizes in the image must be estimated before recognition can take place by a system that can recognize only canonical faces. A face detection task is to report the location, and typically also the size, of all the faces from a given image. Fig. 3.2 gives an example of an image which contains a number of faces. From this figure, we can see that recognition of human faces from an uncontrolled environment is a very complex problem: more than one face may appear in an image; lighting condition may vary tremendously; facial expressions also vary from time to time; faces may appear at different scales, positions and orientations; facial hair, make-up and turbans all obscure facial features which may be useful in localizing and recognizing faces; and a face can be partially occluded. Further, depending on the application, handling facial features over time (e.g., aging) may also be required.

Given a face image to be recognized, the number of individuals to be matched against is an important issue. This brings up the notion of face recognition versus verification: given a face image, a recognition system must provide the correct label (e.g., name label) associated with that face from all the individuals in its database. A face verification system just decides if an input face image is associated with a given face image.

Since face recognition in a general setting is very difficult, an application system typically restricts one of many aspects, including the environment in which the recognition system will take place (fixed location, fixed lighting, uniform background, single face, etc.), the allowable face change (neutral expression, negligible aging, etc.), the number of individuals to be matched against, and the viewing condition (front view, no occlusion, etc.).

Figure 3.2 An image that contains a number of faces. The task of face detection is to determine the position and size (height and width) of a frame in which a face is canonical. Such a frame for a particular face is marked in the image.

2.   The Human Capacity for Face Recognition

Though we currently are unable to endow machines with the capability of the human visual system, it is a good reference point from which to start.   Much research has been done on face recognition, both by machine vision and biological system researchers.   Research issues of interest to neuroscientists and psychophysicists include the human capacity for face recognition [29], the modeling of this capability [35, 42], and the apparent modularity of face recognition [36]; the human facility for learning to recognize faces [7, 21, 11, 32]; the role of distinctive or unusual features in faces for recognition [49]; the degradation of face recognition capability as humans age [2, 30, 50]; and conditions which result in the human inability to recognize faces, such as prosopagnosia [46].

There is evidence to suggest that the human capacity for face recognition is a dedicated process, not merely an application of the general object recognition process [10].   This may have encouraged the views that artificial face recognition systems should also be face-specific.   The issue that which features humans use for face recognition has been subject to much debate and the result of the related studies has been used in the algorithm design of some face recognition systems. Apparently, in human, both global and local features are used in a hierarchical manner, the local features providing a finer classification system for facial recognition [22].   The most difficult faces for humans to recognize are those faces which are considered neither "attractive" nor "unattractive" by the observer.   This gives support to the theories suggesting that distinctive faces are more easily recognized than typical ones [1]. Information contained in low spatial frequency bands is used in order to make the determination of the sex of the individual, while the higher frequency components are used in recognition [40].   Young children typically recognize unfamiliar faces using unrelated cues, such as glasses, clothes, hats, and hair style.   By age twelve, these paraphernalia are usually reliably ignored [8]. Psychosocial conditions also affect the ability to recognize faces. Humans may encode an "average" face; these averages may be different for different races, and recognition may suffer from prejudice or unfamiliarity with the class of faces from another race [20] or gender [19]. "Thatcher's illusion" [47] demonstrates that facial expression is very difficult to recognize if the face is presented at a rarely seen orientation (upside down).

Some recent studies on neural network models in psychology and neuroscience have put innateness into a new perspective [12]. The neural processing modules responsible for face recognition result from extensive interactions between nature (genes) and nurture (learning through experience). Such interactions span the entire life of a human individual. That is why children and adults recognize faces differently, as mentioned above. In terms of the learning mechanism, the face recognition and general object recognition may have a lot in common, although the resulting recognition processes can be very different in terms of what features each uses. This view is related to the appearance based approach discussed in the following section.

The emulation of the human capacity for face recognition is the goal espoused by many computer vision researchers in face recognition. However, systems which are designed for specific environments are very useful for the intended application, even though they do not have the general face recognition capacity of a human being.

3. Approaches

A wide variety of approaches to machine recognition of faces has been published in the literature. Categorization of the approaches may depend on different criteria. In terms of the sensing modality, a system can take 2-D intensity images, color images, infra-red images, 3-D range images, or a combination of them. In terms of viewing angle, a system may be designed for frontal views, profile views, general views, or a combination of them. In terms of temporal component, a system can be designed for a static image or for time-varying image sequences (which may facilitate face segmentation, face tracking, expression identification, and other use of temporal context). In terms of computational tools used, a system can use programmed knowledge rules, statistical decision rules, neural networks, genetic algorithms, etc.

The reader is referred to two excellent surveys for face recognition research. One is the survey written by Samal and Iyengar for research prior to 1991 [39]. The other was authored by Chellappa, Wilson and Sirohey [9] which surveyed research on face recognition up to 1994. A good source for more recent research on face recognition is the series of Proceedings of International Conferences on Automatic Face and Gesture Recognition [14, 15, 16].

The last 10 years have seen very active research on face recognition. Accompanying the increase in research activities is a wave of commercialization for face recognition technology. In a survey by *Biometric Technology Today* [5], a total of 25 commercially available facial systems from 13 companies were listed. The FERET program, a face recognition program administered by US Army Research Laboratory, provided, for the first time, a large face database and conducted a series of blind tests for many face recognition algorithms [34].

Many factors have contributed to the recent increase in these face recognition activities and the successes. One of the major reasons is attributed to a basic, but fundamental change in methodology. That is, manually defining features versus automatically deriving features using statistical methods.

Manually Defining Features

Traditionally, face recognition methods have relied on humans to define geometry-dependent features to be used for recognition. These feature values depend on the detection of geometric facial features, including items such as the distance and angles between geometric points such as eye corners, mouth extremities, nostrils and chin top. The features defined for face profiles (side views) typically include a set of characteristic points on the profile (such as the notch between the brow and the nose or the tip of the nose) and the angles between these points. For example, Kaya and Kobayashi [25] used Euclidean distances between manually identified points in the images to characterize the faces. Kanade [24] used the distances and angles between eye corners, ends of the mouth, nostrils, and top of the chin, but the location of those facial features were extracted automatically by a program. More resent work used a combination of distance and angle measurements with local intensity patches. For example, Campbell, *et al.* [6] utilized hair and cheek intensity values coupled with subimage patches for eye regions to recognize faces.

Automatically Deriving Features

Manually defined features are intuitively understandable. However, methods based on this approach have run into basic problems. First, automatic detection of these features is not reliable due to various variations. Second, the number of features measurable is small. Third, the reliability of each feature measurement is difficult to estimate accurately. Thus, the subsequent classification method, even if itself is optimal, does not result in a reliable overall system. An important advance is brought about by neural networks which implicitly but automatically derive features.

Nonstatistical methods. Using a neural network, humans do not need to define facial features for face recognition. Kohonen [27] demonstrated the use of a self-organizing map for face recollection applications. Even when the input images were very noisy or had portions missing, an accurate recall capability was achieved on a small set of face images. Multilayer perceptron neural networks and radial basis function networks have also been used for face recognition. A back-propagation training algorithm for multi-layer perceptron may be sufficient for a low dimensionality feature vector with a small number of classes. For example, for face detection (a two-class problem) [37] used low resolution images and have successfully tested a multi-layer feedforward network with back-propagation training with momentum. More sophisticated and more powerful training methods (such as the statistical methods) have been used when the input dimension is high and the number of classes is large.

A different example of automatic feature derivation for face recognition is the Cresceptron [53] which was tested for general object segmentation and recognition, including faces. The method uses multilevel retinotopic layers of neurons to automatically determine the configuration of its network in the training phase. Unlike most neural networks, the Cresceptron does not use back propagation for learning. Instead, it analyzes the structure of an object in a bottom-up manner. Although this method allows incremental learning for general objects, the network grows quickly and suffers from speed and performance problems when the number of faces is large. The network structure can be predesigned and fixed if only faces need to be recognized, which allows effective size control. Lawrence et al. [28] used a 5-layer self-organization feature map for face recognition.

Although the neural network based methods can alleviate the problems with manually defined features, an efficient and scalable framework is required in order not only to use the information in the image as much as possible, but also use it efficiently.

Statistical methods. This type of approach originated from an image representation task. Kirby and Sirovich [26] treated a face image as a high dimensional vector, each pixel being mapped to a component in that vector. They used the Karhunen-Loève projection to the corresponding vector space for face image *characterization*. Although they did not use it originally for face recognition [26], their idea of representing the intensity image of a face by a linear combination of the principle component vectors can be used for recognition as well. Turk and Pentland used this technique for face recognition problem [48]. This statistical approach was extended later for 3-D object recognition [33]. Using this image vector representation, the linear discriminant analysis (LDA) has been independently used for face recognition by several research groups, including [13, 3, 51, 45, 44], among many other groups. It has been proposed that this type of approaches be called appearance-based approach, in order to distinguish other view-based approaches (e.g., aspect graph based). For distinguishing it from neural-network based approaches that use intensities directly, we call them appearance-based statistical approaches. Appearance-based statistical methods derive features directly from intensity images, using statistical techniques. They do not require humans to write explicit procedures to detect facial features, such as eyes, nose, and mouth.

It should be noted that neural network and statistical methods are not incompatible. In fact, a significant amount of recent research on neural networks uses statistical methods in combination with a network computational structure. For example, [43] used a combination of statistical measures and a multi-layer perceptron network for face detection. A major limitation of the appearance-based statistical methods is that they are not invariant to the position and size of the face. They require the input face images to be canonical. To deal with variation in the position and size of the faces in an input image, a pixel-based scan window has been used. The size of the window changes within an expected range. For each size, the scan window scans the input image by centering it at each pixel (or along a subsampled pixel grid for efficiency). Each position with each size of the scan window determines a subimage. Such a subimage is scaled to the standard input size for face recognition. Many appearance-based statistical methods use such a scan method to deal with position and size variation of face in a static input image. If an image sequence is available, motion information can be used to roughly locate a moving face [48].

Major Algorithmic Considerations

The appearance-based statistical methods are well understood and easy to implement. Various versions of this class of method have been implemented by many research groups and have been tested extensively in the blind FERET tests [34] with a large number of images. A large number of commercial systems are based on this class of algorithms. Therefore, in the remainder of this chapter, we concentrate on this class of methods and discuss some major algorithmic considerations when using this class of methods.

Face space. First, we describe a now well-used vector representation for an image: A digital image with r rows and c pixel columns can be denoted by a vector X in a d-dimensional space S, where $d = rc$. Suppose the pixel intensity at the i-th row and j-th column of the image is denoted by f_{ij}, $0 \leq i < r$ and $0 \leq j < c$. The vector X representing the image can be defined as a d-dimensional vector $X = (g_0, g_1, ..., g_{d-1})$, where $g_{ic+j} = f_{ij}$, $0 \leq i < r$ and $0 \leq j < c$. It is worth noting that this vector representation is just a notation change. It does not lose any 2-D neighborhood information among pixels. In the following, a face image is a canonical image unless stated otherwise.

Statistically, face images can be considered as random samples in the corresponding space S. From the sample covariation matrix Γ computed from all the training face images $G = \{X_1, X_2, ..., X_s\}$, the principal component analysis (PCA) [18] computes the basis vectors of a linear subspace S' ($S' \subseteq S$) that contains the centered face images $G_s = \{X_1 - \overline{X}, X_2 - \overline{X}, ..., X_S - \overline{X}\}$, where $\overline{X}$ is the sample mean vector of the face images in G. These basis vectors are the eigenvectors of Γ and are mutually orthogonal. We rank them according to the decreasing eigenvalues. The top k eigenvectors span a subspace S'' that minimizes mean squared errors between the original samples and their projections onto the subspace S'' [18, 23]. In this sense, they are suited for image reconstruction [26]. We call them the *Most Expressive Features* in contrast to the Most Discriminating Features described below[1]. Fig. 3.3 shows the mean face and the top 8 MEFs computed from images in the Weizmann face database.

Given any face image X, the Karhunen-Loève projection projects $X - \overline{X}$ onto the subspace S'' ($S'' \subseteq S'$) spanned by the top k MEFs. If k is large enough to include all the nonzero eigenvalues, S'' is the same as S'. However, for practical applications, the number k can be much smaller than the number of training images [48] due to the fact that the regions representing face images can be roughly contained in a relatively lower dimensional space S''. The number k is determined so that the sample variation represented by the sum of the eigenvalues of MEFs not retained, is small enough compared to the total sample variation represented by the sum of all the eigenvalues.

A simple PCA-based face recognition algorithm is as follows. In the training phase, every training face image is projected onto space S'' and its projection is represented by a k-dimensional vector. Every projection of training samples is

---

[1] It is also called eigenfaces by [48] in the context of face recognition.

associated with the name of the corresponding person. In the performance phase, an unknown face image X is given. It is then projected onto space S''. The nearest neighbor, in the Euclidean distance sense, among all the training samples in S'' is considered the best match of X. The class label of the nearest neighbor is assigned to the input image X, if it is known or assumed that X belongs to a person in the database. For face detection, the distance between an input image X and that of the nearest neighbor can be used to decide if X is a face or not [31].

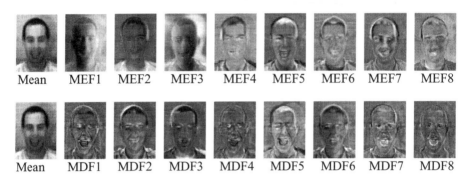

Mean MEF1 MEF2 MEF3 MEF4 MEF5 MEF6 MEF7 MEF8

Mean MDF1 MDF2 MDF3 MDF4 MDF5 MDF6 MDF7 MDF8

Figure 3.3 MEFs and MDFs. The first row shows the mean image followed by MEFs. The second row shows the mean image followed by MDFs.

The most discriminating feature space. As can be seen from Fig. 3, the first several MEFs characterize lighting variation, since lighting variation typically accounts for a major variation from the mean face. The MEFs capture the direction of major variations in the training images, such as those due to lighting direction; but these variations may well be irrelevant to how the classes are divided.

If class labels of the training images are available (i.e., all the images of the same person have the same class label), Fisher's linear discriminant analysis (LDA)[17, 18] can be performed. LDA determines a subspace in which the between-class scatter is as large as possible while keeping the within-class scatter constant. In this sense, the subspace obtained using the LDA optimally discriminates classes represented in the training set, among all the linear projections [17, 55]. The basis of this optimal subspace is the eigenvectors of $W^{-1}B$ associated with the largest eigenvalues, where W and B are the within-class scatter and the between-class scatter matrices, respectively. Ranking these eigenvectors according to the decreasing eigenvalues, we call them the *Most Discriminating Features* (MDF)[2].

The discriminant analysis procedure breaks down, however, when the number of classes is smaller than the dimensionality of the input image (i.e., W is not invertible). This problem can be resolved by performing LDA in the full MEF space, which is represented by a sufficient number, m, of MEFs [45]. Thus, MDFs are a set of basis vectors, each of which is a linear combination of MEF vectors. Therefore, LDA actually computes optimal matrix weights (linear combinations) for MEFs so that the resulting MDF subspace maximizes the ratio of the between-class scatter over the

---

[2] It is also called Fisherface by [3] in the context of face recognition

within-class scatter. Fig. 3.3 shows the first 8 MDFs. As can be seen in the figure, MDFs show patterns consisting of higher order frequency than MDFs. In some sense, MDFs tend to capture a combination of edge locations.

It is common that some classes have more samples for training and others have few or even just one. Since each pixel has at least some digitization noise, we can add a base scatter matrix $\sigma^2 I$ to the within-class scatter matrix of each class, including classes that have only one sample for training, where σ is the expected standard deviation of pixel-value noise. Examining the detail of the computational procedure described in [45], we realize that the (average) within-class scatter matrix of the training face images in MDF space is a unit matrix. This means that the average shape of all the class clusters is a unit ball in the MDF subspace. Although this does not mean that the Mahalanobis distance is degenerated into the Euclidean distance (unless all the covariance matrices of all the classes are the same), it does support the use of Euclidean distance in the nearest neighbor search if there are not enough samples in each class to estimate the class-specific distribution. For example, if a class has only one training sample available and its within-class scatter matrix is thus estimated by the base scatter matrix $\sigma^2 I$. The Euclidean distance in the MDF subspace then corresponds to a matrix weighted distance in the original MEF subspace using the within-class scatter and between-class scatter information of other classes, many of which have more training samples. Such a cross-class distribution generalization property of LDA is not used if one estimates the distribution of every class using only the samples from the class.

A special case of the above discussion is worth mentioning. MDFs weight input components automatically according to the discrimination power of each. For example, if a component corresponds to pure random noise, its contribution in the MDF subspace will be nearly zero. However, in the MEF subspace, the amount of contribution of such a noise component will be roughly proportional to the variance of the noise. The larger the variance, the more weight it has in the more significant MEFs. That explains why we have the well-known curse-of-dimensionality problem in MEF space (i.e., more features do not necessarily give a better recognition rate) but such a problem does not exist in the MDF space — it is true that larger the number of MDF features the better the recognition rate. In our experiments, we have observed that this is indeed the case, and we may use all the MDFs although the lower order MDFs contribute less to the performance improvement.

PCA vs. LDA. From the above discussion, we expect that LDA can give a better recognition rate than PCA. In our own test using Weizmman face database (29 individuals each with different lightings and different expressions), LDA producing MDF) produced significantly better face recognition result than PCA (producing MEF), as shown in Fig. 3.4.

In the training phase, different images from the same person under different lightings and expressions do have the same label. As shown in the figure, when recognition based on the nearest neighbor rule in the MDF space has reached 100% correct recognition rate using 14 features, the corresponding result in MEF space reached a maximum recognition rate of about 89%.

In practice, one can use MDFs when the class labels of the training images are available. Otherwise, MDF degenerates into MEF.

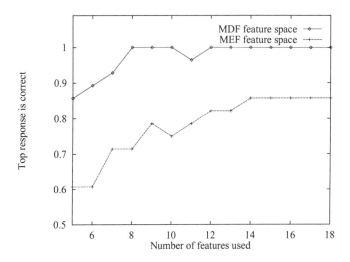

Figure 3.4 The performance of the system for different numbers of MEF/MDF features. The number of features from the subspace used was varied to show how the MDF subspace outperforms the MEF subspace. 95% of the variance for the MDF subspace was attained when fifteen features were used; 95% of the variance for the MEF subspace did not occur until 37 features were used. Using 95% of the MEF variance resulted in an 89% recognition rate, and that rate was not improved using more MEF features.

Hierarchical spaces vs. a flat space. The advantage of using a hierarchical space is twofold: speed and generalization. The hierarchical space allows the search for top matches to be much faster. The time complexity for searching a roughly balanced classification tree is $O(\log(n))$ instead of $O(n)$. This is a very important property for a large face database. In contrast, other flat-space methods have a time complexity of $O(n)$. The speedup due to the hierarchical space (represented by a tree) is $O(n/\log(n))$. Only when n is large, the speed up is large.

The hierarchical space makes features more effective in separating different classes and thus allows better generalization with a given set of training samples. When many samples from many classes are put together into a single set, it is not always possible to find a linear subspace (or a set of features) in which all different classes are separated well. This is because the boundary between classes are typically very nonlinear (i.e., curved hyper-surfaces). However, when the training samples are broken into smaller and smaller sets through a hierarchical space partition scheme, each smaller set typically contains fewer number of classes and fewer number of samples. Thus, at deep levels of the tree, it is much more likely to be able to produce a good set of linear features that separate the samples into different classes. The hierarchical space partition can use piecewise linear boundary (linear features MEFs and MDFs) to approximate curved, nonlinear decision boundaries needed to separate classes in the training set.

Among the existing appearance-based facial recognition methods, the SHOSLIF to be explained later in this chapter is among the very few that uses a hierarchical space.

Incremental learning vs. batch learning. Learning methods fall into two categories, batch and incremental. A batch learning method requires that all the training face images are available at a fixed training time. It is difficult to determine *a priori* how many and what kinds of training images are needed in order to reach a required performance level. Thus, a batch learning method requires multiple cycles of collecting data, training, and testing. The limited space available to store training images and the need for more images for better performance are two conflicting factors. Therefore, if a batch learning method is used, the task of collecting a sufficiently good set of training samples is very tedious in practice. Further, each batch training session takes a significant amount of time to learn the entire batch of the training data.

With an incremental learning method, training samples are available only one (or a small set) at a time. Each training sample is discarded as soon as it has been incorporated into the system. If the output result from the current system is not correct (or with a large error), the current sample is used to update the system [54]. Otherwise, the current training image is rejected. This selective learning mechanism effectively prevents redundant learning in order to keep the size of the face-image database relatively small. Using this incremental learning mode, updating the system is convenient. We do not need to load all the old images to re-learn when new images are added. All we need to do is to run an update algorithm using only the new images.

Built-in deformation models. The appearance-based method directly uses intensity patterns of a face image. Given a face, some parts of the face may deform more than other parts. For example, the mouth region may deform more than the nose region. There are two types of approaches in dealing with such deformations. (1) Let the system learn the deformation, (2) hand build a deformation model into the system.

For the first approach, one can use the techniques discussed earlier. However, one must collect enough samples that sufficiently cover the observable deformations. For example, for deformations caused by expression changes, all the expressions that the system must deal with should be contained in the training samples.

Following the second approach, one needs to design a deformation model. Moghaddam et al. [31] used a method that treats an image $I(x, y)$ as a 3-D surface $(x, y, I(x, y))$, where $I(x, y)$ is the height of the 3-D surface at position (x, y). A 3-D deformable surface model is applied to this 3-D surface, which includes the mass and the stiffness of the surface. Thus the deformation is allowed in both position (x, y) and intensity $I(x, y)$. When matching a face image I_1 with a reference image I_2, the external force at each 3-D point on the surface I_1 is the 3-D vector from the point to the closest point on surface I_2. The 3-D deformation thus estimated is used to estimate the intrapersonal deformation and extrapersonal deformation. The maximum a posterior (MAP) rule is used to decide if the two images arise from the same person. [56] used a type of elastic graph matching based on position deformation and the intensity pattern information computed using a set of Gabor filters.

The second approach may impose some assumptions. For example, the statistics of a face deformation is based on all the faces instead of a particular face or a particular group of faces. This treatment is effective when the number of training samples is limited and the deformation present between an unknown image and the trained images is large.

In the following, we present a specific face recognition algorithm as an example.

4. The SHOSLIF

The Self-Organizing Hierarchical Optimal Subspace Learning and Inference Framework (SHOSLIF) [52] is a framework that aims to provide a unified methodology for visual learning. The SHOSLIF uses PCA and LDA to generate a hierarchical tessellation of a space defined by the training images. The incremental version of the SHOSLIF learns incrementally [54]. Each query to the SHOSLIF takes $O(\log n)$ time.

System overview

The method generates a *Space-Tesselation Tree* (STT) during the training phase. An example of such a tree is shown in Figure 3.5.

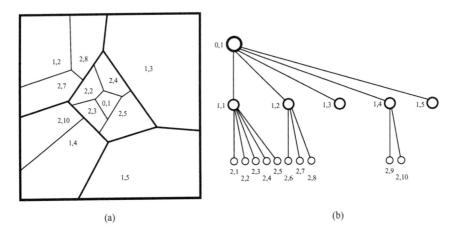

(a)   (b)

Figure 3.5 (a) A sample partition of the face image space generated from the training samples. (b) The tree structure associated with the partition shown in (a). Each cell in (a), which corresponds to a node in (b) does not need to cover a meaningful class. Each cell operates in a different subspace, and the leaf nodes give the final tessellation, which can approximate virtually any complex decision region.   The hierarchy provides a logarithmic retrieval complexity.

As the processing moves down from the root node of the tree, the Space-Tessellation Tree recursively subdivides the training samples into smaller problems until a manageable problem size is achieved. When a face image is presented to a node, a distance measure from each of the node's children is computed to determine the most likely child to which the face image belongs. At each level of the tree, the node that best captures the features of the face image is used as the root of the subtree for further refinement, thereby greatly reducing the search space for the best matches. This scheme is very similar to other works on classification trees used in statistics [4]. Two major differences are (a) that SHOSLIF uses LDA to automatically generate the splitter at each internal node and (b) the input to SHOSLIF is the high-dimensional image vector instead of a number of feature values defined by system designer.

Figure 3.6 explains how additional training images are generated at the root node and the processing performed at each node of the tree.

We want to allow for some variations in the position, scale, and orientation of the faces in the canonical face image to be recognized. This can be accomplished either through more image acquisition, but that is expensive in terms of time, storage, and cost. Each canonical image this system receives provides an attention point and scale to be used to extract a fovea image of the object of interest. Rather than extracting just a single fovea image from this attention point and scale, a family of fovea images are generated by varying the attention point and scale from the supplied points, as shown in Fig. 6. This allows the system to learn some measure of positional and scale variation in the training set.

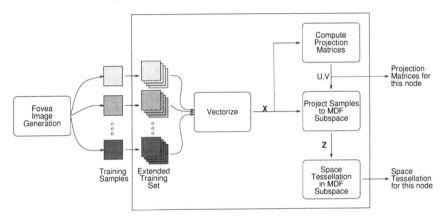

Figure 3.6 A top-level flow of the processing in the Space-Tessellation Tree during the training phase. The steps after the vectorization are performed also at every internal node of the tree.

Hierarchical Space Tessellation

As we discussed earlier, the hierarchical space tessellation allows one to recursively decompose a large and complex problem into smaller and simpler problems. Figure 3.7 shows an example of the difference in the complexity of the class separation problem for the root node and an internal node of the tree. A child node contains fewer samples than its parent does, and the MDF vectors can therefore be optimized to the smaller set of samples in the child node. The finest tessellation level in a hierarchical space tessellation tree allows the linear features such as the MEFs and MDF to approximate any smooth decision regions to a desired accuracy using piecewise hyper-planes, as shown in Fig. 6, as long as the number of training samples is sufficiently large.

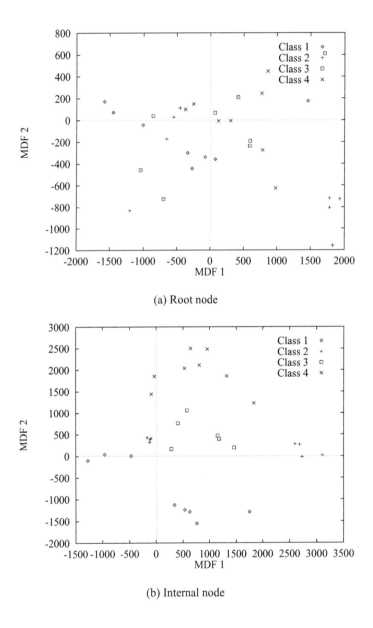

(a) Root node

(b) Internal node

Figure 3.7 An example showing the complexity of the class separation problem at two different levels of the tree. Samples from the same classes are given for both the graphs. This figure shows a more effective clustering in (b) than in (a) because the number of samples and classes in (b) is smaller than in (a).

Definition 1 Given n leaves, a Bounded Unbalanced Tree with Unbalance Bound $0 < a < 1$ (a constant) is a tree such that for any node N containing $n_1 + n_2 + \cdots + n_k$ leaves, where N has k children with n_i leaves assigned to node i and $n_1 \geq n_2 \geq \cdots \geq n_k$, $n_1 \leq a(n_1 + n_2 + \cdots n_k)$.

Automatic Tree Construction

Each level of the tree has an expected radius $r(l)$ to characterize the size of the cells at level l. Note that $r(l)$ is a decreasing positive function based on the level. Let $d(X, A)$ be the distance measure between node N with center vector A and a sample vector X. The tree is built one level at a time. The algorithm is summarized in Figure 3.9.

Lemma 1 The number of levels in a Bounded Unbalanced Tree with n leaves is bounded above by $\log_{(1/a)} n$, where a is the Unbalance Bound of the tree.

Proof 1 Each node N of the tree is assigned with $n_1 + n_2 + \cdots + n_k$ leaves, where n_i is the number of leaves assigned to the ith child of N. Rank these n_i's so that $n_1 \geq n_2 \geq \cdots \geq n_k$. Because the tree is a Bounded Unbalanced Tree, we know that $n_1 \leq a (n_1 + n_2 + \cdots + n_k)$, and this is true for all nodes N of the tree; a is a constant. Each deeper level of the tree will reduce the number of leaves by a factor of at least a. The lth level down the tree will receive na^l leaves. At tree height h, we have just a single sample by Algorithm 1. So, $na^h = 1$, and $a^h = (1/n)$, or $(1/a)^h = n$. Then the height of the tree $h = \log_{(1/a)}n = (\log(n)/\log(1/a))$.

Figure 3.8 Lemma 1.

Algorithm 1 The Hierarchial Quasi-Voronoi Tessellation Algorithm

Input: Node N at level $l - 1$, list of samples X to add.

Output: A tessellation of N based on the new samples.

1. Compute the projection matrices V and W for the MEF and MDF subspaces for this node.
2. For each sample X_i:
   (a) Project X_i to the MEF space to get Y_i.
   (b) Project Y_i to the MDF space to get Z_i.
   (c) If $d(Z_i, C_j) > r(l)$ for all C_j children of N, add Z_i as the center vector for a new child of N.
3. For each feature vector Z_i, add Z_i to the child C_j with the nearest center vector.
4. For each child C_j of N, perform the space tessellation.

Figure 3.9 The Hierarchial Quasi-Voronoi Tessellation Algorithm.

Recognition as Tree Retrieval

The retrieval algorithm is given by Figure 3.10, which provides the top k matches for human examination. Given an unknown face image, the nearest neighbor in the Euclidean distance in image space S is typically not the best match because the class variation is not taken into account. For the same reason, the nearest neighbor in the MEF space is not either, as shown in Figure 3.11.

We do not want the STT to give the nearest neighbor in any single space and it will not either, since each internal node uses its own subspace.

Algorithm 2 *The Image Retrieval Algorithm*

Input: *Probe X, level l, and a list of at most constant k nodes which were explored at level l.*

Output: *A list of nodes explored at level l+1.*

1. *For each node N_i in the list explored at level l:*
       (a) If N_i is not a leaf node:
           i.   Project X to the MDF subspace of node N_i, producing Z .
           ii. Compute $d(C_j, Z)$ for all children j of N_i with center vectors C_j.
       iii.   Transfer at most constant k of the children of N_i to the output list such that those transferred are the k nearest neighbors of Z .
2. *Truncate the output list to hold at most constant k nodes to explore at the next level.*

This algorithm is repeated for all levels of the Space-Tessellation tree until k leaves are found.

Figure 3.10 The image retrieval algorithm.

Distance Measure

In the SHOSLIF tree, every node has its own different MEF and MDF spaces. Given an input, we must compare its match with all the competitive nodes. The Distance from Subspace (DFS) distance measure takes into account the distance from the *projection space* in addition to the distance of the projection to the node centers. The DFS distance measure is given by

$$d(X, A) = \sqrt{\| X - VV'X \|^2 + \| MM'X - MM'A \|^2}$$

where $M = VW$, X is the test probe, A is the center vector, V is the projection matrix to the MEF space, and W is the projection matrix to the MDF space. Figure 3.12 gives a geometric illustration.

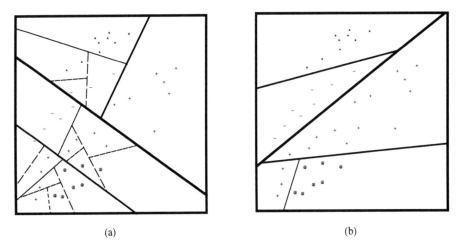

(a)   (b)

Figure 3.11 (a) The binary tree built without class information taken into account, as would be built using the MEF space. (b) The binary tree built optimized to separate classes, as would be built using the MDF space. The MDF typically yields a smaller tree than the MEF space provides. The MDF is effective if the samples cover all the within-class variations. The samples of a class are denoted by a single type of character.

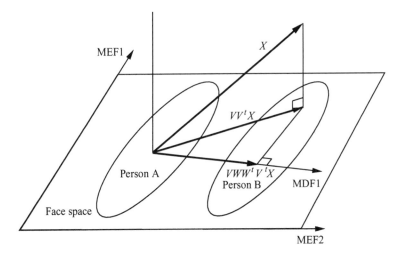

Figure 3.12 Distance from subspace description for 3D.

5. Experimental Results

In the following, we present some experimental results which provide quantitative examples for some design options.

Feature Spaces and the Tree

We examined the effects of MEF and MDF flat subspaces, a single-space tree in which a tree is built in a single subspace, and the multi-subspace tree (SHOSLIF) as described above. The training images come from a set of real-world objects in natural settings. At least two training images from each of 36 object classes were provided; a disjoint set of test images were used in all of the tests. The results of the studies are summarized in Table 3.1. Top one means that the top one choice retrieved is correct. Top 15 means that the correct one is among the top 15 retrieved from the tree.

		Flat	Single-space tree	SHOSLIF tree
Image space	Top one	90%	86%	N/A
	Top 15	100%	90%	N/A
MEF space	Top one	90%	76%	86%
	Top 15	100%	83%	90%
MDF space	Top one	90%	89%	97%
	Top 15	100%	100%	100%

Table 3.1 Results of subspace comparison study.

Face Recognition

To test face recognition, a face database of individuals was organized; each individual had a pool of images which are divided into disjoint training and test data sets. Each individual had at least two images for training with a change of expression. The images of 38 individuals (182 images) came from the Michigan State University Pattern Recognition and Image Processing laboratory. Images of individuals in this set were taken under uncontrolled conditions, over several days, and under different lighting conditions. Another 303 classes (654 images) came from the FERET database. All of these classes had at least two images of an individual taken under controlled lighting, with a change of expression; 24 of these classes had additional images taken of the subjects on a different day with very poor contrast. Sixteen classes (144 images) came from the MIT Media lab under identical lighting conditions (ambient laboratory light). Twenty-nine classes (174 images) came from the Weizmann Institute, and are images with three very controlled lighting conditions for each of two different expressions. Table 2 shows a summary of the results obtained both by re-substituting the training samples as test probes and by using a disjoint set of images for testing.

Handling Variations

In the experiments further conducted, the training images include 1316 images from 526 classes of human faces and other objects. We trained the system using training samples artificially generated from the original training samples to vary in (a) 30% of size, (b) positional shift of 20% of size and 20% of size; (c) 3D face orientation by about 45 degrees and testing with 22.5 degrees. A total of 298 images from 298 classes, all disjoint from the training images, were used for test. The top 1 and top 10 correct recognition rates were, respectively, (a) 93.3% and 98.9%, (b) 93.1% and 96.6%, (c) 78.9% and 89.4%.

No. of training images	1042 of 384 individuals
No. of nodes in the tree	1761
No. of explored paths	10
Re-substitution:	
Top one	100.0%
Disjoint test set:	
No. of test images	246 of 246 individuals
Top one	95.5%
Top 10	97.6%

Table 3.2 Summary of experiment on a face database of 1042 images (384 individuals). The number of explored paths equals the number of top matches explored in the level-by-level tree retrieval.

Timing

Table 3.3 shows how the tree structure speeded up the retrieval process. The test was done on a Sun SPARC-20. A total of 2,850 training images were used in the learning. Three schemes are compared in the table. The flat image space scheme uses a linear search for the nearest neighbor in the original image space S. The flat MEF space scheme uses a linear search for the nearest neighbor in the MEF subspace S'' and the projection time for the input image is included. The SHOSLIF tree scheme is a real-time version using a binary tree of the SHOSLIF [54]. The speed-up of the tree is more drastic when n is larger.

Method	Flat image space	Flat MEF space	SHOSLIF tree
Time	2.854 s	0.738 s	0.027 s

Table 3.3 Average computer time per test probe.

6. Conclusions

Face recognition research has been very active and has made tremendous progress over the past 10 years. Face recognition systems have a lot of immediate applications as long as the environment can be controlled appropriately.

An important lesson that we can draw from the history of face recognition is as follows. For very complex perception tasks, such as face recognition, it is not the most productive way to develop a system according to manually developed content-level rules, rules that are directly related to the object to be perceived. For face recognition, such rules include what facial features should be used, how face images may change if the lighting is changed, what facial changes one may see when one smiles, which part of the face is more invariant, etc. A general learning scheme, one that does not depend on the objects to be perceived can turn out to be very effective because it can adapt well to the objects without being impaired by the content-level rules.

Although the current face recognition systems have achieved very good results for faces that are taken in a controlled environment, they perform poorly in less uncontrolled situations (see, e.g. FERET tests [34]. Humans know how to take environmental context into account but our existing systems do not[3]. A fundamental methodology revolution is necessary for a breakthrough advance from the current state of the art. It becomes increasingly evident that breakthrough solutions to tough computer vision problems can probably be found by looking beyond the visual modality.

Acknowledgments

This work was supported in part by National Science Foundation Grant IRI 9410741, Office of Naval Research Grant N00014-95-0637 to Weng, National Science Foundation Grant CDA-9724462 and NASA Grant No. NGT 40046 to Swets.

[1] R. Baron, "Mechanisms of human facial recognition," *International Journal of Man-Machine Studies*, pp. 137-178, 1981.

[2] J. C. Bartlett and A. Fulton, "Familiarity and recognition of faces in old age," *Memory and Cognition*, Vol. 19, No. 3, pp. 229-238, 1991.

[3] P. N. Belhumeur, J. P. Hespanha, and D. J. Kriegman, "Eigenfaces vs fisherfaces: Recognition using class specific linear projection," *IEEE Trans. Pattern Analysis and Machine Intelligence*, Vol. 19, No. 7, pp. 711-720, 1997

[4] L. Breiman, J. Friedman, R. Olshen, and C. Stone, *Classification and Regression Trees*. Chapman and Hall, New York, 1993.

[5] C. Bunney, "Survey: Face recognition systems," *Biometric Technology Today*, Vol. 5, No. 4, pp. 8-12, 1997.

[6] R. A. Campbell, S. Cannon, G. Jones, and N. Morgan, "Individual face classification by computer vision," In *Proceedings Conference Modeling Simulation Microcomputer*, pp. 62-63, 1987.

[7] S. Carey, "A case study: Face recognition," In *Explorations in the Biological Language*. Bradford Books, New York, 1987.

[8] S. Carey, R. Diamond, and B. Woods, "The development of face recognition-a maturational component?" *Developmental Psychology*, No. 16, pp. 257-269, 1980.

[9] R. Chellappa, C. L. Wilson, and S. Sirohey, "Human and machine recognition of faces: A survey," *Proceedings of the IEEE*, Vol. 83, No. , pp. 705-740, 1995.

[10] H. D. Ellis, "Introduction to aspects of face processing: Ten questions in need of answers," In H. Ellis, M. Jeeves, F. Newcombe, and A. Young, editors, *Aspects of Face Processing*, pp. 3-13. Nijhoff, 1996.

[11] H. D. Ellis, D. M. Ellis, and J. A. Hosie, "Priming effects in children's face recognition," *British Journal of Psychology*, Vol. 84, No. 1, pp. 101-110, 1993.

---

[3] See, e.g., the Clinton and Gore example designed by Sinha and Poggio [41]

[12] J. Elman, E. A. Bates, M. H. Johnson, A. Karmiloff-Smith, D. Parisi, and K. Plunkett, *Rethinking Innateness: A connectionist perspective on development.* MIT Press, Cambridge, MA, 1997.

[13] K. Etemad and R. Chellappa, "Discriminant analysis for recognition of human face images," In *Proceedings International Conference Acousics, Speech, Signal Processing,* pp. 2148-2151, Atlanta, Georgia, 1994.

[14] FG1, *Proceedings of International Workshop on Automatic Face- and Gesture-Recognition.* Mutimedia lab. Department of Computer Sciece, University of Zurich, Zurich, Switzerland, 1995.

[15] FG2, *Proc. 2nd International Conference on Automatic Face and Gesture Recognition.* IEEE Computer Society Press, Los Alamitos, CA, 1996.

[16] FG3, *Proceedings 3rd International Conference on Automatic Face and Gesture Recognition.* IEEE Computer Society Press, Los Alamitos, CA, 1998.

[17] R. A. Fisher, "The statistical utilization of multiple measurements," *Annals of Eugenics,* Vol. 8, pp. 376-386, 1938.

[18] K. Fukunaga, *Introduction to Statistical Pattern Recognition.* Academic Press, New York, NY, second edition, 1990.

[19] A. G. Goldstein, Facial feature variation: Anthropometric data II. *Bulletin of the Psychonomic Society,* Vol. 13, pp. 191-193, 1979.

[20] A. G. Goldstein, "Race-related variation of facial features: Anthropometric data I," *Bulletin of the Psychonomic Society,* Vol. 13, pp. 187-190, 1979.

[21] P. Green, "Biology and cognitive development: The case of face recognition," *Animal Behaviour,* Vol. 43, No. 3, pp. 526-527, 1992.

[22] D. C. Hay and A. W. Young, "The human face," In H. D. Ellis, editor, *Normality and Pathology in Cognitive Function,* pp. 173-202. Academic Press, London, 1982.

[23] I. T. Jolliffe, *Principal Component Analysis.* Springer-Verlag, New York, 1986.

[24] T. Kanade, *Computer Recognition of Human Faces.* Birkhauser, Basel and Stuttgart, 1977.

[25] Y. Kaya and K. Kobayashi, "A basic study on human face recognition," In Wantanabe, S., editor, *Frontiers of Pattern Recognition,* pp. 265-289. Academic Press, New York, 1972.

[26] M. Kirby and L. Sirovich, "Application of the karhunen-loève procedure for the characterization of human faces," *IEEE Trans. Pattern Analysis and Machine Intelligence,* Vol. 12, No. 1, pp. 103-108.

[27] T. Kohonen, *Self-Organization and Associative Memory.* Springer-Verlag, Berlin, 1988.

[28] S. Lawrence, C. L. Giles, A. C. Tsoi, and A. D. Back, "Face recognition: A convolutional-network approach," *IEEE Trans. Neural Networks,* Vol. 8, No. 1, pp. 98-113, 1997.

[29] R. Mauro and M. Kubovy, "Caricature and face recognition," *Memory and Cognition,* Vol. 20, No. 4, pp. 433-441, 1992.

[30] E. A. Maylor, "Recognizing and naming faces: aging, memory retrieval, and the tip of the tongue state," *Journal of Gerontology,* Vol. 45, No. 6, pp. 215-226, 1990.

[31] B. Moghaddam, C. Nastar, and A. Pentland, "A Bayesian similarity measure for direct image matching," In *Proceedings International Conferecne Pattern Recognition,* pp. 350-357, Vienna, Austria, 1996.

[32] J. Morton and M. H. Johnson, "Conspec and conlern: a two-process theory of infant face recognition," *Psychological Review,* Vol. 98, No. 2, pp. 164-181, 1991.

[33] H. Murase and S. K. Nayar, "Visual learning and recognition of 3-D objects from appearance," *Internationa Journal of Computer Vision,* Vol. 14, No. 1, pp. 5-24, 1995.

[34] P. J. Phillips, H. Moon, P. Rauss, and S. A. Rizvi, "The FERET evaluation methodology for face-recognition algorithms," In *Proceedings IEEE Conf. Computer Vision and Pattern Recognition,* pp. 137-143, Puerto Rico, 1997.

[35] S. S. Rakover and B. Cahlon, "To catch a thief with a recognition test: the model and some empirical results," *Cognitive Psychology,* Vol. 21, No. 4, 423-468, 1989.

[36] G. Rhodes and T. Tremewan, "The simon then garfunkel effect: semantic priming, sensitivity, and the modularity of face recognition," *Cognitive Psychology*, Vol. 25, No. 2, pp. 147-187, 1993.

[37] H. A. Rowley, S. Baluja, and T. Kanade, "Neural network-based face detection," *IEEE Trans. Pattern Analysis and Machine Intelligence*, Vol. 20, No. 1, 23-38, 1998.

[38] S. K. S. Lin and L. Lin, "Face recognition/detection by probabilistic decision-based neural network," *IEEE Trans. Neural Networks*, Vol. 8, No. 1, pp. 114-132, 1997.

[39] Samal and P. Iyengar, "Automatic recognition and analysis of human faces and facial expressions: A survey," *Pattern Recognition*, Vol. 25, pp. 65-77, 1992.

[40] J. Sergent, "Microgenesis of face perception," In *Aspects of Face Processing*. Nijhoff, Dordrecht, 1986.

[41] P. Sinha and T. Poggio, " I think I know that face ...", *Nature*, Vol. 384, pp. 404, 1996.

[42] W. Sjoberg, B. Gruber, and C. Swatloski, "Visual-field asymmetry in face recognition as a function of face discriminability and interstimulus interval," *Perceptual and Motor Skills*, Vol. 72, No. 3, pp. 1267-1271, 1991.

[43] K. K. Sung and T. Poggio, "Example-based learning for view-based human face detection," *IEEE Trans. Pattern Analysis and Machine Intelligence*, Vol. 20, No. 1, pp. 39-51, 1998.

[44] D. Swets and J. Weng, "Discriminant analysis and eigenspace partition tree for face and object recognition from views," In *Proc. Int'l Conference on Automatic Face- and Gesture-Recognition*, pages 192-197, Killington, Vermont, 1996.

[45] D. L. Swets and J. Weng, "Using discriminant eigenfeatures for image retrieval," *IEEE Trans. Pattern Analysis and Machine Intelligence*, Vol. 18, No. 8, pp. 831-836, 1996.

[46] M. Szpir, "Accustomed to your face," *American Scientist*, Vol. 80, No. 6, 537-540, 1992.

[47] P. Thompson, "Margaret Thatcher: a new illusion," *Perception*, Vol. 9, pp. 483-484, 1980.

[48] M. Turk and A. Pentland, "Eigenfaces for recognition," *Journal of Cognitive Neuroscience*, Vol. 3, No. 1, pp. 71-86, 1991.

[49] T. Valentine and A. Ferrara, "Typicality in categorization, recognition and identification: evidence from face recognition," *British Journal of Psychology*, Vol. 82, No. 1), pp. 87-102, 1991.

[50] A. Wahlin, L. Backman, T. Mantyla, A. Herlitz, M. Viitanen, and B. Winbald, "Prior knowledge and face recognition in a community-based sample of healthy, very old adults," *Journals of Gerontology*, Vol. 48, No. 2, pp. 54, 1993.

[51] J. Weng, "On comprehensive visual learning," In *Proceedings NSF/ARPA Workshop on Performance vs. Methodology in Computer Vision*, pp. 152-166, Seattle, WA, 1994.

[52] J. Weng, "Cresceptron and SHOSLIF: Toward comprehensive visual learning," In S. K. Nayar and T. Poggio, editors, *Early Visual Learning*. Oxford University Press, New York, 1996.

[53] J. Weng, N. Ahuja, and T. S. Huang, "Learning recognition and segmentation using the Cresceptron," In *Proceedings International Conference on Computer Vision*, pp. 121-128. Berlin, Germany, 1993.

[54] J. Weng and S. Chen, "Incremental learning for vision-based navigation," In *Proceedings Internatinal Conference Pattern Recognition*, volume IV, pp. 45-49, Vienna, Austria, 1996.

[55] S. S. Wilks, *Mathematical Statistics*. Wiley, New York, NY, 1963.

[56] L. Wiskott, J. M. Fellous, N. Kruger, and C. von der Malsburg, "Face recognition and gender determination," In *Proceedings International Workshop on Automatic Face- and Gesture-Recognition*, pp. 92-97, Zurich, Switzerland, 1995.

4 HAND GEOMETRY BASED VERIFICATION

Richard L. Zunkel
Recognition Systems Inc
Campbell, CA
DickZunkel@recogsys.com

Abstract *The physical dimensions of a human hand contain information that is capable of authenticating the identity of an individual. This information has been popularly known as hand or palm geometry; the hand geometry based identity verification systems are being widely used in a number of access control, time and attendance, and point-of-sale applications. This chapter introduces hand geometry based identity authentication system. It provides an overview of operation of a specific system and its performance. Finally, it discusses some emerging applications of hand geometry based authentication systems.*

Keywords: *hand geometry, palm geometry, finger geometry, access control.*

1. Introduction

Anthropologists suggest that humankind survived and evolved due to our large brains and opposing thumbs. The versatile human hand allows us to grasp, throw, and make tools. Today, the human hand has another use, a media to verify identity. Ancient Egyptians used body measurements to classify and identify people. Today's hand geometry scanners use infrared optics and microprocessor technology to quickly and accurately record and compare hand dimensions. Several hand geometry verification technologies have evolved during this century. They range from electro-mechanical devices to the solid state electronic scanners being manufactured today. The U.S. Patent office issued patents to Robert P. Miller in the late 1960's and early 1970's for a device that measures hand characteristics, and records unique features for comparison and ID verification (e.g., [1]). Miller's machines were highly mechanical and manufactured under the name "Identimation." Several other companies launched development and manufacturing efforts during the 70's and early 80's (e.g., [2,3,7,8]). In the mid-1980's, David Sidlauskas developed and patented an electronic hand scanning device (e.g., [5]) and established the Recognition Systems, Inc. of Campbell, California in 1986.

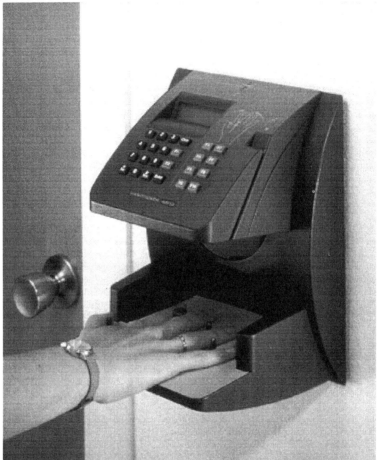

Figure 4.1 Hand geometry biometrics for access control.

The first applications for hand scanners were as access control components. Government and nuclear facilities used them to protect their facilities [6]. The availability of low cost, high speed processors and solid state electronics made it possible to produce hand scanners at a cost that made them affordable in the commercial access control market. At first, systems providers installed hand scanners in the stand-alone mode. The products contained basic access control functions such as time zones, alarm inputs and outputs, duress, and request for exit functions. Sensor Engineering's Wiegand format became widely used to interface hand scanners into existing access control systems. Hand scanners have a twenty-six bit Wiegand output format as one of the menu choices. As the access control market expanded, manufacturers introduced other data formats. Systems integrators demanded, and got alternative protocols. End users became more sophisticated and demanded more elaborate engineered versions of the hand scanner. Today, hand scanners perform a variety of functions including access control, employee time recording and point-of sale applications (Figure 4.1).

2. System Operation

Each human hand is unique. Finger length, width, thickness, curvatures and relative location of these features distinguish every human being from every other person (Figure 4.2). The hand geometry scanner uses a charge coupled device (CCD) camera, infrared light emitting diodes (LEDs) with mirrors and reflectors to capture black and white images of the human hand silhouetted against a thirty-two thousand pixel field (Figure 4.3). The scanner records no surface details, ignoring fingerprints, lines, scars and color. The process is much like placing a hand on a beaded projector screen. The hand scanner reads the hand shape by recording the silhouette of the hand. In combination with a side mirror and reflector, the optics produce two distinct images, one from the top and one from the side. This method is known as orthographic scanning.

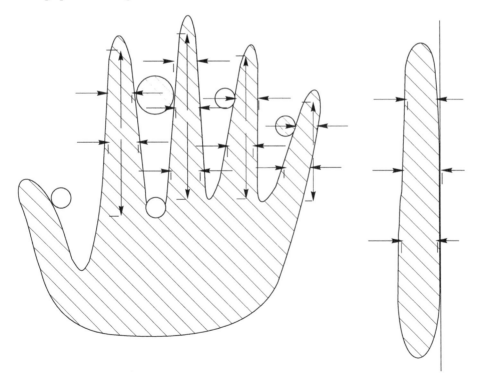

Figure 4.2 Typical hand geometry measurements.

Figure 4.3 Input image for extraction of hand geometry features. The smaller inset image shows a view of the thickness of the hand.

Scanners typically use an optical path approximately 11 inches (28 cm) between the camera and the platen. An optical path folded with mirrors reduces the space required to half the original length. Enclosing the optical path in a structure results in the typical hand geometry scanner that is approximately 8-1/2 inches (22 cm) square by 10" (25 cm) high (Figure 4.4) The scanner takes ninety-six measurements of the user's hand. A microprocessor and internal software convert the measurements to a nine-byte "template" that it stores for later comparison. The process of recording a user's hand template is known as enrollment. During the enrollment session, the scanner prompts the enrollee to place his or her hand on the scanner platen three consecutive times. The platen is the highly reflective surface that projects the silhouetted hand image. Pins projecting from the platen surface position the enrollee's fingers to assure accurate image capture. The hand geometry scanner mathematically averages the three templates and generates an accurate template that the scanner stores in resident memory. To verify, the user enters a personal identification number (PIN) in the scanner through the use of a keypad or other data entry device. The scanner retrieves his or her individual template for comparison. The user places his or her hand on the scanner. The hand image is captured and a representation is derived using the same steps as those used for generating the template at the time of enrollment. The representation thus derived is compared to the stored template. The comparison may involve, for instance, accumulation of absolute differences in the individual features in the input representation and the stored template. The comparison typically results in a single number indicating the strength of the similarity (score) or their difference (distance). A predetermined threshold determines whether the score/distance is acceptable to consider the input representation and stored templates are "matched". The match/no-match decision controls the output of the scanner.

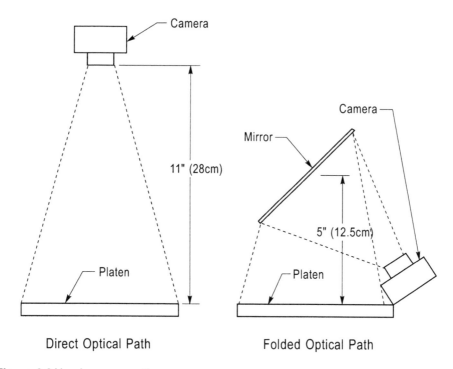

Direct Optical Path   **Folded Optical Path**

Figure 4.4 Hand scanner optics.

Enrollment

The quality of enrollment affects the false reject rate, especially during the first few weeks of use. Quality enrollment depends on several factors. Different platen heights change the relative position of the body and hand, and affect the hand shape. Enrolling at one height and verifying at another can cause enough difference in the hand shape to reject the user. The enrollment scanner should be of the same height as the verification scanner. If persons must stand to use a verification scanner, they should also stand to enroll. Likewise, if the user normally sits to use the verification scanner, (such as using a scanner to enter a password into a computer), they should also enroll in the sitting position.

Enrollers should train users during the enrollment process and correct potential errors before they occur. Successful and correct enrollment may require demonstration and user training. Hand placement training greatly reduces verification problems; it involves enrollees learning the "feel" of the hand geometry scanner platen and finger pins. The popular methods of user training rely on visualization techniques to teach the correct hand placement. For example, it is found that "landing an airplane" scenario is an effective teaching aid, where the enrollee is told to touch the platen with the ends of the index and middle fingers, then slide the hand onto the platen in a motion much like a landing airplane. Learning (e.g., to keep the hand flat, to feel the surface of the platen with the second knuckle of each finger, and to squeeze the fingers against the finger pins) improves   the performance of the system.

Therefore, user training is recommended *before* enrollment. When the user shows that he or she can place the hand properly, the user is enrolled. The enroller enters the user's ID number and prompts the user to place his or her hand on the platen. The hand geometry scanner may prompt the user to place the hand multiple (e.g., three) times. Multiple placements allow the scanner to capture images of the hand in slightly different positions. The scanner processor may use a statistic of the multiple measurements (e.g., average) to form the template, which constitutes the mathematical representation of the hand.

At birth, human hands are nearly symmetrical. As the body ages, the hands change due to natural and environmental changes. Most people become either right or left handed, causing one hand to be slightly larger than the other. The "favored" hand tends to be more susceptible to injury from sports or work activities. Young peoples' hands change rapidly as they mature. Older peoples' hands change with the natural aging process or the onset of arthritis. All these factors necessitate that practical hand geometry scanners "learn" minor hand shape changes and continually update templates as users are verified by the system. This process is known as template averaging. Template averaging updates the mathematical description of the user's hand. It occurs when the differences between user's hand and the template stored during enrollment reach a predetermined limit.

One particular installation, a large metropolitan housing project, uses hand geometry scanners for access by tenants and their children. Many of the children first used the scanners at about the age eight. Now in their mid-teens, their hands have changed considerably, but they continue to use the scanners without difficulty. Hand geometry scanners will work reliably for children above the age of seven or eight, depending on the size of their hands. The user's fingers must be of sufficient length to reach the platen finger pins. Children of some ethnic groups are smaller in stature, and may have to be older before they can reliably use hand geometry scanners. Likewise, some population groups are larger in stature and their fingers will reach the platen pins at a younger age. The hand geometry scanners update their templates as children use them and mature.

Ideally, the placements of the hand on the platen at enrollment and verficiation need to be identical. This process is called registration. Present hand geometry technology accomplishes registration by requiring the use of locator pins on the platen to position fingers. As mentioned, children's and small adults' fingers must be long enough to reach the pins for them to verify. Likewise, persons with amputations or birth defects that prevent them from contacting each of the finger pins may have to use alternative enrollment and verification methods. Standard production hand scanners are designed to accept right hands, palms down. However, because they capture shape information, and not surface details, it is possible to enroll a user's left hand with the palm facing upward. While this placement feels unnatural at first, left-hand verification becomes easier with practice. Other factors affect enrollment and verification. Hand geometry scanners accept non-biological shapes such as rings and small bandages during the enrollment process. To ensure that the contaminants in the optical path do not affect performance, it is advisable to occasionally wipe exposed platens and mirror surfaces with non-abrasive window cleaner on a cloth.

Verification

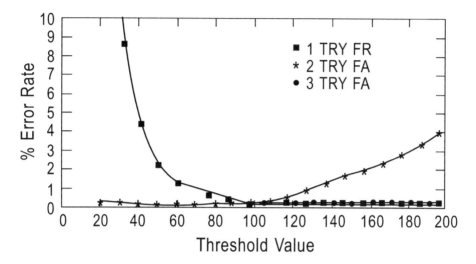

Figure 4.5 Sandia Labs evaluation results for ID3D System for authenticating identity based on hand geometry [4].

Verifying the user identity is the primary purpose of a hand geometry scanner. There are several outputs from the typical scanner. In the stand-alone mode, the system can enable a locking device directly. In this mode, the scanner lock output operates a control relay upon verification of the user. The relay enables the lock for a pre-determined period, then releases it. Direct lock control upon verification is the most basic access control function. Door contacts connected to scanner input terminals sense door position and cause the scanner to shunt the lock time to reduce the chance of people following the user through an unlocked door.

The FRR/FAR crossover error rates for Recognition System's hand geometry scanner was 0.1% at the manufacturer's default threshold as documented by Sandia Laboratories, in 1991 [4] (Figure 4.5). The results were based on a two-try false reject rate. The one-try false reject rate was 0.2%. Field results taken from actual sites with over 100,000 events confirm the controlled test false reject rate.

3. Implementation Issues

Card Reader Emulation

Hand scanner interfaces emulating a card reader are available. In this arrangement, the hand scanner connects to a third-party access control terminal and acts as a card reader. Three common card protocols emulated by hand scanners are Wiegand, magnetic stripe, and barcode. The Wiegand protocol is a stream of binary data sent over two opposing channels commonly known as "data 0" and "data 1" that share a common ground. The data stream has a field of "bits" that contains the card number, another that has a facility or "site" code and two parity bits for error correction. The user enters a PIN number in the scanner keypad or presents a card to an auxiliary card reader connected to the scanner. The scanner reads the ID field from the Wiegand data stream and recalls the user's template from memory for comparison. Upon verification, the card number including the facility code and parity bits are output to the access control terminal in Wiegand format. The third party access control makes the decision to admit or deny the user.

The magnetic card emulation mode operates very much the same as the Wiegand output. A field of data and additional characters are transmitted over the data line with markers or "sentinel" characters that the card access system recognizes. Another channel transmits clocking information and a third conductor is common.

The implementations of hand scanner interfaced as barcode emulators are also available. In this emulation mode, information is transmitted to a receiver over two wires, data and ground. The operation is much the same as the other card emulation modes, where the user enters data in the scanner. Upon verification, the scanner sends the emulated barcode information to a third party device.

The systems integrator has several ID input choices when integrating hand scanners into card reader systems. The typical hand scanner has an integral numerical keypad with which the user can enter a PIN number. In this application, the scanner forwards the user's ID number and a pre-programmed facility code (where applicable) to the third party device once the user's identity has been verified by the scanner. Optionally, a card reader can be attached to the hand geometry scanner thus allowing the user to simply present or swipe a card to start the verification process. Integrating the card reader and hand scanner reduces the total verification time and greatly reduces user errors in entering PIN numbers.

The typical user template in a hand scanner is a 9 byte or 72 bit string of data. The template is small enough to be written on other media, such as a card. In this scenario, the user swipes a card, which sends his or her ID number and the actual hand geometry template to the hand scanner. The scanner makes the comparison based on the template received and sends the ID number on to a third party system that makes the access decision. The advantage of keeping the template on the card is that biometric information does not have to be in the resident memory of the hand scanner. Therefore, large networks and user databases are easily managed using only ID numbers. However, the human hand changes over time and there is a risk of the template becoming obsolete. The evolution of "smart" cards easily solves this problem. Upon enrollment, the chip within the card receives the user's template from the hand scanner where it remains in memory. The user inserts or presents the card at a smart card reader that forwards the user's template to a hand scanner for comparison. The scanner verifies the user, records the image of the user's hand and returns an updated template to the smart card. The smart card retains the user's latest template information.

Stand Alone Access Control

The simplest implementations of hand geometry based authentication operate in stand alone mode. In the stand-alone mode, a hand geometry scanner is more than a comparison device. It is a complete single door access control system. When configured as an unlocking device, the hand scanner has two control outputs, a lock output and an auxiliary output that controls other devices. The lock output energizes whenever a user successfully verifies, assuming there are no time or location restrictions in the user's database file. Typically, a relay connected to the lock output controls an electric locking device. A built-in timer keeps the lock output energized for a pre-programmed period after successful verification.

Other inputs to the scanner can also energize the lock output. A door position switch connected to one of the scanner inputs will cause the lock output to immediately time out when it senses an opened door. The time shunt causes the lock to re-secure the moment it closes thus reducing the opportunity for a follow-on unauthorized entry.

An internal tamper switch energizes the auxiliary output if the scanner is dislodged from its mounting. Operation of a "request to exit" switch or motion sensor on the secure side of an entrance energizes the lock output unlocking the door to allow persons to exit. The request to exit can also be controlled by a numerical keypad on the secure side. This requires the entry of an ID number stored in the scanner's database to energize the lock output. The lock output can also be controlled from a programmed time zone within the scanner. This function causes the automatic unlocking and relocking of doors during business hours or at any time determined by the system administrator. Using external logic components, it is also possible to devise a "first in" function, wherein the door remains locked until the scanner processes a valid entry. The door remains unlocked throughout the balance of the time zone.

Privacy Issues

Hand geometry measurments have information to verify a person's identity but the templates cannot be accurately "reverse engineered" to identify users. Consequently, hand geometry based authentication protects privacy of the users better than other biometrics, say, fingerprints.

Operation by Disabled People

In general, hand scanners could be enabled for blind persons to use. Keys on the ID entry keypad are raised and arranged in the typical "telephone" format, with three keys across and four keys down. An audio feedback can be provided to guide the blind user.

Outdoor Conditions

Typical hand scanner electronics and scanning components will work in extreme cold conditions, however, cold can hamper performance due to naturally occurring phenomena when a human hand contacts a cold surface. The moisture of the human body forms an "aura" of vapor when exposed to cold temperatures and may adversely affect imaging optics. For instance, when the hand comes in contact with the hand scanner platen, the vapor condenses, causing a visible "halo" to form on the platen surface. If the halo is sufficiently dense, the scanner's optics interprets it as being part of the hand and will likely reject it. The use of integral platen and LCD heaters allows the use of hand scanners in sub-freezing environments.

Typical hand geometry sensing employs infra-red imaging. The infrared content of sunlight may "blind" scanner optics. Stray infrared light (or the lack of it) does not affect the false accept rate but may adversely affect false reject rate. When possible, systems designers should shield hand scanners from direct sunlight. When sun shielding is not practical, the users should be trained to block stray sunlights with their bodies.

4. Applications

Parking Lot Application

It would seem that hand scanners would make an ideal access control for parking lots. Users need not carry cards and facility managers will be assured that only authorized users can enter the lot. However, there are human engineering factors to consider. Hand scanners presently on the market are designed for use with the right hand. Left-hand drive automobiles make it difficult for users to reach a scanner platen. It is possible to enroll users with the left hand palm facing up, but it still requires extra effort to verify. Further, sports utility vehicles and sports cars may require the platen height to vary by as much as three feet (.9 meter). In general, hand scanners are not recommended for ordinary parking lot use, however, they are in use in some special circumstances. Some truck terminals, especially those at trade free zones require drivers to log in using biometric verification. In this application, all truck cabs are the same height. Specially built left-hand scanners allow drivers to verify without leaving their cabs. In entrances where both car and truck drivers must verify, it may be necessary for them to leave their vehicle to use the hand scanners.

Cash Vault Applications

A particularly interesting hand scanner application is in use within a cash vault mantrap. The trap has two doors, an entry and an exit, and a hand scanner inside the trap to verify entrants. A personnel counter records the number of people entering from the public side of the mantrap. A programmable logic controller (PLC) reports the "count" to the hand scanner. The number of different people using the scanner must match the count. For example, if three people enter the trap, the scanner requires three verifications from three different people before the inner door unlocks. If the count is under or over, the inner door will not unlock and a supervisor manually processes the users.

Dual Custody Applications

Dual custody access control is common in physical security. Two persons, each with a key must operate two separate locks to gain access. This concept translates easily to electronic access control. In hand scanner dual custody applications, two different people must verify before the scanner sends an output. There are several variations on this method. The most common requires any two people from the database. A variation of dual custody requires one person from one database and one person from the other. Another variation requires two persons from one database and one person from another for a total of three different hands to produce an output.

Anti-passback

Anti-passback is a common access control function. To prevent a user from "passing" a card to an accomplice, the access control requires an exit before it will grant the card an access, or a time interval must pass before the card can be used again. At first, anti-passback would seem redundant for hand scanner applications. It would be difficult at best to pass back a hand. However, another possible scenario makes hand scanner anti-passback a viable application. Again, it assumes an accomplice. The authorized enrollee verifies, but sends the accomplice through the turnstile or revolving door. Then he or she verifies again and also passes through the security barrier. A special hand scanner option prevents the ID from being used twice in succession without either an exit or a time delay after verification. A higher security application places the hand scanner inside a revolving door. The user enters his or her unique PIN number to start the door in motion and the user steps in. Meanwhile, the scanner retrieves the user's template from resident memory or a host computer. The door rotates one-quarter turn and stops, making the hand scanner accessible to the user, now inside the door quadrant. The user places his or her hand to verify and continue through the door. There is no opportunity to enter a different ID number in the hand scanner as it has no keypad or data entry device. It only receives information from the outside keypad or card reader. Ultrasonic or weight sensors in the door assembly detect two persons attempting to pass as one. Upon successful verification, the door rotates another one-quarter turn allowing the user to exit on the secure side. If the user fails to verify, the door reverses returning the user to the public side.

Time and Attendance

The first hand scanner applications were for security access control. They were used like card readers. Users enter data and place their hands. In the mid-1990's a new market for hand scanners appeared, time and attendance. The market evolved from the security applications. Payroll and human relations managers saw employees using hand scanners to enter their buildings. They realized these devices worked on peoples' identity, not what they were carrying. It became apparent that hand geometry technology could take the place of a traditional punch clock. Business owners asked systems integrators to adapt hand geometry technology to time and attendance use.

The first hand geometry time and attendance installations used hand scanners connected to a printer or access control software to record users' arrival and departure. This required manual sorting of the event data, though some "computer savvy" managers exported event data files to spreadsheet programs where they could sort and calculate the data. When it appeared there was a market emerging, hand scanner firmware was changed to include specific time and attendance functions. During verification, workers could enter employment specific data such as departmental or job codes in the scanner keyboard. Access control software stored this information for later retrieval by a host computer. Hand scanner time and attendance functions were limited and data exportation was cumbersome compared to the functions of automated time and attendance systems and software. End users wanted all the time and attendance functions with biometric verification. As a result, time and attendance providers connected hand scanners to electronic time clocks. The scanners function as "dumb terminals," using magnetic stripe or barcode outputs to send employees' ID numbers to the clocks. Users "punch" at the hand scanner, then enter job specific data in the electronic time clock. This interface method is still very much in use today.

The market for electronic time keeping products continues to grow. This growth has resulted in an increasing demand for fully integrated biometric terminals to reduce system cost and simplify employee data entry. The result is a hand scanner dedicated to time and attendance use. It includes many of the time and attendance functions of an electronic time clock. Employees enter their ID numbers in the scanner, then select from a menu of work related functions. These include:

☐   Explicit punch. The scanner prompts the employee to select from a menu to record if he or she is reporting to work, leaving, taking a break, lunch, etc.

☐   Departmental transfer. The employee enters his or her department or job code.

Supervisors have specific functions as well. These may include:

☐   Supervisor override. Supervisors can edit employee punch information to authorize overtime or pay employees for work related activities prior to coming on site.

☐   Bell schedules. The scanner can operate bells or other annunicating devices at specific times programmed into time zone menus.

Time and attendance hand scanners store employee and supervisor events in a memory buffer. A host computer polls the scanner on a regular basis to collect data and leave new employee hand templates or change existing employee status.

Point of Sale Applications

The ability to verify identity has many other uses. Among them is a growing market, point of sale applications. Debit systems are becoming more common in our everyday lives as we move toward being a cashless society. The ATM (automatic teller machine) card is no longer restricted to cash withdrawals. It is rapidly becoming a universal payment card. However, there are many debit systems that are not tied to ATM cards. They are closed systems specifically for a defined group of people such as students of a high school or college, health care recipients or members of an athletic club. Members present identification, select a commodity or service, and the service provider deducts payment for goods and services from the member's account. Many service providers use hand scanners as the means to collect their members' transactions. The following are three examples of hand scanners used in point of sale applications.

A school district administers a government subsidized lunch service. The program requires accurate reporting of the number of students receiving lunches. Some students purchase the lunch program by paying a monthly fee. The school deducts the cost of each lunch served from the student's account. Other students are on public assistance and a social service agency pays the school for each lunch served to students receiving benefits. Previously, the school depended on the staff verifying student identity by examining each participating student's ID card. The staff entered each transaction on a log sheet. An administrator collected the information and manually settled accounts. The accounting process alone was a nightmare. Students frequently lost their ID cards or damaged them to the point where they were unreadable. Rather than deny students their lunches, the staff often allowed them access to the lunch program without identification. There were other problems. The number of students being served often exceeded the number enrolled in the program. To solve the problem, the school district installed an automated point of sale system with two hand scanners in each of eleven schools. Students no longer present ID cards to receive lunches. Each participating student enters his or her ID number in a hand scanner. The automated cash register and point of sale system records the transaction to be later processed by a computer running accounting software. The cashier permits verified students to proceed through the lunch line and collects cash from students who do not participate in the program. The lunch program runs smoothly and the savings paid for the hand scanners and software in one school term.

A resort hotel in the Crimea, the "Riviera" of the former Soviet Union uses hand scanners to debit guests' accounts. The hotel debits the account of guests with funds in a sponsoring bank for hotel services such as bars and restaurants. Guests who do not have an account with the sponsoring bank simply leave a deposit with the hotel when they check in. To pay a bill, the guest enters his or her room number in a hand scanner at the cashier's station and verifies. The point of sale terminal automatically debits the guest's account for the cost of the goods or service. If the guest has insufficient funds or is otherwise denied access to accounts, the cashier tactfully records the transaction and advises the guest to check with the hotel accounting staff. The biometric debit system assures that only registered guests can use hotel services. It is customary for European Hotels to require guests to leave their keys with the front desk when they are not in their rooms. With the hand scanner interface to the debit system, they need not carry identification to purchase hotel goods and services.

In the U.S., athletic clubs use a similar system. Members pay a monthly or annual fee to use the club's facilities. They can access the club without a key or identification using a hand scanner. The automated point of sale terminal logs each user's arrival time for accounting and marketing purposes. The scanner may unlock the door or release a turnstile, allowing verified members into the club. The administrative staff places members whose dues are in arrears into a null time zone within the scanner. Scanners deny access or alert staff to any attempted entry by non-active members. When the member settles his or her account, club administrators restore his or her ID and template to an active time zone.

Software used in this application can enhance biometric access by displaying the member's name and file information on a monitor screen when the member enters. The supervising staff member can greet the member by name, even though the greeter has never seen the member before. The monitor can display messages concerning account status, or special discounts available. Because the point of sale transaction is biometric, the staff administrator can be assured that the file belongs to the member who just verified.

Hand scanners adapt to many point-of-sale terminals through the use of an intermediate decoder, also known as a "keyboard wedge." The hand scanner attaches to the decoder "wand" port. The computer keyboard, usually located away from the hand scanner, also attaches to the decoder, which connects to the keyboard port on the computer. When an enrolled user verifies, the scanner sends the user ID number to the decoder, which forwards the number to the computer with a "return." Therefore, the user enters his or her ID number in the computer the same way as if using the computer keyboard. However, programmers can add characters to the decoder output. A password or keyboard entry contains information from both the hand scanner and decoder thus making unauthorized data entry difficult. The decoder application is not intended as a high security means to protect computer programs. It is intended for unsupervised data entry in point-of-sale applications.

Interactive Kiosks

Hand scanners have found broad applications in the interactive kiosks. A host computer maintains user files and interacts with the user through a touch screen monitor or keyboard. When the user verifies, the monitor displays a menu of choices from which the user may select. Similar to modern automatic teller machines, the interactive kiosk communicates with the user after ID entry and verification. The interactive biometric kiosk is in use in automated border crossings.

5. Conclusions

Current technology hand scanners have been in use for over ten years. Their applications are limited only by imagination. They offer fast, reliable ID verification in applications from nuclear power plants to day care centers. Future hand scanners will be smaller and faster as the "silicon revolution" produces faster and smaller processors. The day will come when hand scanners look just like the in movies, where the hero just waves his hand in front of a small dot on the wall and the door opens, automatically, of course.

References

[1] R. P. Miller, "Finger dimension comparison identification system," *US Patent No. 3576538*, 1971.
[2] R. H. Ernst, "Hand ID system", *US Patent No. 3576537*, 1971.
[3] I. H. Jacoby, A. J. Giordano, and W. H. Fioretti, "Personnel Identification Apparatus," *US Patent No. 3648240*, 1972.
[4] *A Performance Evaluation of Biometric Identification Devices*, Technical Report SAND91-0276, UC-906, Sandia National Laboratories, Albuquerque, NM and Livermore, CA for the United States Department of Energy under Contract DE-AC04-76DP00789, 1991.

[5] D. P. Sidlauskas, "3D hand profile identification apparatus," *US Patent No. 4736203*, 1988.

[6] J. R. Young and R. W. Hammon, "Automatic Palmprint Verification Study," Rome Air Development Center, Report No. RADC-TR-81-161, Griffith AF Base, New York, June 1981.

[7] H. Yoshikawa and S. Ikebata, "A Microcomputer-Based Personal Identification System," *Proc. of International Conference on Industrial Electronics, Control and Instrumentation (Cat. No. 84CH1991-9)*, Vol. 1, pp. 105-109,October 1984.

[8] J. Svigals, "Low Cost Personal Identification Verification Device Based on Finger Dimensions," *IBM Technical Disclosure Bulletin*, Vol. 25, No. 4, September 1982.

5 RECOGNIZING PERSONS BY THEIR IRIS PATTERNS

John Daugman
Cambridge University
Cambridge, UK
john.daugman@cl.cam.ac.uk

Abstract *Performance in biometric identification is determined by two kinds of variability among the acquired biometric templates: (1) within-Subject variability, which sets a minimum False Reject rate; and (2) between-Subject variability, whose lower limit sets a minimum False Match or False Accept rate. Clearly, it is desirable for a biometric to have maximal between-Subject variability but minimal within-Subject variability. It is also desirable for recognition decisions to be based upon features which have very little genetic penetrance (so that genetically identical or related individuals would still be distinguishable), yet high complexity or randomness, and stability over the life of the individual. A phenotypic facial feature with exactly these properties is the iris pattern within either eye. When imaged at distances up to a meter, the population entropy (information density) of iris patterns is about 3.4 bits per square millimeter, and their complexity spans about 266 independent degrees-of-freedom. The resulting decision environment for recognizing persons by their iris patterns has a decidability index of about $d' = 11$. Quantitative decision metrics such as these, resulting from 223,000 comparisons between IrisCodes in tests published by British Telecom, may be used to compare the intrinsic decision-making power of different biometrics.*

Keywords: *Phenotype, genotype, monozygotic twins, iris, randomness, complexity, degrees-of-freedom, 2D Gabor wavelets, IrisCode, demodulation, decidability.*

1. Introduction: Biological Variability, Genotype, and Phenotype

The central issue in pattern recognition is the relation between within-class variability and between-class variability. These are determined by the number of degrees-of-freedom (forms of variation) spanned by the pattern classes. Ideally the within-class variability should be small and the between-class variability large, so that decisions about "same" versus "different" can easily and reliably be made. In the case of biometric identification of persons, this basic principle implies that an optimal biometric measurement should have maximal variation across individuals, but minimal variation for any given person across time or conditions. These two dimensions of biological variation, in turn, reflect the genetic, developmental, and environmental influences associated with the concepts of genotype, phenotype, and genetic penetrance.

Genotypic and Phenotypic Limitations on Biometric Performance

Genotype refers to a genetic constitution, or a group sharing it, and *phenotype* refers to the actual expression of a feature through the interaction of genotype, development, and environment. *Genetic penetrance* describes the heritability of factors or the extent to which the features expressed are genetically determined. Those that are (such as blood group or DNA sequence) I will here call *genotypic features*, and those that are not (such as fingerprints or iris patterns, as shown below) I will call *phenotypic features*. (This terminology is somewhat idiosyncratic, but it serves to capture the intended distinction.)

Persons who are genetically identical share all their genotypic features, such as gender, blood group, race, and DNA sequence. All biological characteristics of individuals can be placed somewhere along this "genotypic-phenotypic" continuum of genetic determination, with many features (e.g., gender, fingerprints) placed firmly at either endpoint. Some features such as overall facial appearance reveal both a genotypic factor (hence identical twins "look identical") and a phenotypic factor (hence everyone's facial appearance changes over time).

The importance of these properties of biometric features, especially in the context of forensics, is that they directly influence the two basic error rates. Nearly one percent of persons have an identical twin, with whom they share all genotypic features such as their entire DNA sequence; this creates a minimum False Match rate (for a population) which we may call the *genotypic error rate*. Similarly, the tendency for some biometric features to change over time (such as facial appearance) creates some minimum rate of False Rejections which we may call a *phenotypic error rate*.

Roughly one in 80 births are twins, and about a third of these are "identical" (monozygotic). So a representative sample of 240 births (counting twin births as one event) usually yields 243 persons, among whom there are three pairs of twins; one such pair of persons are genetically identical. Thus the chances are roughly one in 121 (or $2/243 = 0.82\%$) that any person selected at random has an identical twin. Biometrics dependent upon genotypic features thus must have a minimal False Match rate of 0.82% due to this birth rate alone. Exactly the same argument, and numbers, pertain to the maximum possible confidence levels that can theoretically be achieved by DNA tests. This minimum error rate for genotypic features will worsen proportionally if human cloning, now on the horizon, becomes a future reality.

Using the example of overall facial appearance and variation, it is interesting to consider these factors especially for the case of identical (monozygotic) twins. Obviously any pair of twins are always matched in age. Each twin's appearance changes over time in the normal dramatic way, yet the pair usually remain strikingly similar to each other in appearance at any age. Nobody would deny that identical twins look much more similar to each other than do unrelated persons. Since such twins are genetically identical, their similarity in appearance serves to calibrate the extent of genetic determination of facial structure. A further, but secondary, calibration for this factor is provided by persons who share only 50% rather than 100% of their genes. These include fraternal twins, full siblings, double cousins, and a given parent and offspring. Occasionally the latter pairings have virtually indistinguishable appearance at a similar age, such as Robert F. Kennedy and his son Michael in adulthood.

It is clear that the phenotypic variation over time of any biometric such as facial appearance imposes one limit on its performance (not necessarily the lowest limit), since when such variation is great enough, it causes a False Reject. Likewise, the high genetic penetrance for facial appearance imposes a different limit on performance (again, not necessarily the lowest such limit), since persons with identical genetic constitution look so similar and so would be susceptible to a shared-genotype error, namely a False Match. These two performance limitations are summarized in the following general table:

Type of Feature	Performance Limitation
"Genotypic"	False Match Rate $\geq$ birth rate of identical twins
"Phenotypic"	False Reject Rate $\geq$ feature variability over time

2. Iris Patterns: Complex Phenotypic Features

The most numerous and dense degrees-of-freedom (forms of variability across individuals), which are both stable over time and easily imaged, are found in the complex texture of the iris of either eye. This protected internal organ, whose pattern can be encoded from distances of up to almost a meter, reveals about 266 independent degrees-of-freedom of textural variation across individuals. One way to calibrate the "information density" of the iris is by its human-population entropy per unit area. As we will see, this works out to 3.4 bits per square millimeter on the iris, based upon 222,743 IrisCode comparisons recently reported by the British Telecom Research Laboratories using the algorithms for iris encoding and recognition to be described here.

Properties of the Iris

The iris is composed of elastic connective tissue, the trabecular meshwork, whose prenatal morphogenesis is completed during the 8th month of gestation. It consists of pectinate ligaments adhering into a tangled mesh revealing striations, ciliary processes, crypts, rings, furrows, a corona, sometimes freckles, vasculature, and other features. During the first year of life a blanket of chromatophore cells often changes the colour of the iris, but the available clinical evidence indicates that the trabecular pattern itself is stable throughout the lifespan. Because the iris is a protected internal organ of the eye, behind the cornea and the aqueous humour, it is immune to the environment except for its pupillary reflex to light. (The elastic deformations that occur with pupillary dilation and constriction are readily reversed mathematically by the algorithms for localizing the inner and outer boundaries of the iris.) Pupillary motion, even in the absence of illumination changes (termed *hippus*), and the associated elastic deformations in the iris texture, provide one test against photographic or other simulacra of a living iris in high security applications. There are few systematic variations in the amount of detectable iris detail as a function of ethnic identity or eye colour; even visibly dark-eyed persons reveal plenty of iris detail when imaged with infrared light. Further discussion of anatomy, physiology, and clinical aspects of the iris may be found in Adler [2].

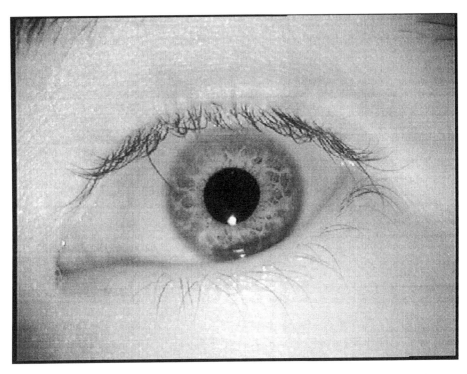

Figure 5.1 Example of an iris pattern, imaged in infrared at a distance of about 50 cm.

Localizing Irises and Analyzing Their Patterns

The two-dimensional modulations which create iris patterns are extracted by *demodulation* [9] with complex-valued 2D wavelets (Figure 5.2).

First, it is necessary to localize precisely the inner and outer boundaries of the iris, and to detect and exclude eyelids if they intrude. These detection operations are accomplished by integro-differential operators of the form

$$\max_{(r, x_0, y_0)} \left| G_\sigma(r) * \frac{\partial}{\partial r} \oint_{r, x_0, y_0} \frac{I(x, y)}{2\pi r} ds \right| \tag{5.1}$$

where contour integration parameterized for size and location coordinates r, x_0, y_0 at a scale of analysis σ set by $G_\sigma(r)$ is performed over image data $I(x,y)$.

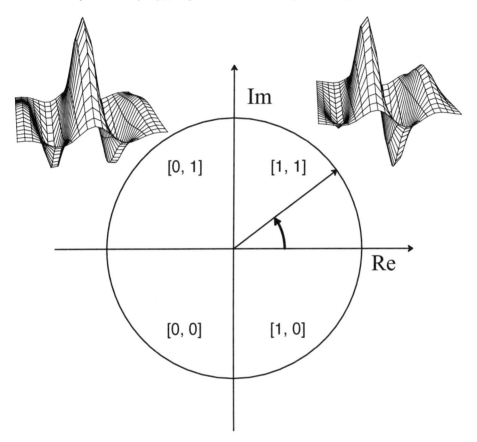

Figure 5.2 Pattern encoding by phase demodulation using complex-valued 2D wavelets.

Then, a doubly-dimensionless coordinate system is defined which maps the tissue in a manner that is invariant to changes in pupillary constriction and overall iris image size, and hence also invariant to camera zoom factor and distance to the eye. This coordinate system is pseudo-polar, although it does not assume concentricity of the inner and outer boundaries of the iris since the pupil is normally somewhat nasal, and inferior, in the iris. The coordinate system compensates automatically for the stretching of the iris tissue as the pupil dilates. It is illustrated graphically in Figure 5.3, together with a phase-demodulation IrisCode indicated in the top left as a bit stream.

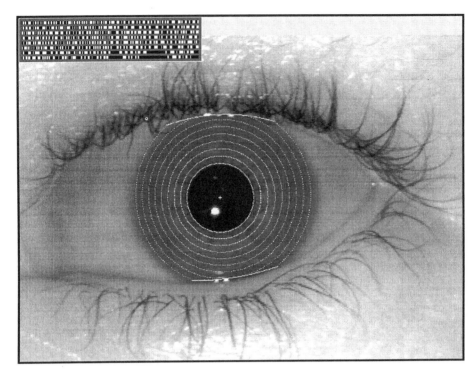

Figure 5.3 Isolation of an iris for encoding, and its resulting "IrisCode."

The detailed iris pattern is encoded into a 256-byte "IrisCode" by demodulating it with 2D Gabor wavelets [5,6], which represent the texture by phasors in the complex plane. Each phasor angle (Figure 5.2) is quantized into just the quadrant in which it lies for each local element of the iris pattern, and this operation is repeated all across the iris, at many different scales of analysis. Such local phase quantization is described by the following conditional integral equations, in which each code bit h is represented as having both a "real part" h_{Re} and an "imaginary part" h_{Im}, with $h = h_{Re} + i\, h_{Im}$, and the raw image data is given in a pseudo-polar coordinate system $I(\rho,\phi)$:

$$h_{\text{Re}} = 1 \text{ if } \text{Re} \int_\rho \int_\phi e^{-i\omega(\theta_0-\phi)} e^{-(r_0-\rho)^2/\alpha^2} e^{-(\theta_0-\phi)^2/\beta^2} I(\rho,\phi)\rho d\rho d\phi \geq 0 \quad (5.2)$$

$$h_{\text{Re}} = 0 \text{ if } \text{Re} \int_\rho \int_\phi e^{-i\omega(\theta_0-\phi)} e^{-(r_0-\rho)^2/\alpha^2} e^{-(\theta_0-\phi)^2/\beta^2} I(\rho,\phi)\rho d\rho d\phi < 0 \quad (5.3)$$

$$h_{\text{Im}} = 1 \text{ if } \text{Im} \int_\rho \int_\phi e^{-i\omega(\theta_0-\phi)} e^{-(r_0-\rho)^2/\alpha^2} e^{-(\theta_0-\phi)^2/\beta^2} I(\rho,\phi)\rho d\rho d\phi \geq 0 \quad (5.4)$$

$$h_{\text{Im}} = 0 \text{ if } \text{Im} \int_\rho \int_\phi e^{-i\omega(\theta_0-\phi)} e^{-(r_0-\rho)^2/\alpha^2} e^{-(\theta_0-\phi)^2/\beta^2} I(\rho,\phi)\rho d\rho d\phi < 0 \quad (5.5)$$

Independence and the Degrees-of-Freedom in IrisCodes

It is important to establish that there exists independent variation in iris patterns, across populations and across positions in the iris. This is confirmed by tracking the probability of a bit being set, as shown in Figure 5.4. If there were any systematic correlations among irises, this plot would not be flat. The fact that it is flat at a value of 0.5 means that any given bit in an IrisCode is equally likely to be set or cleared, and so IrisCodes are maximum entropy codes [7] in a bit-wise sense.

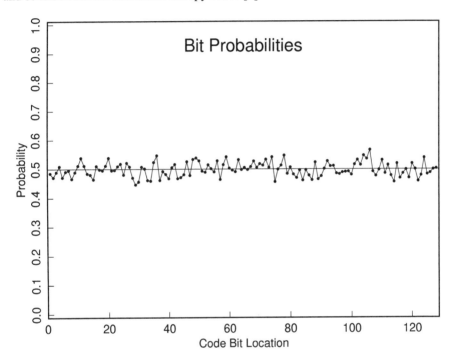

Figure 5.4 Test for independence of code bits across a population of IrisCodes.

The histogram in Figure 5.5 compares different eyes' IrisCodes by vector Exclusive-OR'ing them in order to detect the fraction of their bits that disagree. Since any given bit is equally likely to be set or cleared, an average Hamming Distance fraction of 0.5 would be expected. The observed mean was 0.498 in comparisons between 222,743 different pairings of IrisCodes enrolled by British Telecom. The standard deviation of this distribution, 0.0306, indicates that the underlying number of degrees-of-freedom in such comparisons is $N = pq/\sigma^2 = 266$. This indicates that within any given IrisCode, only a small subset of the 2,048 bits computed are independent of each other, due to the large correlations (mainly radial) that exist within any given iris pattern. (If every bit in an IrisCode were independent, then the distribution in Figure 5.5 would be very much sharper, with an expected standard

deviation of only $\sqrt{pq/N} = 0.011$; thus the Hamming Distance interval between

0.49 and 0.51 would contain most of its area.) The solid curve fitted to the data is a binomial distribution with 266 degrees-of-freedom; this is the expected distribution from tossing a fair coin 266 times in a row, and tallying up the fraction of heads in each such run. The factorials which dominate the tails of such a distribution make it astronomically improbable that two different IrisCodes having these many degrees-of-freedom could accidentally disagree in much fewer than half their bits. For example, the chances of disagreeing in only 25% or fewer of their bits (achieving a Hamming Distance below 0.25, or equivalently the chances of getting fewer than 25% heads in 266 coin tosses) are less than one in 10^{16}. Thus the observation of a match even of such poor quality (25% bits incorrect) is extraordinarily compelling evidence of identity.

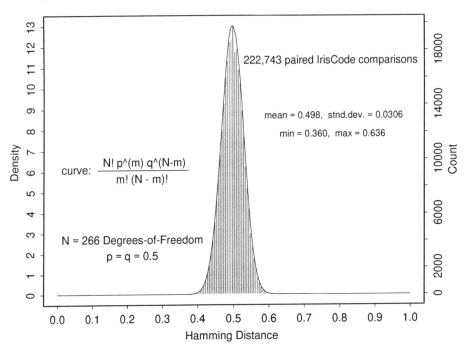

Figure 5.5 Histogram of raw Hamming Distances between 222,743 pairs of unrelated IrisCodes. The fitted curve is a binomial distribution with 266 degrees-of-freedom.

3. Genetically Identical Irises

Just as the striking visual similarity of identical twins reveals the genetic penetrance of overall facial appearance, a comparison of genetically identical irises reveals that iris texture is a phenotypic feature, not a genotypic feature. A convenient source of genetically identical irises are the right and left pair from any given person. Such pairs have the same genetic relationship as the four irises of two identical twins, or indeed in the probable future, the *2N* irises of *N* human clones. Eye colour of course has high genetic penetrance, as does the overall statistical quality of the iris texture, but the textural details are uncorrelated and independent even in genetically identical pairs. This is shown in Figure 5.6, comparing 648 right/left iris pairs from 324 persons.

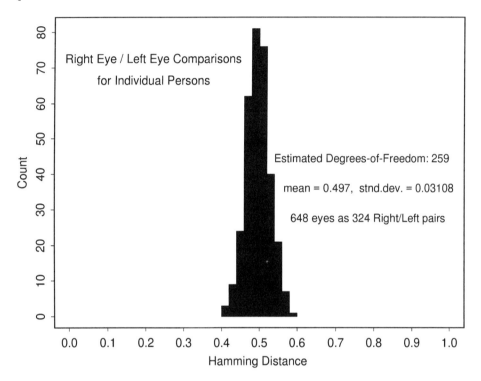

Figure 5.6 Histogram of raw Hamming Distances between IrisCodes computed from 324 pairs of genetically identical irises (648 eyes in right/left pairs). This distribution is statistically indistinguishable from Figure 5.5, which compared unrelated irises.

The mean Hamming Distance is 0.497 with standard deviation 0.031, indicating 259 degrees-of-freedom between genetically identical irises. These results are statistically indistinguishable from those shown in Figure 5.5 for genetically unrelated irises. This shows that the detailed phase structure extracted from irises by the phasor demodulation process is purely phenotypic, so performance is not impaired (as it is for faces) by the birth rate of identical twins.

4. Statistical Recognition Principle

The principle of operation underlying this approach to pattern recognition is the *failure* of a test of statistical independence. Samples from stochastic sequences with sufficient complexity need reveal only a little unexpected agreement, in order to reject the hypothesis that they are independent. For example, in two runs of 1,000 coin tosses, agreement rates between their paired outcomes higher than 56% or lower than

44% are extremely improbable: the odds against a higher or lower rate of agreement are roughly 10,000 to 1. The failure of a statistical test of independence can thereby serve as a basis for recognizing patterns with very high confidence, provided they possess enough degrees-of-freedom. Combinatorial complexity of random patterns generates similarity metrics having binomial-class distributions, even when the underlying Bernoulli trials are correlated [13] as they are in IrisCode comparisons. With so many degrees-of-freedom, the binomial-class distributions have tails that are dominated by large factorials. For this key reason, iris patterns allow recognition decisions about personal identity to be made with astronomic confidence levels. The practical importance of such astronomic odds against any False Match arising by chance is that it permits huge databases (even of "planetary" size) to be searched exhaustively, without diluting down the odds to unacceptable levels as a consequence of allowing so many opportunities for a False Match by performing (say) 10^8 or 10^9 comparison tests. Biometrics that lack so many degrees-of-freedom and therefore lack this huge combinatorial property would be limited to use in one-to-one verification mode, or to searching only rather small databases.

Extreme-Value Distribution for Rotated IrisCodes

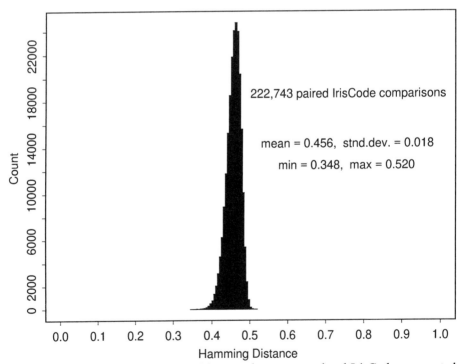

Figure 5.7 Histogram of Hamming Distances between unrelated IrisCodes computed after comparisons in multiple (n=7) relative rotations, keeping only each best match.

The computed IrisCode for any eye is invariant under translations and dilations (size change), including changes in the pupil diameter relative to the iris diameter.

However, the phasor information scrolls in phase as the iris is rotated, due to tilt of the head or camera or due to torsional rotation of the eye in its socket. Therefore all iris comparisons need to be repeated over a range of relative rotations, keeping only the best match. This amounts to sampling the distribution of Figure 5.5 many times and keeping only the smallest value, which leads to the extreme-value distribution given in Figure 5.7.

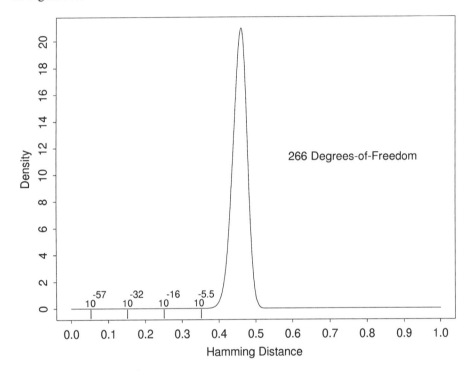

Figure 5.8 Theoretical density function for the derived binomial distribution for best IrisCode matches after multiple (n=7) relative rotations.

The raw binomial distribution shown earlier in Figure 5.5 had the form:

$$f(x) = \frac{N!}{m!(N-m)!} p^m q^{(N-m)} \tag{5.6}$$

where $N = 266$, $p = q = 0.5$, and $x = m/N$ is the Hamming Distance. Let $F_0(x)$ be its cumulative from the left, up to x: $F_0(x) = \int_o^x f(x)dx$. When only the smallest of n samples from such a distribution is kept, the resulting extreme-value distribution (derived in [7]) has density $f_n(x)$:

$$f_n(x) = nf(x)[1 - F_0(x)]^{n-1} \tag{5.7}$$

as plotted in Figure 5.8 (fit to the data will be seen later in Figure 5.9.) The areas under the tail of this probability density function are shown marked off at various points, illustrating that, for example, finding accidental agreement of two unrelated IrisCodes in even 65% or more of their bits (a Hamming Distance of 0.35 or lower) has very small probability (1 in 295,000). This illustrates that we can tolerate a huge amount of corruption in iris images due to poor resolution, poor focus, occluding eyelashes and eyelids, contact lenses, specular reflections from the cornea or from eyeglasses, camera noise, etc. We can accept matches of very poor quality, say up to 30% of the bits being wrong, and still make decisions about personal identity with very high confidence.

5. Decidability of Iris-Based Personal Identification

The overall *decidability* of the task of recognizing persons by their iris patterns is revealed by comparing the Hamming Distance distributions for "same" versus "different" irises. To the degree that one can confidently decide whether an observed sample belongs to the left or the right distribution in Figure 5.9, this recognition task can be successfully performed.

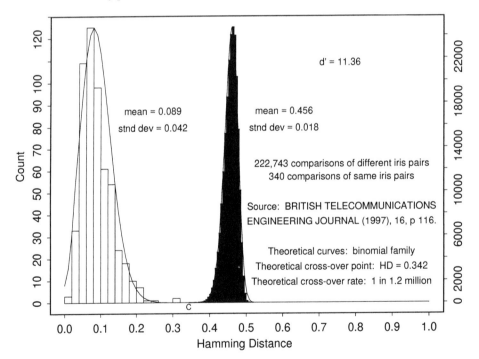

Figure 5.9 Decision environment for personal identification based on iris patterns.

For such a decision task, the Decidability Index d' measures how well separated the two distributions are, since recognition errors are caused by their overlap. If their two means are μ_1 and μ_2, and their two standard deviations are σ_1 and σ_2, then d' is defined as

$$d' = \frac{|\mu_1 - \mu_2|}{\sqrt{(\sigma_1^2 + \sigma_2^2)/2}} \tag{5.8}$$

This measure of decidability (or detectability) is independent of how liberal or conservative is the acceptance threshold used. Instead it reflects the degree to which any improvement in (say) the False Accept error rate must be paid for by a worsening of the False Reject error rate. The measured decidability is $d' = 11.36$ for iris recognition, which is much higher than that reported for any other biometric or inferred from their corresponding dual histogram plots showing "same" versus "different" template comparisons as in Figure 5.9.

By calculating the areas under the curves fitted to the observed distributions of Hamming Distances, we can compute the theoretical error rates as a function of the decision criterion employed. These are provided in the following Table, for various Hamming Distance acceptance thresholds. The cross-over point is at 0.342, at which fraction of disagreeing bits the odds of either type of error are equal to 1 in 1.2 million for the fitted pair of distributions in Figure 5.9.

	Error Probabilities	
HD Criterion	*Odds of False Accept*	*Odds of False Reject*
0.28	1 in 10^{12}	1 in 11,400
0.29	1 in 10^{11}	1 in 22,700
0.30	1 in 6.2 billion	1 in 46,000
0.31	1 in 665 million	1 in 95,000
0.32	1 in 81 million	1 in 201,000
0.33	1 in 11.1 million	1 in 433,000
0.34	1 in 1.7 million	1 in 950,000
0.342 Cross-over	**1 in 1.2 million**	**1 in 1.2 million**
0.35	1 in 295,000	1 in 2.12 million
0.36	1 in 57,000	1 in 4.84 million
0.37	1 in 12,300	1 in 11.3 million

6. Identification versus Verification

Because the probabilities of False Accepts are so low even at rather high Hamming Distances, as shown in the Table above, it is possible (indeed routine) with this approach to perform exhaustive searches through very large databases for *identification* of a presenting iris pattern, rather than merely a one-to-one comparison for *verification*. Clearly, exhaustive search identifications are far more demanding than mere verifications, since the probabilities of a False Accept in any single comparison are increased proportionately with the size of the exhaustive search database. More precisely, if P_1 is the probability of a False Accept in a single (one-to-one) verification trial with an impostor, then P_N, the probability of getting any False

Matches in identification trials after searching exhaustively through a database of N different impostors, is:

$$P_N = 1 - (1 - P_1)^N \tag{5.9}$$

This is a terribly demanding relationship. For example, even if P_1 were 0.001 (better than any published test results for overall face recognition, or most other non-iris biometrics), then even after searching through a database of merely N = 200 impostors, the probability of getting one or more False Matches among these impostors is P_N = 0.181. When the database of impostors has grown merely to N = 2,000 the probability of a False Match among them will have grown to P_N = 0.86. However, with iris recognition, the confidence levels against a False Match are so high that one can afford to search even a national or a planetary database exhaustively, and still suffer only minuscule chances of a False Match despite so many opportunities. The above Table of cumulatives under the fitted British Telecom distributions indicates that if we use an acceptance Hamming Distance criterion of 0.28 (i.e. allowing up to 28% of the bits in two IrisCodes to disagree while still accepting them as a match), the False Accept probability in single trials is 10^{-12}. Even after diluting down these odds by performing an exhaustive search over the total number of human irises on the planet, roughly 10^{10}, the chances of any False Match among them would still be only about 1%. This is an extraordinary statistical situation for a recognition system, and it reveals the power of combinatorics to solve pattern recognition problems by reducing them to the detection of the failure of a test of statistical independence, when there are enough degrees-of-freedom.

7. Stability of Iris Patterns Over Time

The within-class variability for a biometric (variation in a given person's template over time or conditions) is the source of False Reject decisions. We saw in Figure 5.9 that typically about 10% of the bits in an IrisCode disagree when the enrolled and presenting patterns are compared, due to factors such as inadequate imaging resolution, poor focus, motion blur, occlusion by eyelashes, artifacts from contact lenses, corneal reflections, scattering from scratches or dust on eyeglasses, CCD camera noise, etc. Because of non-uniform thickness of the iris, its elasticity is not uniform, and hence the first-order ("rubber sheet") model for inverting the deformations in the iris pattern as the pupil undergoes large dilation or constriction is not completely accurate. Moreover, at extreme dilations an iris may display radial folding rather than elastic deformation. All of these sources of corruption produce non-zero Hamming Distances for genuine matches; the tests by British Telecom (reported in [11] and reproduced above in Figure 5.9) show typical disagreement even under "good" imaging conditions in 8.9% of the bits. The fitted binomial distribution shown at left in Figure 5.9 has 46 degrees-of-freedom: this is the distribution one would expect to obtain from tossing a coin, whose probability of heads is p = 0.089, in runs of N = 46 tosses. We can tolerate this distribution because it remains so well separated from the narrower one having 266 degrees-of-freedom centered on much higher Hamming Distances, that results from comparing different irises.

There is a popular belief in long-term changes in the appearance of the iris, reflecting the state of health of the various organs in the body, one's personality or mood, and indeed one's future. Practitioners skilled in the art of interpreting these aspects of iris patterns and in diagnosing one's health, personality, and mutual compatibilities, are called iridologists. Iridology is popular in Rumania and in California; a Bay-Area directory of practitioners is available at http://www.best.com/~joyful/contactlist.html, and a chart showing how each individual organ of the body is mapped onto a specific region of each iris for diagnosis is given at http://www.best.com/~joyful/charttoiridology.GIF.

There may be a scientific basis for two sorts of changes in iris appearance: (1) soon after birth, a blanket of chromatophore cells establishes eye colour; until this happens, many babies have slate-blue eyes; and (2) some drug treatments for glaucoma involving prostoglandin-analogues are reported anecdotally to change iris colour, but this is not yet documented in the medical literature. In any case, changes in iris colour itself are irrelevant for the method of iris recognition described here, as all imaging is done with monochrome cameras and using primarily infrared illumination, in the 700nm - 900nm band. The clinical database of iris images made available to this author from ophthalmologists' photographs spanning a 25 year period did not reveal any noticeable changes in iris patterns for individual Subjects; some changes of hue in the colour prints were apparent, but these are difficult to disentangle from variations in the colour printing process over such a long time period.

As for the general claim that iris patterns reveal one's state of health and thus may change systematically over time, there have been four reviews published in medical journals reporting various scientific tests of iridology [3,4,10,12], and all dismiss it as a medical fraud. In particular, the review by Berggren [3] concludes: "Good care of patients is inconsistent with deceptive methods, and iridology should be regarded as a medical fraud."

8. Countermeasures Against Subterfuge

There are several ways to confirm that a living iris is being imaged, and not (for example) a photograph, a videotape, or a fake iris printed onto a contact lens, glass eye, or other artifice. One obvious method is to track the ratio of pupil diameter to iris diameter, either when light levels are changing, or even under steady illumination. The pupil can be driven larger or smaller by programmed random changes in light level, with a response time constant of about 250 msec for constriction and about 400 msec for dilation. But even without programmed illumination changes, the disequilibrium between excitatory and inhibitory signals from the brainstem to the enervation of the pupillary sphyncter muscle [2] produces a steady-state small oscillation at about 0.5 Hz termed *hippus*. Since the algorithms must constantly track both the pupil boundary and the iris boundary anyway [7], it is routine to monitor the amount of hippus; its coefficient of variation is normally at least 3%.

Further tests to exclude a photograph of somebody else's iris involve tracking eyelid movements, and indeed examining ocular reflections when simply turning on and off small infrared LEDs in random sequences at various positions on a device face plate in front of the Subject. These should create correspondingly changing reflections from the moist cornea of a living eye, whereas a photograph would not be able to change the locations of specular reflections from the cornea of the photographed eye. (In fact there are four "Purkinje" reflections from a real eye, arising from its four optical surfaces: the front and back of the cornea, and the front and back of the lens. Three of these surfaces are curved outward and produce ipsilateral reflections, whereas the fourth is curved inward and so produces a contralateral reflection. Normally all four reflections can be detected, but the first dominates in intensity.)

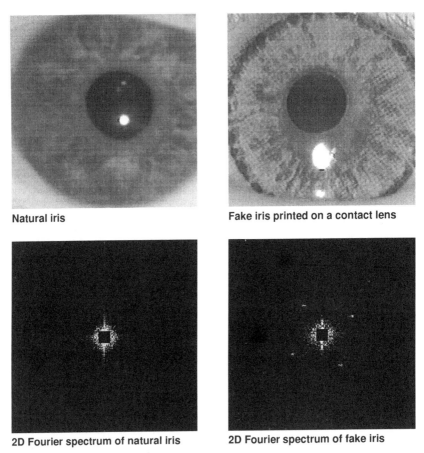

Natural iris   **Fake iris printed on a contact lens**

2D Fourier spectrum of natural iris   **2D Fourier spectrum of fake iris**

Figure 5.10 Illustration of one countermeasure against subterfuge: detecting a printed iris pattern on a contact lens by the 2D Fourier domain artifacts of printing.

Still further tests involve the characteristic spectral signature of living tissue in infrared illumination. Hemaglobin in oxygenated blood has an absorption band in near infrared wavelengths, and in specialized applications it is even possible to distinguish between arterial (oxygenated) versus venous (deoxygenated) blood with such measures. Vein patterns create distinctive silhouettes in infrared wavelengths, and indeed this is the basis of a number of other biometrics. In contrast, printers' dyes and emulsions and the reflectance properties of photographic papers are often completely ineffective for infrared light.

Finally, certain vanity contact lenses are available in the USA with fake iris patterns printed onto them (purely for the purpose of changing one's apparent eye colour). The fact that such a fake "iris" is floating on the spherical, external surface of the cornea, rather than lying in an internal plane within the eye, lends itself to optical detection; likewise the fact that the printed iris pattern does not undergo any distortions when the pupil changes in size, as a living iris pattern does. Moreover, the printing process itself creates a characteristic signature that can be detected, as illustrated in Figure 5.10. The panels show a natural iris, and a fake one printed onto a contact lens, together with their 2D Fourier power spectra. (The central square of each Fourier spectrum has been blanked out to prevent its domination.) The dot matrix printing process generates four points of spurious energy in the Fourier plane, corresponding to the directions and periodicities of coherence in the printer's dot matrix, whereas a natural iris does not have these spurious coherences.

9. Execution Speeds

On an embedded Intel 486DX66 processor (66 MHz), the execution times for the critical steps in iris recognition are as follows, with optimized integer code:

Operation	Execution Time
Assessing image focus	28 milliseconds
Localizing the eye and iris	408 milliseconds
Fitting the pupillary boundary	76 milliseconds
Detecting and fitting the eyelids	93 milliseconds
Demodulation and IrisCode creation	102 milliseconds
XOR comparison of two IrisCodes	10 microseconds

Once an IrisCode has been computed, it is compared exhaustively against all enrolled IrisCodes in the database, in search of a match. This search process is facilitated by vectorizing the Exclusive-OR comparisons to the word-length of the machine, since two integers of such length (say 32 bits) can have all of their bits XOR'd at once in a single machine instruction. Thus the elementary integer XOR instruction is an extremely efficient way to detect and tally up the total number of bits that disagree (i.e. the Hamming Distance) between two IrisCodes. Ergodicity (representativeness of subsamples) and commensurability (universal format of IrisCodes) facilitate extremely rapid comparisons in searches through large databases. On a 486DX66 processor the rate of raw comparisons approaches 100,000 IrisCodes per second, and this rate could be increased using dedicated PLA hardware to many millions of persons per second if such large databases of IrisCodes are ever enrolled.

10. Current Usage of this Technique

All current commercially available systems for iris recognition are based on the algorithms described here, by software license of the executable binary code. These include systems made or under research by: British Telecom (UK); Sensar/Sarnoff Inc. (USA); NCR Corp. (UK); Oki Electric Co. (Japan); NTT Data (Japan); LG Electronics (Korea); Garny AG (Germany); GTE Corp. (USA); Electronic Data Systems (USA); Spring Technologies (USA); and IriScan Inc. (USA) to which the author's Patent 5,291,560 has been assigned. Current applications include: bank automatic teller machines (ATMs); telecommunications and Internet security; portal entry control; nuclear power station security; computer login validation; prison controls; electronic commerce security; and various government applications.

Acknowledgments

Supported by grants from British Telecom and The Gatsby Foundation. The large database of iris images from which IrisCodes were computed and compared were acquired and made available by British Telecom. Their cooperation and support for this research are gratefully acknowledged.

References

[1] Y. Adini, Y. Moses, and S. Ullman, "Face recognition: the problem of compensating for changes in illumination direction," *IEEE Trans. Pattern Analysis and Machine Intelligence*, Vol. 19, No. 7, pp.721-732, 1997.

[2] F. H. Adler, *Physiology of the Eye: Clinical Application (fourth edition)*. London: The C.V. Mosby Company, 1965.

[3] L. Berggren, "Iridology: A critical review," *Acta Ophthalmologica*, 63(1), pp. 1-8, 1985.

[4] D. M. Cockburn, "A study of the validity of iris diagnosis." *Australian Journal of Optometry*, Vol. 64, pp. 154-157, 1981.

[5] J. G. Daugman, "Two-dimensional spectral analysis of cortical receptive field profiles," *Vision Research*, Vol. 20, No. 10, pp. 847-856, 1980.

[6] J. G. Daugman, "Complete discrete 2D Gabor transforms by neural networks for image analysis and compression," *IEEE Trans. Acoustics, Speech, and Signal Processing*, Vol. 36, No. 7, pp. 1169-1179, 1988.

[7] J. G. Daugman, "High confidence visual recognition of persons by a test of statistical independence," *IEEE Trans. Pattern Analysis and Machine Intelligence*, Vol. 15, No. 11, 1148-1161, 1993.

[8] J. G. Daugman, *United States Patent No. 5,291,560* (issued March 1994). *Biometric Personal Identification System Based on Iris Analysis*, Washington DC: U.S. Government Printing Office, 1994.

[9] J. G. Daugman and C. J. Downing, "Demodulation, predictive coding, and spatial vision," *Journal of the Optical Society of America A,* 12(4), pp. 641-660, 1995.

[10] P. Knipschild, "Looking for gall bladder disease in the patient's iris," *British Medical Journal*, Vol. 297, pp. 1578-1581, 1988.

[11] C. Seal, M. Gifford, and D. McCartney, "Iris recognition for user validation," *British Telecommunications Engineering Journal*, Vol. 16, No. 7, pp. 113-117, 1997.

[12] A. Simon, D. M. Worthen, and J. A. Mitas, "An evaluation of iridology," *Journal of the American Medical Association*, Vol. 242, pp. 1385-1387, 1979.

[13] R. Viveros, K. Balasubramanian, and N. Balakrishnan, "Binomial and negative binomial analogues under correlated Bernoulli trials," *The American Statistician*, Vol. 48, No. 3, pp. 243-247, 1984.

6 RETINA IDENTIFICATION

Robert "Buzz" Hill[1]
Portland, OR
buzzhill@rain.com

Abstract *Retina based identification is perceived as the most secure method of authenticating an identity. This chapter traces the basis of retina based identification and overviews evolution of retina based identification technology. It presents details of the innovations involved in overcoming the challenges related to imaging retina and user interface. The retinal information used for distinguishing individuals and a processing method for extracting an invariant representation of such information from an image of retina are also discussed. The issues involved in verifying and identifying an individual identity are presented. The chapter describes performance of retina based identification and the source of inaccuracies thereof. The limitations of the retina based technology are enumerated. Finally, the chapter attempts to speculate on the future of the technology and potential applications.*

Keywords: *Fundus, choroid, fundus camera, astigmatism, ergonomics, infrared imaging, fixation.*

1. Introduction

Identification of a given person is often an essential part of transactions on a network. While this is the goal, the fact is we often are left with substitutes for true personal identification in such transactions such as something the person knows (password) or has (a card, key, etc.). Retinal identification (RI) is an automatic method that provides true identification of the person by acquiring an internal body image, the retina/choroid of a willing person who must cooperate in a way that would be difficult to counterfeit.

RI has found application in very high security environments (nuclear research and weapons sites, communications control facilities and a very large transaction-

---

[1] The author of this chapter is the original RI inventor and the founder of EyeDentify, Inc. (1976). Although, he no longer owns stock or otherwise has an interest in EyeDentify, ha has, at various times since 1987, served as its consultant.

processing center). The installed base is a testament to the confidence in its accuracy and invulnerability. Its small user base and lack of penetration into high-volume price-sensitive applications is indicative of its historically high price and its unfriendly perception.

2.   Retina/Choroid as Human Descriptor

Awareness of the uniqueness of the retinal vascular pattern dates back to 1935 when two ophthalmologists, Drs. Carleton Simon and Isodore Goldstein, while studying eye disease, made a startling discovery: every eye has its own totally unique pattern of blood vessels. They subsequently published a paper on the use of retinal photographs for identifying people based on blood vessel patterns [7].

Later in the 1950's, their conclusions were supported by Dr. Paul Tower in the course of his study of identical twins [8]. He noted that, of any two persons, identical twins would be the most likely to have similar retinal vascular patterns. However, Tower's study showed that of all the factors compared between twins, retinal vascular patterns showed the least similarity.

The eye shares the same stable environment as the brain and among physical features unique to individuals, none is more stable than the retinal vascular pattern.

Because of its internal location, the retina/choroid is protected from variations caused by exposure to the external environment (as in the case of fingerprints, palmprints etc.).

Referring to Figure 6.1, the retina is to the eye as film is to camera. Both detect incident light in the form of an image that is focused by a lens. The amount of light reaching the retina (or film) is a function of the iris (f-stop). The retina is located on the back inside of the eyeball. Blood reaches the retina through vessels that come from the optic nerve. Just behind the retina is a matting of vessels called the choroidal vasculature.

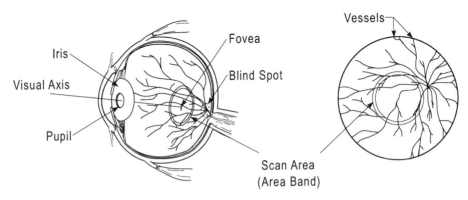

Figure 6.1 Eye and scan circle (area band).

The products of EyeDentify, Inc. have always used infrared light to illuminate the retina as will be discussed later. The retina is essentially transparent to this wavelength of light. The mat of vessels of the choroid just behind the retina reflect most of the useful information used to identify individuals, so the term "retinal identification" is a bit of a misnomer but nevertheless useful because the term is familiar. RI in this chapter will be used interchangeably to mean retina/choroid identification. This area of the eye is also referred to by medical doctors as the eye fundus.

It might seem that corrective error changes (such as becoming more near-sighted over time) might change the image of this very stable structure. In fact, the low resolution required to acquire adequate identification information masks any effect the focus errors might have. The RI products of EyeDentify, Inc. take advantage of this fact. No focusing of the RI system optics is necessary reducing cost and making the unit easier to use.

The operational rule-of-thumb for the circular scan RI systems described here is as follows: If the person to be identified can see well enough to drive with at least one eye, it is highly likely that he/she can use RI successfully.

Children as young as four years of age have been taught how to use RI. Once learned, RI is simple to use for the vast majority of the human population.

3. Background

The concept of a simple device for identifying individuals with RI was conceived in 1975. A practical implementation of this concept did not emerge for several years. The author formed a corporation, EyeDentify, Inc. in 1976 and a full time effort began to research and develop RI. In the late 1970s several different brands of ophthalmic instruments called fundus cameras were modified in an attempt to obtain live images of the retina suitable for personal identification [9,10]. Using then available fundus cameras for the optics portion of RI had at least three major disadvantages:

1. Critical alignment was necessary requiring either extraordinary expertise or the assistance of an operator.

2. A bright illumination light was necessary.

3. They were too complex and therefore too expensive.

The early RI experiments used visible light to illuminate the retina. This proved undesirable since the amount of light required for a sufficient signal-to-noise ratio was often uncomfortable to the user. An experiment was tried using near infrared as the illuminating source. This wavelength is invisible to the human eye and eliminates the bright illuminating light that can be annoying to the subject and cause his/her pupils to constrict (lowering the detected light). Inexpensive light sources and detectors existed for the near IR providing a cost saving advantage as well.

The first practical working prototype of RI was built in 1981. An RI camera using an infrared light was connected to a general-purpose desktop computer for analyzing the reflected light waveforms. Several forms of feature- extraction algorithms were evaluated. Simple correlation proved to be a superior matching technique however.

Four years of refinement led to the first production RI system built by EyeDentify, Inc. (then of Portland, OR). It was called the EyeDentification System 7.5 and performed three basic functions:

1. Enrollment - where a person's reference eye signature is built and a PIN number and text (such as the person's name) is associated with it.

2. Verification - a person previously enrolled claims an identity by entering a PIN number. The RI scans the ID subject's eye and compares it with the reference eye signature associated with the entered PIN. If a match occurs, access is allowed.

3. Recognition - RI scans the ID subject's eye and "looks-up" the correct, if any, reference eye signature. If a match occurs, access is allowed.

System 7.5 performed a circular scan of the retina, reducing the circular fundus image composed of 256 twelve bit logarithmic samples into a reference signature for each eye of 40 bytes. The contrast pattern was coded in the frequency domain. An additional 32 bytes per eye of time-domain information was added to speed up the Recognition mode.

Patents

State-of-the-art RI is covered by at least nine active U.S. patents. The RI implementation described here is covered by at least four major U.S. patents dating back to 1978 [2,3,4,5]. The patent with broad first claim to the method of RI [2] expired in 1995 and is thought to have discouraged others from developing RI technology. Now that the method of identifying individuals by their retinal patterns is no longer protected (as opposed to the apparatus to identify), we may see more interest by others in developing RI technology that is not protected by the active patents whose claims are less general than the expired patent. EyeDentify, Inc. either owns or has exclusive license to the three aforementioned patents that have not, as yet, expired. These patents deal with the alignment/fixation and user interface subsystems of the RI technology.

4. Technology

The three major subsystems of the RI technology are:
- **Imaging, signal acquisition, and signal processing**: An RI Camera that translates a circular scan of the retina/choroid into a digital waveform.

- **Matching**: A computer that verifies or recognizes the acquired eye pattern with a stored template.

- **Representation**: The eye (retina) signature reference templates with the corresponding identification information; storage issues.

Sections 6, 7, 8, 9, and 10 describe in more detail the entire RI system. Section 5 discusses representation issues.

5.   Eye Signature (Reference Template)

The representation of retina is derived from a retina scan composed of an annular region of retina, scan circle (Figure 6.1). The spot size (width of annular band) and scan circle size are chosen to return sufficient light and contrast detail in the worst case (very small eye pupil) to support the performance specification of the RI.

Two major representations for the RI eye signature have emerged. The original representation consisted of 40 bytes of contrast information encoded as real and imaginary coordinates in the frequency domain and was generated with the fast fourier transform.

The second representation, while slightly larger at 48 bytes, leaves the contrast data in the time domain. The primary advantage of the time domain representation of the eye signature is computing efficiency resulting in lower computer cost and/or higher processing speed.

Taking the ratio of the brightness at any point to the average regional brightness removes artifacts that are due to non-uniformity of the beam at the point where it enters the eye. This also normalizes the identifying signal for varying pupil sizes that greatly influence the total light returned to the detector.

The fully processed digital eye signature can be described as a normalized contrast waveform of the entire scan circle. Average RMS contrast averages approximately 1.5 to 4% of the total light detected. The contrast maximum is the brightest reflection from the scan circle and the contrast minimum is the darkest reflection from the scan circle. The waveform is normalized so that the maximum or the minimum is at the 4 bit limit (either +7 or –8, respectively) to fully utilize digital dynamic range.

The simplest form of the RI reference eye signature is an array of 96 four-bit contrast numbers for each of 96 equally spaced scan circle positions for a time-domain pattern of 48 bytes per eye. An optional 49th byte carries the AC RMS value of the waveform to be used for equalizing the RMS values of the acquired and reference waveforms in the correlation (match) routine.

6.   RI Camera

Most of us, at one time or another, have gone to an optometrist or ophthalmologist to have our eyes examined. As part of the exam, the doctor uses an instrument called a retinascope. The RI camera accomplishes the same thing as a retinascope. Its light source is projected onto the subject's retina and (the doctor in case of the retinascope) detects the return light. The light coming from the retinascope is in a collimated beam so that the eye lens focuses it to a spot on the retina.

The retina reflects some of the light back towards the eye lens, which once again collimates the light. This light leaves the eye at the same angle that it enters the eye, a process called retro-reflection. The light reflected from the retina is observed by the examining doctor who holds the instrument to his own eye. In the case of RI, a light detector replaces the examining doctor's eye.

If the doctor were to examine the eye from a number of points 10 degrees off the visual axis of the patient's eye, it would simulate the action of the fovea centered RI scan we will discuss here.

Old Camera

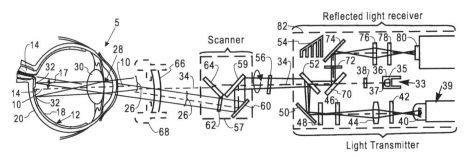

Figure 6.2 Old camera.

The first products of EyeDentify used a camera disclosed in detail in US Patent #4620318 [4] (Figure 6.2). This design used a rotating mirror assembly to generate the scan circle on the retina. Hot mirrors (reflecting infra red while transmitting visible light) are used to combine the Scanner optical path and the align/fixate target optical path. What follows is a description of the operation of the relevant portions of the camera described in the patent as they relate to the EyeDentification System 7.5.

A fixation target (33) allows the RI subject to properly focus his/her eye (5) and align its visual axis (10) with an optical axis (34) of the scanner. Fixation target (33) includes a visible light emitting diode (35) positioned in a mounting structure (36) having a pinhole (37). LED (35) illuminates the fixation reticle (38).

US Patent#4923297 [1] describes an improved fixation targeting system that replaced the system described above in production 7.5's. This patent describes the 7.5 fixation target system as a quasi-reticle that generates enhanced multiple ghost reflections of a single pinhole. It is a simple plate of glass with a partially silvered mirror on one surface and an opaque mirror surface with a pinhole on the other illuminated by a light emitting diode.

Alignment is a critical requirement of RI and this so-called "ghosticle" alignment/fixation system accomplishes its function elegantly. It is simple and intuitive - just line up the dots - and both alignment and fixation are assured. Yet it is inexpensive and easier to align in production than previous RI alignment/fixation systems.

Once alignment and fixation are accomplished the scan can be initiated either manually by pressing a button or automatically when the RI is placed in the Auto-Acquire mode (a feature introduced in the model 8.5 product).

An IR source (39) provides a beam of IR radiation for scanning fundus (12) of eye (5). IR source (39) includes an infrared light emitting diode (the drawing shows a tungsten bulb (40) as the light source) that produces light that passes through a spatial filter (42) and is refracted by a lens (44). An IR filter (46) (not used in the IR LED version) passes only the IR wavelength portion of the beam, which then passes through a pinhole (48). The beam is then reflected by a mirror (50) onto a beam splitter (52) that is mounted to coincide with the fixation target optical axis (34).

· The scanner directs the beam into the fixated eye from an angle of 10 degrees offset from the optical axis. The scanner includes a rotatable housing (57) and scanner optics that rotate with the housing as indicated by a circular arrow (58). As the scanner rotates, the 10 degree offset beam rotates about the optical axis.

The scanner optics include a hot mirror (59) (one that reflects IR radiation while passing visible light), located in the path of the source beam and the fixation beam. The visible wavelength fixation beam is passed through hot mirror (59), while the IR source beam is reflected away from the housing (57) at a point spaced apart from optical axis (34) and is oriented to direct the IR beam through an IR filter (62) and into the eye (5) as housing (57) rotates. Hot mirror (59) causes a displacement of the fixation beam so an offset plate (64) is positioned to compensate for the displacement.

When housing (57) rotates, the IR beam is directed into the eye (5) in an annular scanning pattern centered on the fovea as represented by circular locus of points (32). Light reflected from fundus (12) of the eye (5) varies in intensity depending on the structures encountered by the scan. The reflected light is re-collimated by the lens (30) of the eye (5) directed out pupil (28), back through objective lens (66) and IR filter (62), and reflected off scanner mirror (60) and hot mirror (59). The reflected beam is then focused by objective lens (56) on to beam splitter (52) which passes a portion of the reflected scanning beam to a hot mirror (70) that reflects the beam through a spatial filter (72). The beam is next reflected by a mirror (74), refracted by a lens (76) and passed through another spatial filter (78) to a detector (80).

New Camera

Current RI camera technology is based on an active US Patent [5]. It is a much simpler design that also takes advantage of the concentric nature of the RI's fixation and scanning to reduce labor intensive alignment of camera parts and the part count.

The current RI camera is shown schematically in Figure 6.3. It includes a rotating scanner disk (116) that integrates a multi-focal Fresnel fixation lens (114), a Fresnel optical scanner (122,124) and an angular position encoder (140) into a unitary, inherently aligned, compression-molded scanner disk. An RI subject views through the multi-focal Fresnel lens, an image of a pinhole (108) illuminated by a krypton bulb (104). The multi-focal lens is centered on the disk and creates a multiple in- and out-of-focus images (180 182, 184, 186) of the pinhole image. By setting the focal distances of these images along a range that includes corrective errors of from -7 diopters (very near sighted) to +3 diopters (very far sighted), at least one of the pinhole images will be in relatively sharp focus for virtually everyone in the RI subject population. The images will appear concentric when the RI subject is properly aligned with the scanner disk and associated optics.

Once aligned, the subject initiates scanning which causes the scanner disk to rotate. The Fresnel optical scanner receives IR light from the krypton bulb light source and creates an IR scanning beam (126). IR light reflected by the eye fundus (12) of the RI subject returns via a reciprocal path, by way of a beam splitter (112) and into a detector (134) to generate eye waveform data. Rotational position information from the encoder instructs the signal acquisition system when to sample the detector's output.

The key feature of this new RI camera design is that it integrates and inherently aligns multiple optical elements greatly reducing both the material and labor costs of the original RI camera. Overall costs of the camera yields to manufacturing economies of scale much more so than with the original RI camera.

Both camera types share the same subsystem functions:

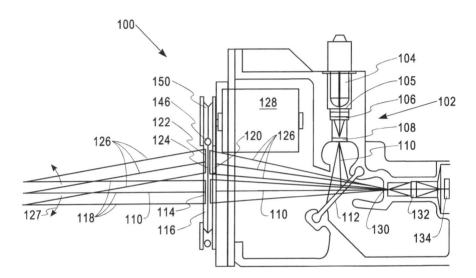

Figure 6.3 New camera.

Target - Align & Fixate

To insure that the circular scan of the RI is centered on the fovea and that the subject is in the scanner beam throughout the scan, an alignment/ fixation target is presented to the ID subject. One form of this target is an optical system that presents say four simple reticles at focal distances of -7, -3, 0 and +3 diopters. For virtually all of the ID population, at least one of the reticles will be in focus regardless of corrective error (near-sightedness through far-sightedness). When the ID subject "focuses" (fixates) on the target, the RI is angularly aligned to subject's eye, centering the RI's scan circle on the eye fundus. When he/she aligns two or more of the reticle patterns nulling their parallax, the RI illumination beam is centered on the eye pupil. Translation along the optical (Z) axis is not critical and is achieved by providing a rest for some part of the face (forehead, eye socket, etc.). Rotation about the Z-axis caused by head tilt is addressed by the Rotator algorithm, discussed later.

It is important to note that Fixation/Alignment is an absolute requirement for this method of RI to work. It would be prohibitively difficult to identify someone using RI without his or her cooperation in performing this function. Depending on one's perspective, this requirement can be seen as a benefit (usually to the ID subject) or a negative (covert ID). This does, however, prevent RI's use in identifying an individual against his or her will which may make RI appear more acceptable to the ID subject population.

Transmitter (Light Source)

The light source is ideally near infrared and is not visible to the identification subject. The illumination spot projected on the retina must be uniform. A suitable diffuser is required to achieve spot uniformity when using a light source that, when projected on the retina/choroid, is not homogenous. This is usually the case with an IR Light Emitting Diode and can also be true of other light sources.

A tungsten lamp is considerably brighter than an IRED and can produce better S/N figures. The disadvantages of such a lamp compared to an IRED is the need for filter, turn-on time and lamp life. Retinal identification systems have been proposed that would use a laser (preferably solid state). The author is not familiar with any commercially available system that uses a laser, however. Further, lasers can be considered dangerous by the RI subject population.

Receiver

The light receiver is composed of a silicon photodetector, a high gain pre-amplifier and a sharp cut-off low-pass filter. The filter is necessary to sharply reduce high frequency noise generated by the detector/preamp that is outside of the useable passband which is determined by the spot and scan circle sizes and the scanner speed. With the selected parameters a good choice is an 8th order switched capacitor elliptic filter with a corner frequency of approximately 220 Hz.

Scanner

The scanner has to deal with the light noise arising from (i) corneal reflections, (ii) other scattered light sources, and (iii) ambient light. Reflected noise in the RI comes from essentially four sources, the front and back surfaces of the cornea, and the front and back surfaces of crystalline lens. Extensive spatial filtering that is conjugate to the retina and the scan angle reduces the light noise to insignificance.

Corneal reflections of the scanner light source is one of the primary reasons for using a circular rather than a raster scan of the eye fundus. The reflections would render the center pixels of a raster scan of the retina useless unless an annular illumination requiring very critical alignment is used. The scanner consists of the following components comprising the signal acquisition and processing subsystem:

7.   Signal Acquisition Subsystem

The signal acquisition subsystem consists of the following components:

Detector/Preamplifier

The silicon photodetector operating in the photo-ampuric mode receives the light collected from the RI camera. It is converted to a voltage by a low noise op-amp configured as a trans-impedance amplifier. With a carefully selected op-amp, the primary sources of electrical noise are the thermal noise of the feedback resistor and quantum noise. A second op-amp brings the signal level up to a level sufficient to drive the contrast processor.

A/D Conversion

The raw unprocessed analog signal derived from the camera photodetector can span at least two orders of magnitude due to the range of pupil sizes encountered in normal operation of the RI. Performing the conversion at this point requires close to 16 bits of resolution to accommodate absolute signal variations, contrast figures and sufficient resolution left to quantize the "contrast" portion of the signal. A more economical scheme is to perform the contrast processing function ahead of the conversion. An 8 bit analog-to-digital converter is all that is required in this case

Contrast Processor

8.   The function of this stage in the signal chain is to reduce the raw camera signal into salient contrast information that has both human descriptor qualities of invariance and discrimination. It can be done in hardware or software and both methods have been used successfully in EyeDentify's commercial products. The far less expensive modality is hardware because it dramatically reduces the resolution required of the analog /digital converter. The contrast processor removes the redundant or variable content from the acquired scanner waveform while retaining sufficient information to yield a unique eye signature.

8.   Computing Subsystem

The computing subsystem could be explained in terms of its hardware and software components.

Hardware

EyeDentify's System 7.5, the first widely available RI, used a Motorola 68000 microprocessor as both the controller and signal processor. By moving contrast processing to hardware and coding correlation in the time domain in the late 1980s, it was possible to move to a 68HC11 micro-controller to replace most of the functionality of the 68000 based System 7.5. The cost of the computing elements of RI have been and currently are insignificant compared to the opto-mechanical portion of the system.

Software

The software performs the following two functions: phase correction and matching.

Phase Correction

Each time the RI subject looks into the RI camera, his or her head may be slightly tilted (rotated) from the position it was before. The rotator algorithm (phase corrector) shifts the acquired waveform through the equivalent of several degrees of rotation or head tilt. This is done while correlating it with the stationary reference eye signature to find the best match (highest correlation).

Matching

Comparison of the acquired contrast waveform is done with a routine that performs the following steps:

1. Sample rate converts the reference eye signature into an array with the same number of elements as the acquired array.

2. Normalize both arrays to have a RMS value of 1.0.

3. Correlate arrays using the time domain equivalent of Fourier-based correlation.

The quality of match is indicated by the correlation value, where the time shift is zero. It ranges from +1.0, a perfect match, to -1.0, a perfect mismatch. Experience has shown that scores above 0.7 can be considered a match (see Performance discussion below).

8. System Operation

Taking an Eye Reading

Central to every RI transaction is the process wherein the camera scans the RI subject's eye. We present here the detailed user instructions below to give an idea of the user involvement and training needed for retina based identification. The subject is instructed as follows:

☐   If you wear glasses, take them off (does not apply to contacts).

☐   If the system requires PIN (Personal Identification Number), enter it (recognition does not require a PIN).

☐   Position camera at eye level (or eye to camera)

☐   The target consists of a number of softly illuminated dots. Moving the head in relation to the eye lens opening, without tilting or skewing the head centers the target. Do this until all of the dots move one behind the other. The smaller dots will then appear inside the larger dots.

☐   Both eyes should be wide open. Squinting or closing one eye can cause eyelashes to be included in the reading.

☐   Be sure that your eye is about three-quarters of inch from the eye lens.

☐   Press the scan button (or wait for scanner to stop if in the Auto-Acquire mode).

☐   Hold your head steady during the reading.

☐   Although it is important to fixate on the center of the target during the reading, you should not fixate for more than a couple of seconds before pressing the button. Otherwise, the eye may drift.

Various incarnations of RI cameras and systems have different user requirements, but the steps above apply generally to all of them.

Alignment/Fixation

To use RI, it is important for the subject to be aligned with the RI camera and fixated on its target. After peering into the camera, the subject achieves alignment by lining up the dots of the target so they appear as one. At that point fixation is also accomplished since the virtual dot image is then focused on the fovea of the subject's eye (Figure 6.1). This process assures that the subject's eye pupil is within the "acceptance diamond" which is the cross sectional shape of the volume where the entire scan's beam will fill the eye pupil. This volume is essentially like two cones placed together at their bases with their centers along the eye's optical axis (the Z axis). A larger volume means less critical alignment. The volume is a function of the exit/entrance aperture, which is determined by the size of the RI camera's objective lens.

Scanning

Eyeglasses must be removed for the RI camera to work reliably. There are two reasons: 1) Reflections from the lens surface may interfere with the scanner signal, and 2) Distortion of the retina/choroid image may occur if eye glasses are not in the same position on the face from use to use such as when they slide down the ID subject's nose. If an attempt is made to enroll an individual with eyeglasses, it is possible that the eye glass reflection will be enrolled, not the retina/choroid, resulting in a very simple eye signature that might be duplicated.

Contacts do not need to be removed. Certain types of contacts can prove problematical. Lenses can cause improper signatures if any part of the edge of the lens is inside the eye pupil while the eye is being scanned. Generally, the effect of contacts on eye signatures is so slight that it is not necessary to enroll a given eye both with and without them, except possibly in cases of severe or unusual correction (extreme near- or far-sightedness and/or astigmatism).

RI at a Distance

Just as the eye doctor can use a retinascope at a distance from the patient, a suitably designed RI can be used at a distance from the ID subject. However, the size of the RI must increase proportionate to the scan distance in order to support the RI's scan circle diameter. Working RI systems with an operating distance of 12 inches have been demonstrated in the laboratory. Other considerations in such systems include ambient light conditions and Fixation/Alignment issues. Light shields sizes have to grow in proportion to the operating distance. A long distance universal focus target's requirements change when the operating distance exceeds a certain threshold. Scanner beam size will need to be larger as well.

Enrollment

RI enrollment is the process of acquiring the reference eye signature. Each eye signature is built from several eye readings. The person responsible for enrolling a new RI subject, the enroller guides that person through the following steps:

☐ Instruction on camera use. Enroller instructs the enrollee on correct RI camera use. Enroller usually demonstrates this by scanning his/her own eye and then describes what the Enrollee must do to align the camera. Fixation is automatic when the enrollee achieves alignment.

☐ Several scans until correct fixation/alignment is verified. Out of beam condition (meaning the subject has not achieved alignment) is detected when the raw signal drops off in some part of the waveform. Both manual and automatic modes have proven effective for this purpose. Fixation can only be verified when scans are compared. Correlation scores of a scan with the reference eye signature greater than, say, 0.75 to 0.8 indicate that correct fixation has been achieved.

☐ Several scans averaged. Scans that have a correlation within a certain range (such as 0.75 to 0.8 are added to a waveform average. The impact on correlation scores of variant features such as a choroidal vessel that is substantially tangent to the scan circle is reduced with averaging.

☐ Optional Recognize - verifies whether or not the new enrollment eye signature already exists in the database either because the new enrollee has already enrolled the eye or the database is large enough to include a sufficiently similar eye signature to cause a false accept error in the Recognition mode. This step assures the new enrollment eye signature is unique to the database.

☐ Assignment of linked data (Name, Pin #, etc.).

☐ Store enrollment data.

Automatic RI enrollment techniques have been studied wherein eye scans that do not match any eye signature in the database (using the Recognition algorithm described below) but appear repeatable are given a label indicating such and stored to indicate intrusion attempts. The RI system can alert an administrator when unrecognized but repeatable eye signatures occur by displaying/printing that label.

It is important to note that enrolling is somewhat of an art as well as a science. The enroller, through experience, learns how best to train each new enrollee and to interpret correlation scores during the enrollment process. An enroller should remember several key points. Correlation scores should get progressively better as enrolment progresses. It is important for the enrollee to look away between scans to insure that the averaging process creates a true average of variations in head position. A person's "dominant eye" can be easier to enroll. If one eye is difficult to enroll, try the other.

Verify

Subsequent to enrollment an enrollee can authenticate his/her identity by entering a code (such as a PIN number). An eye scan is taken and compared with the eye signature associated with the PIN number. If the eye scan matched against the eye signature produces a correlation score above the match threshold (typically a correlation score of 0.7), the person scanned is said to be the person enrolled with the given PIN and an appropriate action is taken.

Recognize

A scan is taken and using the recognize algorithm, a match to the entire database above the correlation threshold identifies the person requesting access. Any eye signature recognize algorithm is considerably more compute intensive than verify algorithm. The simplest form of recognize would take the verify time and multiply it by the number of people in the eye signature database. Several multi-level techniques have been developed that reduce the time it takes for the recognition mode on given computing resources. In some cases, execution time has been reduced as much as two orders of magnitude. The down side of the methods tried is that they sometimes eliminated good candidates, producing false reject errors.

Today's fast microprocessors and time domain correlators have nearly eliminated the need for multi-step recognition routines for databases of medium size (hundreds to tens of thousands) and, in the process, have virtually eliminated false reject errors produced by multi-level recognition algorithms.

Large Database Recognition

Recognition is identification where the ID subject does not claim an identity (with a PIN number, card etc.) as part of the process. The acquired scan waveform is compared to an entire database of eye signatures to find the best match.

Currently available parallel processing computers can perform high accuracy RI recognition of databases composed of millions of eye signatures at very low relative cost. Indeed, some RI recognition mode feasibility has been studied on massively parallel processing supercomputers with very promising results. Simply dividing up the database and having each processor correlate the acquired scan with its portion of the database is the simplest method. RI's small signature size, uniqueness and small variance gives it a significant competitive advantage in terms of cost, speed and reliability for large database recognition mode over other biometric ID methods.

Counterfeiting the Scan

A counterfeit eye must have the following characteristics.
☐   The same optical system to simulate retina/choroid reflectivity.

☐   A lens to substantially focus the incoming collimated beam and to re-collimate the reflected beam.

☐ An alignment/fixation system that angulary orients the counterfeit eye about it's X and Y axes, translates the counterfeit eye along its X and Y axes, positions the eye at the correct distance from RI camera (translation along Z) and rotates the eye about its Z axis within the tilt range of the Rotation algorithm.

The last item is the most difficult to counterfeit. A well-designed RI provides as little information about the correct alignment/fixation as possible to the would-be counterfeiter. Variable fixation displays could also require a counterfeiter to perform an interpretation of the target in order to correctly interact with the acquisition process. For example, the ID subject could be instructed to remember a random three digit sequence that is displayed in the RI fixation target and to key it in later. This would force the counterfeiting system to see, interpret and output to the RI keypad the three digit sequence.

9.   Performance

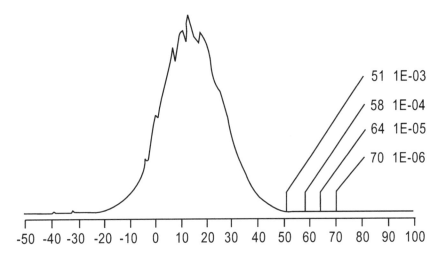

Figure 6.4 Mismatch distribution in retina based identification.

Many tests of performance of the retina/choroid scanning technology described have taken place, some with databases of several hundred individual eyes. Sandia National Laboratory has tested the products of EyeDentify and reported no false accepts and a three-attempt false reject error rate of less than 1.0% [11].

Mismatch Frequency Distribution

A frequency distribution of each eye signature compared against all others matches very closely with an ideal guassian distribution with a mean of 0.144 and a standard deviation of 0.117 as shown in Figure 6.4. The corresponding right tail probability of guassian distribution with this mean and standard deviation at a threshold score of 0.7 is approximately one in one million.

Source of Errors

The retina/choroid contrast waveform has a low variability when acquired under correct conditions. The conditions under which the variability could increase and cause false reject (Type II) errors are:

☐ Lack of Fixation or sustained fixation

☐ Out of scanner beam condition

☐ Incorrect eye distance to RI camera lens

☐ Insufficient pupil size

☐ Obstruction and distortion of the optical path from:

> dirty camera window
>
> contact lens edges
>
> subject neglects to remove eyeglasses

☐ Ambient light interference

Small pupils can cause false rejects. The purpose of the eye pupil is to regulate the amount of light reaching the retina. Bright environments such as those encountered outdoors in the daytime can cause pupils to constrict to a very small size compared to, say, indoor lighting conditions. Because light must pass the eye pupil twice (once entering and once exiting the eye), the return light to the RI camera varies inversely with the fourth power of the pupil diameter. In the worst case (smallest pupil size), resulting retina/choroid signals can be attenuated by as much as four orders of magnitude. The signal can be so low that system noise swamps the acquired eye signature data, lowering correlation scores.

Outdoor environments can also be less conducive to reliable RI performance because of the potential for high ambient light noise entering the camera and interfering with the scanned waveform. Because of the uniqueness of the retina/choroid contrast circle characteristic, false accept (Type I) errors tend to be limited to large database recognition.

10. RI Subject Motivation

An important and enduring observation of the use of RI to enroll and identify several thousand individuals over a period of two decades is the importance of motivation to have the enrollment and identification transactions succeed. Many of these observations can be said, in a general sense, of other biometrics as well.

Enrollment is the subject's first hands-on use of RI. The subject should not fear harm from the RI camera before using it the first time. Learning to align and fixate the RI camera, while a simple process, can be impeded willfully or subconsciously by a suspicious subject. Several scans are necessary and depending on the quality of the scans, this procedure can take several minutes. Because RI requires cooperation from the subject, sabotage at this stage is very easy. If a subject is difficult to enroll, the subject's motivation can deteriorate as time passes during the enrollment process.

The identification transaction (verify or recognize) is less susceptible to fear based motivational problems simply because if a subject has been successfully enrolled he/she must have overcome considerable fear or reluctance already.   But subtle sabotage can be a factor here as well. Deliberate false reject transactions with an accompanying complaint such as "it gives me a headache" can diminish confidence in the system. It is very difficult to ascertain whether subjective comments of this kind are truthful yet the result is the same - RI is less attractive.

Many user's have naively assumed that the lack of negative consequences (I can't work here because I can't/won't use RI) is sufficient to gain acceptance of a RI system by the ID subject population. Experience has taught users of RI that a perceived personal benefit (I am better off than people who can't/won't use RI) to the ID subject has a dramatically positive effect on RI enrollment and identification speed and acceptance.

11. Limitations

Perceived Health Threat

While the low light level is harmless to the eye, there is a widely held perception that retinal identification can hurt the retina. This appears to be less true in information access applications since ID subjects are generally less fearful of new technology.

Outdoors vs. Indoors

Small pupils can reduce the Type II (False Reject) performance. Because light must pass the eye pupil twice (once entering and once exiting the eye), the return light can be attenuated by as much as four orders of magnitude when the ID subjects pupils are small. The signal can be so low that quantum and feedback resistor noise swamp the eye signature data lowering correlation scores. Further, outdoor environments are less conducive to reliable RI performance than indoor environments because of ambient light conditions.

Ergonomics

The need to bring the RI device to an eye or the eye to the device makes the RI more difficult to use in some applications than other biometric identification technologies. For instance, it is quite easy for a subject, regardless of his height to reach a hand to a fingerprint or hand geometry. The eye is much less easily manipulated. Bringing the RI camera to the eye seems more practical in "workstation" applications while the opposite is true in physical access control applications.

Severe Astigmatism

Because eyeglasses must be removed in order to use RI systems reliably, people with severe astigmatism may have trouble aligning the dots in the camera's align/fixate target. To these individuals, what appears to them can be quite different than dots. This can result in ambiguous feedback during the alignment step of RI camera use, causing the eye pupil to be outside the "acceptance diamond" for part of the scan. That part of the scan will therefore be invalid.

High Sensor Cost

The camera requirement of RI puts a lower limit on the cost of the system. Manufacturing economies of scale can mitigate this problem, but RI is likely to always be more expensive than some other biometrics such as fingerprint (using chips) or speaker recognition (telephone hand-set as sensor).

12. Future

The inherent simplicity of the RI means that in mass production the cost of the entire unit could come below, say, $100. This is still considerably more expensive than some competing technologies which have a much cheaper scan component (such as fingerprint chips). The trade off is accuracy. If accuracy is important to the ID application, perhaps the additional cost of RI can be justified.

With the proliferation of e-commerce applications, RI might reach a critical mass. Because of the RI's accuracy and small signature size it fits more naturally with the encryption that is needed for e-commerce security than competing biometric ID technologies. Public key encryption systems are only as secure as their private keys and a high performance biometric identifier like RI is ideal for keeping private keys secret.

13. Conclusions

RI is a highly accurate and secure biometric identification method. The example RI system presented utilizes a small reference data size that makes it attractive in large population networked systems in both verification and recognition modes. RI, currently, is both image and performance based. The performance aspect restricts successful use to those who are motivated to see the ID transaction successful.

RI's weakness are:

☐   The cost of the signal acquisition hardware

☐   ID subject's unfounded fear that it is harmful

☐   Unfriendly access

The future will likely bring the cost of RI down dramatically if a sufficiently large demand is created to achieve manufacturing economies of scale especially as it applies to the RI camera optics and mechanics. Fear becomes less of an issue as the computer/internet age expands and raises the level of technological awareness and acceptance. Lack of a sufficient level of friendly access may prevent RI from becoming a truly ubiquitous method of identification.

References

[1] J. H. Arndt, "Optical alignment System," *US Patent No. 4923297*, 1990.
[2] R. B. Hill, "Apparatus and method for identifying individuals through their retinal vasculature patterns," *US Patent No. 4109237*, 1978.
[3] R. B. Hill, "Rotating beam ocular identification apparatus and method," *US Patent No. 4393366*, 1983.
[4] R. B. Hill, "Fovea-centered eye fundus scanner," *US Patent No. 4620318*, 1986.
[5] J. C. Johnson and R. B. Hill, "Eye fundus optical scanner system and method," *US Patent No. 5532771*, 1990.
[6] J. R. Samples and R. V. Hill, "Use of infrared fundus reflection for an identification device," *American Journal of Ophthalmology*, Vol. 98, No. 5, pp. 636-640, 1984.
[7] C. Simon and I. Goldstein, "A New Scientific Method of Identification," *New York State Journal of Medicine*, Vol. 35, No. 18, pp. 901-906, September, 1935.
[8] P. Tower, "The fundus Oculi in monozygotic twins: Report of six pairs of identical twins," *Archives of Ophthalmology*, Vol. 54, pp. 225-239, 1955.
[9] S. Yamamoto, H. Yokohuchi, and T. Suzuki, "Image Processing and Automatic Diagnosis of Color Fundus Photographs," *Proceedings 2nd International Joint Conference on Pattern Recognition,* Copenhagen, pp. 268-269, August 13-15, 1974.
[10] H. Yokouchi, S. Yamamoto, T. Suzuki, M. Matsui, and K. Kato, "Fundus pattern recognition," *Japanese Journal of Medical Electronics and Biological Engineering*, Vol. 12, No. 3, pp. 123-130, June 1974.
[11] *A Performance Evaluation of Biometric Identification Devices*, Technical Report SAND91-0276, UC-906, Sandia National Laboratories, Albuquerque, NM 87185 and Livermore, CA 94550 for the United States Department of Energy under Contract DE-AC04-76DP00789, 1991.

7 AUTOMATIC ON-LINE SIGNATURE VERIFICATION

Vishvjit S. Nalwa
Bell Laboratories
Holmdel, NJ
vic@bell-labs.com

Abstract *Automatic on-line signature verification is an intriguing intellectual challenge with many practical applications. I review the context of this problem and then describe my own approach to it. My approach breaks with tradition by relying primarily on the detailed shape of a signature for its automatic verification, rather than relying primarily on the pen dynamics. I propose a robust, reliable, and elastic local-shape-based model for handwritten on-line curves. Further, I suggest the weighted and biased harmonic mean as a graceful mechanism of combining errors from multiple models of which at least one model is applicable but not necessarily more than one model is applicable, recommending that each signature be represented by multiple models. Finally, I outline a signature-verification algorithm that I have tested successfully.*

Keywords: *Signature verification, on-line, shape based, performance evaluation, harmonic mean, jitter, aspect, sliding window, warping, saturation, cross-correlation, characteristic function, center of mass, torque, moments of inertia.*

1. Introduction[1]

Signature verification is an art: Whereas we may bring objective measures to bear on the problem, in the final analysis, the problem remains subjective. This art is both well studied and well documented as it applies to the verification by humans of signatures whose only records are visual [13,5,10] — that is, as it applies to signatures during whose production no measurement is made of the pen trajectory or dynamics. Let us call such signatures, for which we have only a static visual record, **off-line**, and let us call signatures during whose production the pen trajectory or dynamics is

---

[1] This chapter is condensed from V. S. Nalwa, "Automatic On-Line Signature Verification," *Proceedings of IEEE*, pp. 215-239, Feb. 1997.

captured, **on-line**.   Whereas attempts to automate the verification of off-line signatures have fallen well short of human performance to this point, I shall demonstrate that automatic on-line signature verification is feasible.

In a break with tradition, I challenge the notion that the success of automatic on-line signature verification hinges on the capture of velocities or forces during signature production. Whereas velocities and forces can assist us in automatic on-line signature verification, I contend that we should not depend on them solely, or even primarily. *If we were indeed unavoidably consistent over the dimensions of time and force when we signed, the use of pen dynamics during signature production — over and above that of signature shape — would be very useful in detecting forgeries, as dynamic information pertinent to a signature is not as readily available to a potential forger as is the shape of the signature, given just the signature's off-line specimens.* However, I have seen no substantive evidence to the effect that our pen dynamics is as consistent as, or more consistent than, our final signature shape when we sign. My own informal experiments indicate that we typically exhibit similar temporal variations over the production of similar handwritten curves: In general, our speed along high-curvature curve segments is low relative to our speed along low-curvature curve segments, our average overall speed varying greatly from one instance of a pattern to another irrespective of whether we are producing our own pattern or forging someone else's. This observation suggests that at least the requirement of consistency over time during signature production is of limited value beyond that of consistency over shape. At any rate, irrespective of the velocities and forces generated during the production of a signature, for us to declare two signatures to be produced by the same individual, clearly, it is necessary that the shapes of the signatures match closely.

I have organized this chapter into eight sections, describing the fundamental concepts that underlie my approach in Section 4. In Section 2, I describe what constitutes successful signature verification. In Section 3, I summarize the state of the art of automatic on-line signature verification as recorded in the published literature. In Section 5, I outline my algorithm, which I have implemented and tested both on databases and in live experiments. In Section 6, I illustrate my algorithm with a detailed example. In Section 7, I describe the performance of my implementation on three databases created by Bell Laboratories. I conclude with Section 8, where I list some of the outstanding issues in automatic on-line signature verification.

2.   Evaluating Performance

For a signature-verification system to be useful, the system must commit few errors in practice. The strategy often adopted to obtain an indication of a system's error rates, without actually introducing the system into the marketplace, is to **field test** the system on a limited scale; but even a limited field test can be expensive and time consuming. Hence, it is useful to devise criteria to help us decide whether to field test a system. I have come up with the **following two criteria to evaluate a signature-verification system** that is yet to be field tested.

Criterion 1: When you try the system in person, it must work.

Criterion 2: When you test the system on large databases, it must exhibit low statistical error rates.

Neither criterion is sufficient, and both are necessary.

Criterion 1, of course, begs the issue unless we can reach an agreement on what *work* means. I can think of at least three conditions that must be met for us to declare a system to *work* when it is tried in person:

- The system must recognize your visually similar scribbles consistently.

- You must find it difficult, if not impossible, to forge someone else's signature successfully.

- You must not be able to generate a scribble that is visually disparate from your signature and is yet accepted by the signature-verification system as your signature.

The first condition enables genuine transactions. The second condition hinders **second-party fraud** — that is, fraud by an entity other than the genuine signer. The third condition hinders **self fraud** — that is, fraud by the genuine signer. Omission of the third condition would open up the possibility of a genuine signer authorizing a transaction with the a priori intent of later denying this authorization by pointing to the visual discrepancy between the authorizing signature and the expected signature as proof of second-party fraud.

Criterion 2 for evaluating a signature-verification system also warrants some discussion. Now, in any verification task, there are two types of errors we can commit: false rejects and false accepts. In the current context, a **false reject** is a signature that we reject even though the signature is not a forgery, and a **false accept** is a signature that we accept even though the signature is a forgery. Clearly, we can trade off one type of error for the other type of error. In particular, if we accept every signature as a genuine, we shall have 0% false rejects and 100% false accepts, and, if we reject every signature as a forgery, we shall have 100% false rejects and 0% false accepts. Thus, in the statistical evaluation of a verification system, whether on a database or otherwise, we must determine the percentage of false accepts as a function of the percentage of false rejects. The ensuing curve — the **error trade-off curve** — which trades off false accepts for false rejects, is often characterized by its **equal-error rate**, which is the error rate at which the percentage of false accepts is equal to the percentage of false rejects. The equal-error rate, despite its convenience as an indicator of system performance, of course, is no substitute for the actual trade-off curve, especially if we intend to operate the system in a range outside the immediate vicinity of the equal-error rate.

3. Prior Art

My review of the literature, and of the most comprehensive published survey [9] with its accompanying bibliography [14] (see also [7]), indicates the existence of a widely held belief that the temporal characteristics of the production of an on-line signature are key to the signature's verification. I am not sure what the basis for this belief is —

after all, we have for centuries relied on a visual examination of a signature to verify the signature's authenticity. Of the many possible reasons for this belief, two reasons come readily to mind. The first reason is that, in experiments, the temporal characteristics of signature production are seen to provide better system performance than alternate characteristics. The second reason is that the production of a signature is believed to be necessarily a **reflex action**, or a **ballistic action**, rather than a **deliberate action**; see, for instance, [9]. Ballistic handwriting is characterized by a spurt of activity, without positional feedback, whereas deliberate handwriting is characterized by a conscious attempt to produce a visual pattern with the aid of positional feedback.

I challenge, on two counts, the belief that signature production is necessarily ballistic, and also the more widely held notion that the temporal characteristics of signature production are key to signature verification. The first count is that many signers — including most of my acquaintances — can produce their signatures both ballistically and deliberately, the exact mechanism of production in a particular instance depending on the urgency and importance of the task. In general, it is fast handwriting that is ballistic (see [3]) rather than signature production per se, and many of us have and exercise control over the speed with which we sign. The second count is that even if we were to group together all the instances of the ballistic production of a signature, there is no compelling reason why these instances would exhibit temporal consistency. I suspect that the apparent success of the use of the temporal characteristics of on-line signatures in their verification is, at least partially, an artifact of the testing methodology: It is clearly easy to detect forgers on the basis of time when these forgers, being unaware that time is critical to verification, are making every effort to reproduce the shape of the signature they are trying to forge, with little attention to time. I must emphasize that I am not arguing here that the temporal characteristics of signature production are not potentially useful for automatic signature verification, but rather that these characteristics should not be the primary determinants of our decision.

The various strategies reported in the literature for the automatic verification of on-line signatures rely typically either on comparing specific features of signatures or on comparing specific temporal functions captured during signature production, or, perhaps, on both (see [9]). Although the signature features that are compared are typically global — such as the total time taken, or the average or root-mean-square speed, acceleration, force, or pressure (e.g., [2,12]) — these features could be local, such as the starting orientation or speed. Typical signature functions that are compared include pressure versus time, and the horizontal and vertical components of position, velocity, acceleration, and force, each versus time (e.g., [15,16,17]). The more sophisticated among the temporal-function-based approaches allow the horizontal axes of the functions to warp during comparison (e.g., [15,17]), and, approaches that rely on comparing temporal functions reputedly perform better, in general, than approaches that rely solely on comparing features. Barring the straightforward representation of the coordinates, orientation, and curvature of a signature along its length, all as functions of time, few attempts have been made to characterize the local shape of a signature. One exception to this observation is the work of Hastie and his coauthors [6], who match signatures by first segmenting them at places of low speed, and then seeking the optimal affine transformation between each segment and its stored prototype. However, segmentation-based approaches are, in general, not robust owing to their rapid deterioration in the presence of segmentation errors that are bound to occur sooner or later.

I have had the opportunity to try out, in person, only one well-known signature-verification system created outside Bell Laboratories, and the various statistical results reported in the literature are difficult to compare because of the very disparate conditions under which these results were produced. In database testing, we can in practice obtain almost any desired statistical trade off between false rejects and false accepts if we allow ourselves the luxury of suitably pruning or restricting the database on which we test the system. Such pruning is often easy as the performance of a verification system on a database is typically limited by the database's **goats**, a term used to describe the typically few individuals who account for a large majority of the errors — in our case, by producing signatures that are either inconsistent or degenerate; see, for instance, [2], and also [9]. Note, here, also that the false-reject statistics obtained in laboratory settings are likely to be overly optimistic vis-a-vis results that would be obtained in more unregulated settings (see [9]). Further, I point out that the validity of many of the results reported in the literature is suspect in real-world settings, because, in most experiments, forgers are not provided all the knowledge that they could gain over time if the verification system were ever introduced into the marketplace. For instance, it is clearly easy to detect forgers who are making every effort to duplicate a shape while all that the verification system is measuring is the total time taken; under such circumstances, a forger would clearly have greater success by ignoring the shape completely and concentrating on duplicating the total time taken. I believe this artifact of testing to be a significant contributor to the widespread emphasis given to the temporal characteristics of on-line signatures in their automatic verification.

There is a plethora of reasons other than those that I have just mentioned why a direct comparison of the various published statistical results is of little value. Some tests allow each user multiple tries to have a signature validated by the verification system, whereas other tests do not permit multiple tries. In some experiments, the users are highly motivated — for instance, by financial reward [16] — whereas in other experiments, the users are largely unmotivated. In some experiments, the false-accept statistics are based on so-called random forgeries that typically have little or no similarity to the genuine signatures they are supposed to represent. A **random forgery**, as its name suggests, is a pattern that by design is not related to the original signature; such a forgery is to be expected when a forger does not have ready access to the original signature, as might happen, for instance, if a credit card were stolen in transit before a genuine signature could be produced on the card.

All the reasons stated above point to the difficulty of comparing the various published statistical results. Hence, as a practical matter, we have no choice but to take recourse to our *common sense* in judging the quality of the various efforts toward automatic on-line signature verification. My own examination of the various published techniques makes me very skeptical of their efficacy in practice as stand-alone techniques. This skepticism is borne of my conviction that the varying local shape of a signature, as we proceed along the length of the signature, is key to the signature's verification, and, in my judgement, the published techniques are by and large conceptually inadequate at capturing this.

4. Key Concepts

In this section, I briefly describe some of the key ideas that underlie my approach to on-line signature verification. Please see the original paper from which this chapter is condensed for details.

Harmonic Mean

The most popular method of combining two errors is to compute their root weighted-mean square. Given two errors ξ_1 and ξ_2, their **root weighted-mean square** is $\xi = [a_1\xi_1^2 + a_2\xi_2^2]^{1/2}$, whose isocontours are ellipses — that is, each of whose loci in the ξ_1–ξ_2 plane for a particular ξ is an ellipse. An immediate generalization of this error combination is $\xi^n = a_1^n|\xi_1|^n + a_2^n|\xi_2|^n$, where a_1, a_2, and n are all positive. The isocontours of this expression for various n when $\xi = 1$, which are generalizations of ellipses, are called **superellipses** (see, for instance, [11]). A possible drawback of this ubiquitous family of error combinations is that this family takes into account each of the two errors, *irrespective of the other error*. In a sense, this mechanism of combining errors ANDs the errors — assuming here that a low error corresponds to a Boolean 1, and a high error corresponds to a Boolean 0. But, what if we wish to OR the errors? We might want to do this, for instance, if ξ_1 and ξ_2 are derived from two different models of which only one model is applicable.

One mechanism of ORing two errors, if you will, is by constraining n in the superelliptic error expression above to be negative instead of positive. Say, $m = -n$. Then, we get, $1/\xi^m = 1/(a_1^m|\xi1|^m) + 1/(a_2^m|\xi_2|^m)$, where a_1, a_2, and m are all positive. Isocontours of this expression, which are generalizations of hyperbolas, are called **superhyperbolas**. Now, if we put $m = 1$, and assume that both ξ_1 and ξ_2 are positive, then ξ will become the **weighted harmonic mean** of ξ_1 and ξ_2, which is

$$\xi = \frac{a_1 a_2 \xi_1 \xi_2}{a_1 \xi_1 + a_2 \xi_2}. \tag{7.1}$$

The isocontours of this expression are hyperbolas with asymptotes $\xi_1 = \xi / a_1$ and $\xi_2 = \xi / a_2$, as illustrated in Fig. 7.1 for $a_1 = 2$ and $a_2 = 1$. Notice that if we put $a_1 = a_2 = 2$ in the above expression, then ξ will simply become the unweighted harmonic mean of ξ_1 and ξ_2. See [1] for more on the harmonic mean. Finally, we may define the **weighted and biased harmonic mean** ξ of two numbers ξ_1 and ξ_2 to be

$$\xi = \frac{a_1 a_2 (\xi_1 + b_1)(\xi_2 + b_2)}{a_1 (\xi_1 + b_1) + a_2 (\xi_2 + b_2)}. \tag{7.2}$$

Here, ξ_1 is said to be biased by b_1, and ξ_2 by b_2. What such biasing does to the hyperbolic isocontours of the weighted harmonic mean is translate them by $-b_1$ along the ξ_1-axis and by $-b_2$ along the ξ_2-axis.

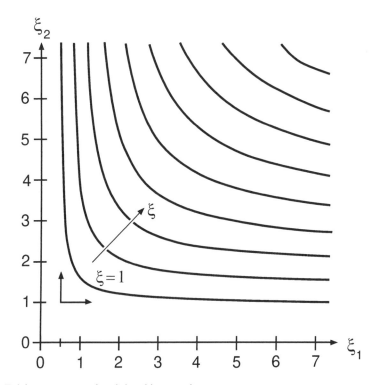

Figure 7.1 Isocontours of weighted harmonic mean.

Jitter

When individuals attempt to copy or trace a preexisting curve closely, as often happens in forgery, they produce a "jitter" owing to the act of constantly correcting the pen trajectory to conform to the a priori curve. This jitter often exceeds the quantization errors that result from the use of a discrete spatial sampling grid to capture on-line signatures — these quantization errors, of course, depend on the rate of pen motion vis-a-vis the temporal sampling rate. A measure of **jitter** that I have found useful is

$$\text{Jitter} = 1 - \frac{\text{length of polygonal (or other) smoothing approximation to data}}{\text{total sum of intersample distances}}. \quad (7.3)$$

Note that $0 \le \text{Jitter} \le 1$.

Aspect Normalization

Individuals do not scale their signatures equally along the horizontal and vertical dimensions when they sign (e.g., see [4]). You might, for instance, make your signature fatter without making it any taller. Hence, before we verify the shape of a signature, we must standardize the signature's ratio of height to width — this ratio called the signature's aspect. A measure of **aspect** that I have found useful is

$$\text{Aspect} = \frac{\text{total sum of vertical displacements}}{\text{total sum of horizontal displacements}}. \qquad (7.4)$$

The displacements in this expression are the unsigned vertical and horizontal components of the arc-lengths of curves fitted to the data.

Parameterization over Normalized Length

The **parameterization** of a curve is the creation of a one-to-one mapping from a subset of the real line onto the curve. The real line here, which is said to **parameterize** the curve, provides an index or **parameter** by which we can conveniently locate any point on the curve. Once we have parameterized a curve, we can describe various properties of the curve as functions of the curve's parameter. Such functions, which we call **characteristic functions**, could provide us with robust descriptions of the local shape of the curve along its length.

One possible parameter of an on-line curve is the time instant(s), relative to an arbitrary fixed time, at which the pen is located at a position along the curve. This particular choice of parameter seems to have been adopted universally for on-line signatures in the past, in part, perhaps, because of the ready availability of the pen trajectory as a function of time, and in part because of the widespread belief we discussed in Section 3 that the temporal characteristics of signature production are key to on-line signature verification. I contend that the parameterization of any handwritten on-line curve, including on-line signatures, should be over a spatial metric, rather than over a temporal metric. I suggest that we parameterize each handwritten on-line signature over its normalized arc-length — that is, over the distance traveled by the pen while the pen is in contact with the writing surface, this distance measured as a fraction of the total distance traveled by the pen while the pen is in contact with the writing surface. Let us denote the normalized arc-length of a signature by l. Parameterization of a curve over its arc-length is typical in differential geometry (e.g., see [8]).

Sliding Computation Window

Once we have parameterized a signature over its normalized arc-length, the question that arises is, what characteristics of the signature do we represent as functions of the signature's normalized arc-length? The characteristics of the signature we shall represent are derived from the center of mass, the torque, and the moments of inertia of the signature computed over a window that is sliding along the length of the signature in unison with the motion of a coordinate frame. Before we discuss each of these characteristics in sequence, let us spend some time on the sliding window over which we shall compute these signature characteristics. Let us call this sliding window the **computation window** to distinguish it from another sliding window that we shall discuss in the context of the moving coordinate frame later in this section.

Two questions that arise immediately in the context of a sliding window are, what is the window's width? and, what is the weighting along the length of the window? With regard to the window's width, the broader we make the sliding window, the more we shall average the signal noise, thus increasingly suppressing the net effect of noise on our computations of signature characteristics. However, a broader window does more than just increasingly smooth out noise: It also increasingly smooths out actual signature variations, making it harder to detect discrepancies between forgeries and genuine signatures. Hence, in choosing the width of our window, we must balance the prospect of undersmoothing the noise against the prospect of oversmoothing the signature. Typically, a reasonable choice for the width of the sliding window is a fraction of the length of an individual "character."

Now, coming to the question of the weighting along the length of the window, let us choose for this task a normalized one-dimensional Gaussian weighting function centered at the center of the sliding window, the σ of this Gaussian satisfying $L \approx 2\sigma$, where L is half the width of the sliding window. That is, let us choose the **window function**

$$g(\lambda) = \frac{\exp(-\lambda^2/(2\sigma^2))}{\int_{-L}^{+L} \exp(-\gamma^2/(2\sigma^2))d\gamma}, \qquad -L \le \lambda \le +L, \qquad (7.5)$$

where $\sigma \approx L/2$, and $g(\lambda) = 0$ outside the range $\pm L$.

Center of Mass

Assume the following: The signature is parametrized over its normalized arc-length l; the signature has a weighted window $g(\lambda)$ of span $\pm L$ sliding over its length; and the signature has unit mass per unit length. Then, the coordinates of the **center of mass** of the signature within the sliding window are

$$\overline{x}(l) = \int_{-L}^{+L} g(\lambda)\, x(l+\lambda)\, d\lambda, \qquad (7.6)$$

$$\overline{y}(l) = \int_{-L}^{+L} g(\lambda)\, y(l+\lambda)\, d\lambda, \qquad (7.7)$$

where $(x(l), y(l))$ are the point coordinates along the length of the signature. The varying coordinates of the center of mass, $\overline{x}(l)$, $\overline{y}(l)$, computed over a window that is sliding along the length of a signature, together provide us with a robust position-dependent description of the shape of the signature.

Torque

The torque **T** exerted by a vector **v**, which is located at position **p** with respect to the point about which we measure the torque, is $\mathbf{T} = \mathbf{v} \times \mathbf{p}$. Now, assume the following: The signature is parameterized over its normalized arc-length l; the signature has a weighted window g (λ) of span $\pm$L sliding over its length; and the signature is decomposed into a series of infinitesimal vectors, each vector with magnitude equal to its length and with direction pointing in the direction of pen motion. Then, we can define the **torque** exerted about the origin by the signature within the sliding window to be

$$\mathbf{T}(l) = \int_{-L}^{+L} g(\lambda)[dx(l + \lambda) \quad dy(l + \lambda)] \times [x(l + \lambda) \quad y(l + \lambda)], \qquad (7.8)$$

where $(x(l), y(l))$ are the point coordinates along the length of the signature. Here, $dx(l + \lambda)$ and $dy(l + \lambda)$ are the differential changes in x and y at the location $(l + \lambda)$ under a $d\lambda$ change in λ. Given that each of our on-line signatures resides in the x-y plane, $\mathbf{T}$ here can point only in a direction orthogonal to the x-y signature plane: $\mathbf{T}$ will point orthogonally out of the x-y plane if the net torque is counterclockwise, and $\mathbf{T}$ will point orthogonally into the x-y plane if the net torque is clockwise. As a result, it suffices for us to consider only the following scalar T, which we obtain by expanding the above vector $\mathbf{T}$:

$$T(l) = \int_{-L}^{+L} g(\lambda) \left(y(l + \lambda) \, dx(l + \lambda) - x(l + \lambda) \, dy(l + \lambda) \right). \qquad (7.9)$$

If we ignore the window function $g(\lambda)$, we can interpret the torque $T(l)$ here to be twice the signed area swept with respect to the origin by the portion of the signature within the sliding window centered at position l, a positive value of $T(l)$ indicating a net counterclockwise sweep and a negative value indicating a net clockwise sweep. The varying torque, $T(l)$, computed over a window that is sliding along the length of a signature, can provide us with a robust position- and orientation-dependent description of the shape of the signature.

Moments of Inertia

Assume the following: The signature is parameterized over its normalized arc-length l; the signature has a weighted window g (λ) of span $\pm$L sliding over its length; and the signature has unit mass per unit length. Then, the **moments of inertia** about the y-axis and the x-axis, respectively, of the signature within the sliding window are

$$\overline{x^2}(l) = \int_{-L}^{+L} g(\lambda) \, x^2(l + \lambda) \, d\lambda, \qquad (7.10)$$

$$\overline{y^2}(l) = \int_{-L}^{+L} g(\lambda) \, y^2(l + \lambda) \, d\lambda, \qquad (7.11)$$

where $(x(l), y(l))$ are the point coordinates along the length of the signature. And, the second-order cross-moment is

$$\overline{xy}(l) = \int_{-L}^{+L} g(\lambda) \, x(l + \lambda) \, y(l + \lambda) \, d\lambda. \qquad (7.12)$$

The varying second-order moments, $\overline{x^2}(l)$, $\overline{y^2}(l)$, and $\overline{xy}(l)$, computed over a window that is sliding along the length of a signature, when expressed in forms $s_1(l)$ and $s_2(l)$ below can together provide us with a robust orientation- and curvature-dependent description of the shape of the signature:

$$s_1(l) = a\frac{\overline{x^2}(l)\,\overline{y^2}(l) - \overline{xy}^2(l)}{[\overline{x^2}(l) + \overline{y^2}(l)]^2} + b\frac{2\overline{xy}(l)}{\sqrt{[\overline{x^2}(l) - \overline{y^2}(l)]^2 + 4\overline{xy}^2(l)}}, \qquad (7.13)$$

$$s_2(l) = a\frac{\overline{x^2}(l)\,\overline{y^2}(l) - \overline{xy}^2(l)}{[\overline{x^2}(l) + \overline{y^2}(l)]^2} - b\frac{2\overline{xy}(l)}{\sqrt{[\overline{x^2}(l) - \overline{y^2}(l)]^2 + 4\overline{xy}^2(l)}}, \qquad (7.14)$$

where a and b are positive weights. Together, $s_1(l)$ and $s_2(l)$ provide us with measures of the curvature and orientation of a curve segment that are independent of scale — that is, that are invariant under uniform magnification or reduction of the curve segment. Let us, for brevity, call s_1 and s_2 **curvature-ellipse measures**.

Moving Coordinate Frame and Saturation

We now have a complete list of the signature characteristics we shall represent as functions of the normalized length l of the signature. These characteristics are $\overline{x}(l)$, $\overline{y}(l)$, T, s_1, and s_2, as defined in equations (7.6), (7.7), (7.9), (7.13), and (7.14). Given the dependence of $\overline{x}(l)$, $\overline{y}(l)$, and T on the location of the origin of the coordinate frame in which we compute them, we are now faced with deciding how to choose our coordinate frame as our computation window slides along the length of the signature. Let us attach our coordinate frame to the center of mass of the signature computed over another window that too is sliding along the length of the signature. However, let us align the axes of this moving coordinate frame permanently with the global axes of maximum and minimum inertia, rather than compute these axes locally.

Now, we have two windows that are sliding over the length of the signature — one window over which we compute the origin of our moving coordinate frame, and the other window over which we measure the center of mass, the torque, and the moments of inertia of the signature. Let us call the window over which we compute the origin of our moving coordinate frame the **coordinate-frame window**; we earlier named the window over which we compute the signature characteristics the computation window. As illustrated in Fig. 7.2, both windows slide in unison over the length of the signature with a fixed — but, in general, nonzero — displacement between their centers. Further, the two windows have fixed — but, in general, unequal — widths. Just as we had earlier weighted the computation window by a Gaussian centered over the window, let us now weight the coordinate-frame window by its own Gaussian centered over the window.

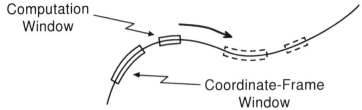

Figure 7.2 Sliding computation and coordinate-frame windows.

In our discussion to this point, we have assumed implicitly that both our sliding windows span continuous curve segments. Gaps along the length of a signature, in themselves, do not pose any conceptual hurdle to the motion of our sliding windows: All we have to do whenever either of our sliding windows spans a gap is split the window, with its (Gaussian) weighting function, across the gap in the curve segment. Although such window splitting suffices to characterize a signature continuously across gaps in the signature, it is not sufficient for our purposes: Our measurements of signature characteristics in the vicinity of gaps in a signature could exhibit unusually large magnitudes when the gaps are large, these large magnitudes posing the threat of disproportionately influencing comparisons of characteristic functions to their models. We can circumvent this threat by **saturating** our measurements of the two center-of-mass coordinates and the torque exerted about the origin, employing, in each case, the following **saturation function**:

$$m_{sat} = m_0 \tanh\left[\frac{m_{unsat}}{m_0}\right], \qquad (7.15)$$

where m_{unsat} is the original unsaturated measurement, m_{sat} is the same measurement after saturation, and m_0 — which is positive and chosen individually for each signature characteristic — determines the degree of saturation. When $|m_{unsat}| \ll |m_0|$, $|m_{sat}| \approx |m_{unsat}|$, but when $|m_{unsat}| \gg |m_0|$, $|m_{sat}| \approx |m_0|$, the signs of m_{sat} and m_{unsat} always being the same.

Weighted Cross-Correlation and Warping

No signer is uniformly consistent along the entire length of the signer's signature. Further, the consistency of a signer at a particular location along a signature depends on the signature characteristic we examine. As a result, whenever we measure a characteristic of a signature along its length, we must also measure, as a function of the normalized length of the signature, the consistency of the characteristic across multiple instances of the signature. Doing so, for every characteristic function of a signature, we shall have a **consistency function** that provides a measure of the consistency of the characteristic function along its length. A natural choice for the consistency function of a characteristic function is the inverse standard deviation of the characteristic function at each point along its length. Let us adopt this choice. Once we have a consistency function to accompany the prototype of a characteristic function, whenever we compare a **characteristic function** to this prototype, we shall weight each of the two functions along its length by the consistency function of the prototype.

The question now is, how do we compare a characteristic function to its prototype? or equivalently, what is our measure of error in comparing a characteristic function to its prototype? Of the several error measures that are possible — for instance, the integral of the difference of squares — I have chosen this: ($1 - cross\text{-}correlation$), where we compute the *cross-correlation* between the characteristic function and its prototype while weighting each function by the consistency function of the prototype. The **weighted cross-correlation** of two functions $f(l)$ and $h(l)$, each function weighted by the function $w(l)$, is, by definition,

$$\text{Cross - Correlation} = \frac{\int w^2(l)\, f(l)\, h(l)\, dl}{\sqrt{\int w^2(l)\, f^2(l)\, dl \int w^2(l)\, h^2(l)\, dl}}. \qquad (7.16)$$

If $f(l)$ is a characteristic function here, $h(l)$ is this characteristic function's prototype, and $w(l)$ is the prototype's consistency function, then, for us, $f(l)$, $h(l)$, and $w(l)$ are related as follows: $h(l) = E[f(l)]$ and $w(l) = 1/(E\,[\,(f(l) - E\,[\,f(l)\,])^2\,])^{1/2}$.

When we compute the various individual weighted cross-correlations between a signature's characteristic functions and their models, we will allow all the characteristic functions of the signature — or, equivalently, all the models of these functions — to **warp** simultaneously along their lengths such that an overall error measure is minimized. This simultaneous warping of the individual functions must, of course, be constrained to be identical at identical abscissae along the lengths of all the functions because the abscissa of each function is the same length parameter l, whose each specific value corresponds to a specific physical location along the signature. Warping allows us to accommodate instances of signatures that deviate from one another with respect to the fractional lengths of their various parts, such deviations being unavoidable even when all the signatures are produced by the original signer.

5. Algorithm

My algorithm has three distinct components — normalization, description, and comparison — each of which I outline broadly next. **Normalization** makes the algorithm largely independent of the orientation and aspect of a signature; the algorithm is inherently independent of the position and size of a signature. **Description** generates the five characteristic functions of the signature. **Comparison** computes a net measure of error between the signature characteristics and their prototypes.

A. Normalization

1. Fit a polygon to the ordered set of samples of the on-line signature, and keep a count of the total number of pen-down samples, a number proportional to the total pen-down time under uniform temporal sampling.

2. Compute the jitter. There is no further need for the original samples.

3. Compute the global axes of maximum and minimum inertia of the signature through the signature's global center of mass, and then rotate the signature to normalize the orientation of these axes.

4. Compute the aspect of the signature from the fitted polygon, and then scale the signature either vertically or horizontally to normalize its aspect.

B. Description

1. Parametrize the signature over its length l, measured along the fitted and normalized polygon as a fraction of the total length.

2. Compute a moving coordinate frame.

3. In the moving coordinate frame, measure, as a function of l, the following signature characteristics over a sliding computation window: the coordinates $\bar{x}(l)$ and $\bar{y}(l)$ of the center of mass, the torque $T(l)$ exerted about the origin, and the curvature-ellipse measures $s_1(l)$ and $s_2(l)$. All these computations can be conveniently performed over a discrete l that is uniformly sampled.

4. Saturate the characteristic functions and normalize each function to have a zero mean.

C. Comparison

1. Simultaneously warp the five characteristic functions to maximize the sum of the weighted cross-correlation of each function with respect to its model, and retain a measure of the total warping performed.

2. Compute the error between each characteristic function and its model by subtracting from 1.0 the weighted cross-correlation between the two functions; then, normalize each such error by first subtracting its mean from it and then dividing the resultant by the error's standard deviation; finally, bias each normalized error and then threshold it so as to make it 0.0 if it is negative.

3. Compute the root mean square of the individual biased and thresholded errors between each of the five characteristic functions and their models (C-2) to arrive at the **net local error**.

4. Compute the root mean square of the differences between the following four global entities and their means, after first normalizing each difference by the entity's standard deviation, to arrive at the **net global error**: jitter (A-2), aspect (A-4), warping (C-1), and the total number of pen-down samples (A-1).

5. Compute the weighted and biased harmonic mean of the net local error (C-3) and the net global error (C-4) — the weights and biases reflecting the overall spatial consistency of the signature across its multiple instances — to arrive at the **net error**, which provides us a measure of the discrepancy between the signature being verified and its model and whose comparison against a threshold determines whether we accept or reject the signature being verified.

6. Example

I now illustrate the algorithm we discussed in Section 5 — specifically, the nature of the characteristic functions that lie at the core of this algorithm — through an example. This pedagogically contrived example, shown in Fig. 7.3, has four signatures, all shown in the left-most column: Proceeding from top to bottom, the first signature is a typical genuine, the second signature has a loop missing from its "w", the third signature has an extra loop in its "w," and the fourth signature is written with a slant. Shown alongside each signature, in a row, are the characteristic functions of that signature: Proceeding from left to right, shown in sequence are $\overline{x}(l)$, $\overline{y}(l)$, $T(l)$, $s_1(l)$, and $s_2(l)$, with the result of the weighted cross-correlation of each characteristic function with its prototype indicated at the lower right of the function. The prototype of each characteristic function is shown immediately above the four corresponding characteristic functions of the four signatures, and immediately above each prototype is shown the consistency function of the signature characteristic, this function bounded below by 0.

As is clear from the figure, local deviations in the shape of a signature from its typical genuine instance lead to locally identifiable deviations in the characteristic functions from their prototypes, and systematic deviations in the shape of a signature from its typical genuine instance lead to distributed deviations in the characteristic functions from their prototypes. In particular, note that the characteristic functions of the two signatures with the extra and missing loop in "w" each differ from its prototype roughly within the interval of l between 0.5 and 0.6; because of the discrepancy between the shapes of the signatures, the characteristic functions of these two signatures are neither aligned with each other along the l axis, nor are they aligned with their prototypes (until we warp the l axis).

7. Database Results

The three databases on which I tested my implementation of the algorithm we discussed in Section 5 were compiled by Bell Laboratories, and are proprietary. Let us call these databases **DB1**, **DB2**, and **DB3**, calling their union **DB**. The details of the creation of these databases are available in the original paper on which this chapter is based. In a nutshell, DB1 was created in the most carefully controlled fashion, and DB3 was created in the least carefully controlled fashion. Owing to the varying circumstances of their creation, I report not only the error trade-off curve (see Section 2) for the three databases collectively, but also the error trade-off curves for the three databases individually.

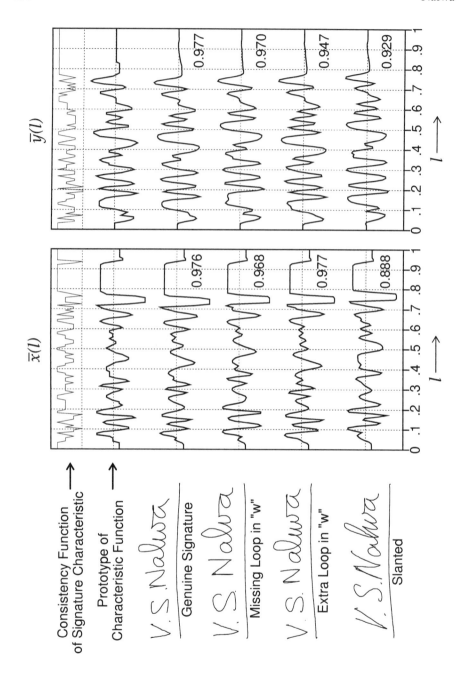

Figure 7.3 Example illustrating characteristic functions of signatures (cont'd...).

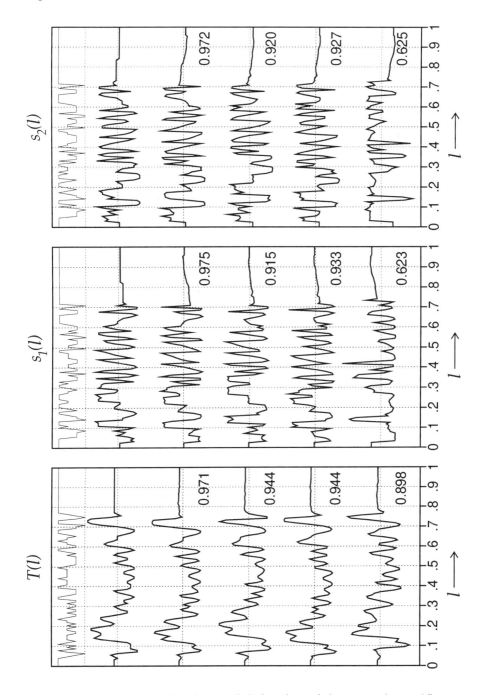

Figure 7.3 Example illustrating characteristic functions of signatures (…cont'd).

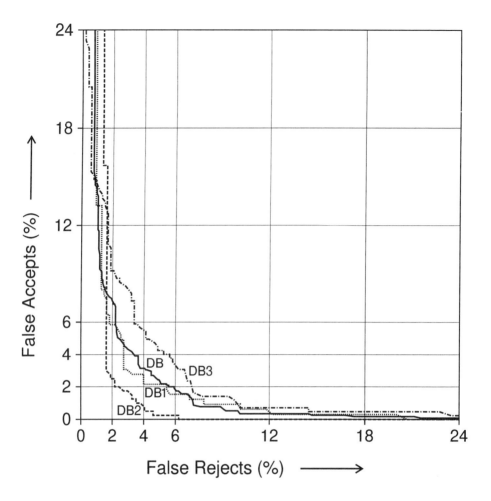

Figure 7.4 Error trade-off curves when modeling with first 6 genuine signatures.

In Fig. 7.4, I show the error trade-off curves for the three databases individually and collectively, labeling the last curve DB. Each curve was generated using the first 6 genuine signatures of each signer to build a model of the signer's signature, this model requiring about 600 bytes of storage after suitable compression. Under these circumstances, DB1 provided a test set of 550 genuines from 59 signers in addition to 325 forgeries, DB2 provided a test set of 370 genuines from 102 signers in addition to 401 forgeries, and DB3 provided a test set of 532 genuines from 43 signers in addition to 424 forgeries. For the curve labeled DB in Fig. 7.4, then, the total number of genuines tested is 1452 from 204 signers, and the total number of forgeries tested is 1150.

I used 6 genuine signatures to build each signature model because, at least on these databases, as I increased the number of signatures for modeling up to 6, there were tangible, albeit increasingly smaller, improvements in the various error trade-off curves. However, I did not see a tangible improvement in the error trade-off curves as I increased the number of signatures for modeling up from 6.

It is clear from the various error trade-off curves that each such curve depends highly on the nature of the database on which the curve was produced. For instance, in Fig. 7.4, the equal-error rates for DB1, DB2, and DB3 are about 3%, 2%, and 5%, respectively. The equal-error rate of my implementation's net error trade-off curve in Fig. 7.4 is about 3.6%. An operating point that I consider reasonable for many credit-card transactions is at about 1% false accepts, which corresponds to about 7% false rejects on the net error trade-off curve in Fig. 7.4. At this operating point, statistically, approximately 1 out of every 100 forgeries will be accepted and approximately 1 out of every 14 genuine signatures will be rejected — an instance of rejection requiring either a fresh signature, or some other action. *In practice, of course, the point on the error trade-off curve at which we operate in a particular situation must depend on the relative penalties we would incur for committing the two types of errors in that situation.*

A final point I would like to make is this: For any given database, perhaps a composite of multiple individual databases, we can always fine tune a signature-verification system to provide the best overall error trade-off curve for that database – for the three databases here, I was able to bring my overall equal-error rate down to about 2.5% — but we must always ask ourselves, does this fine tuning make *common sense* in the real world? If the fine tuning does not make common sense, it is in all likelihood exploiting a peculiarity of the database. Then, if we do plan to introduce the system into the marketplace, we are better off without the fine tuning.

8.   Conclusions

Topics in on-line signature verification that deserve our further attention include the following:

•   Better models for signatures whose instances are not shaped consistently or for which we have fewer than 6 instances to build a model.

•   Acquisition of instances of a signature used to build its model over multiple sessions, rather than over a single session, to obtain a more representative variety of instances of the signature.

•   Invocation of multiple models for individuals with multiple distinct signatures.

•   Statistically well-founded procedures for determining the parameters of a model from the relatively few instances of a signature available to model the signature.

•   Automatic adaptation of models to signatures as they evolve over time.

•   Theoretically sound statistical framework to exploit fully each of the various individual error measures generated from comparing the characteristic functions of a signature to their prototypes.

•   Partial matching of signatures, highlighting discrepancies if they are specific.

- Identification of problem signers, including those who are unusually inconsistent or have signatures that are trivial to forge.

- Comparison of signatures to their models at multiple or personalized resolutions, rather than at a single common resolution.

Over and above these issues, we must also further investigate the usefulness of pen dynamics during on-line signature production in automatic on-line signature verification. Such dynamics might include not only velocities and forces, but also the varying orientation of the pen, and the way in which a signer grasps a pen.

References

[1] G. Chrystal, *Algebra: An Elementary Text-Book for the Higher Classes of Secondary Schools and for Colleges*, Part 1, Seventh Edition, Chelsea Publishing Co., New York, 1964.

[2] H. D. Crane and J. S. Ostrem, "Automatic Signature Verification Using a Three-Axis Force-Sensitive Pen," *IEEE Trans. on Systems, Man, and Cybernetics*, Vol. SMC-13, No. 3, pp. 329-337, May-June 1983.

[3] J. J. Denier van der Gon and J. Ph. Thuring, "The Guiding of Human Writing Movements," *Kybernetik*, Vol. 2, No. 4, pp. 145-148, February 1965.

[4] Evett and R. N. Totty, "A Study of the Variation in the Dimensions of Genuine Signatures," *Journal of the Forensic Science Society*, Vol. 25, pp. 207-215, 1985.

[5] W. R. Harrison, *Suspect Documents: Their Scientific Examination, Praeger*, New York, 1958. Reprinted by Nelson-Hall Publishing Co., Chicago.

[6] T. Hastie, E. Kishon, M. Clark, and J. Fan, "A Model for Signature Verification", in *Proceedings of the 1991 IEEE International Conference on Systems, Man, and Cybernetics*, Vol. 1, Charlottesville, Virginia, pp. 191-196, October 1991.

[7] F. Leclerc and R. Plamondon, "Automatic Signature Verification: The State of the Art — 1989-1993," *International Journal of Pattern Recognition and Artificial Intelligence*, Vol. 8, No. 3, pp. 643-660, June 1994.

[8] M. M. Lipschutz, *Differential Geometry*, McGraw-Hill Book Co., New York, 1969.

[9] G. Lorette and R. Plamondon, "Dynamic Approaches to Handwritten Signature Verification," in *Computer Processing of Handwriting*, R. Plamondon and C. G. Leedham, Eds., World Scientific Publishing Co., Singapore, pp. 21-47, 1990.

[10] J. Mathyer, "The Expert Examination of Signatures" *The Journal of Criminal Law, Criminology and Police Science*, Vol. 52, pp. 122-133, 1961.

[11] V. S. Nalwa, *A Guided Tour of Computer Vision*, Addison-Wesley Publishing Co., Reading, Massachusetts, 1993.

[12] W. Nelson and E. Kishon, Use of Dynamic Features for Signature Verification, in *Proceedings of the 1991 IEEE International Conference on Systems, Man, and Cybernetics*, Vol. 1, Charlottesville, Virginia, pp. 201-205, October 1991.

[13] S. Osborn, *Questioned Documents*, Boyd Printing Co., Albany, New York, 1929. Reprinted by Nelson-Hall Publishing Co., Chicago.

[14] R. Plamondon and G. Lorette, "Identity Verification from Automatic Processing of Signatures: Bibliography," in *Computer Processing of Handwriting*, R. Plamondon and C. G. Leedham, Eds., World Scientific Publishing Co., Singapore, pp. 65-85, 1990.

[15] Y. Sato and K. Kogure, "Online Signature Verification Based on Shape, Motion, and Writing Pressure," in *Proceedings of the Sixth International Conference on Pattern Recognition*, Munich, Germany, pp. 823-826, October 1982.

[16] T. K. Worthington, T. J. Chainer, J. D. Williford, and S. C. Gundersen, "IBM Dynamic Signature Verification," in *Computer Security: The Practical Issues in a Troubled World*, J. B. Grimson and H.-J. Kugler, Eds., North-Holland (Elsevier Science Publishers), Amsterdam, pp. 129-154, 1985.

[17] P. Zhao, A. Higashi, and Y. Sato "On-Line Signature Verification by Adaptively Weighted DP Matching," *IEICE Trans. on Information and Systems*, Vol. E79-D, No. 5, pp. 535-541, May 1996.

8 SPEAKER RECOGNITION

Joseph P. Campbell, Jr.
Department of Defense
Fort Meade, MD
j.campbell@ieee.org

Abstract *A tutorial on the design and development of automatic speaker recognition systems is presented. Automatic speaker recognition is the use of a machine to recognize a person from a spoken phrase. These systems can operate in two modes: to* identify *a particular person or to* verify *a person's claimed identity. Speech processing and the basic components of automatic speaker recognition systems are shown and design tradeoffs are discussed. The performances of various systems are compared.*

Keywords: *Access control, authentication, biometrics, biomedical measurements, biomedical signal processing, biomedical transducers, communication system security, computer network security, computer security, corpus, databases, identification of persons, public safety, site security monitoring, speaker recognition, speech processing, verification.*

1. Introduction

The focus of this chapter is on facilities and network access-control applications of speaker recognition. Speech processing is a diverse field with many applications. Figure 8.1 shows a few of these areas and how speaker recognition relates to the rest of the field. This chapter will emphasize the speaker recognition applications shown in the boxes of Figure 8.1.

Speaker recognition encompasses verification and identification. Automatic speaker verification (ASV) is the use of a machine to verify a person's claimed identity from his voice. The literature abounds with different terms for speaker verification, including voice verification, speaker authentication, voice authentication, talker authentication, and talker verification. In automatic speaker identification (ASI), there is no *a priori* identity claim, and the system decides who the person is, what group the person is a member of, or (in the open-set case) that the person is unknown. General overviews of speaker recognition have been given by Atal, Doddington, Furui, O'Shaughnessy, Rosenberg, Soong, Sutherland, and Jack [2,9,13,28,38,39,46].

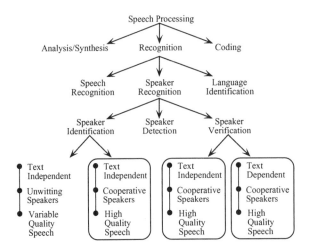

Figure 8.1 Speech processing.

Speaker verification is defined as deciding if a speaker is who he claims to be. This is different than the speaker identification problem, which is deciding if a speaker is a specific person or is among a group of persons. In speaker verification, a person makes an identity claim (e.g., entering an employee number or presenting his smart card). In text-dependent recognition, the phrase is known to the system and it can be fixed or not fixed and prompted (visually or orally). The claimant speaks the phrase into a microphone. This signal is analyzed by a verification system that makes the binary decision to accept or reject the user's identity claim or possibly to report insufficient confidence and request additional input before making the decision.

A typical ASV setup is shown in Figure 8.2. The claimant, who has previously enrolled in the system, presents an encrypted smart card containing his identification information. He then attempts to be authenticated by speaking a prompted phrase(s) into the microphone. There is generally a tradeoff between recognition accuracy and the test-session duration of speech. In addition to his voice, ambient room noise and delayed versions of his voice enter the microphone via reflective acoustic surfaces. Prior to a verification session, users must enroll in the system (typically under supervised conditions). During this enrollment, voice models are generated and stored (possibly on a smart card) for use in later verification sessions. There is also generally a tradeoff between recognition accuracy and the enrollment-session duration of speech and the number of enrollment sessions.

Many factors can contribute to verification and identification errors. Table 8.1 lists some of the human and environmental factors that contribute to these errors, a few of which are shown in Figure 8.2. These factors are generally outside the scope of algorithms or are better corrected by means other than algorithms (e.g., better microphones). However, these factors are important because, no matter how good a speaker recognition algorithm is, human error (e.g., misreading or misspeaking) ultimately limits its performance.

| Misspoken or misread prompted phrases |
| Extreme emotional states (e.g., stress or duress) |
| Time varying (intra- or intersession) microphone placement |
| Poor or inconsistent room acoustics (e.g., multipath and noise) |
| Channel mismatch (e.g., using different microphones for enrollment and verification) |
| Sickness (e.g., head colds can alter the vocal tract) |
| Aging (the vocal tract can drift away from models with age) |

Table 8.1 Sources of verification error.

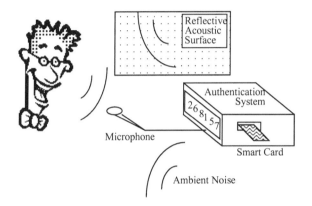

Figure 8.2 Typical speaker-verification setup.

Motivation

ASV and ASI are probably the most natural and economical methods for solving the problems of unauthorized use of computer and communications systems and multilevel access control. With the ubiquitous telephone network and microphones bundled with computers, the cost of a speaker recognition system might only be for the software for the recognition algorithm.

Biometric systems automatically recognize a person using distinguishing traits (a narrow definition). Speaker recognition is a performance biometric; i.e., you perform a task to be recognized. Your voice, like other biometrics, cannot be forgotten or misplaced, unlike knowledge-based (e.g., password) or possession-based (e.g., key) access control methods. Speaker-recognition systems can be made somewhat robust against noise and channel variations [25,36], ordinary human changes (e.g., time-of-day voice changes and minor head colds), and mimicry by humans and tape recorders [18].

Problem Formulation

Speech is a complicated signal produced as a result of several transformations occurring at several different levels: semantic, linguistic, articulatory, and acoustic. Differences in these transformations appear as differences in the acoustic properties of the speech signal. Speaker-related differences are a result of a combination of anatomical differences inherent in the vocal tract and the learned speaking habits of different individuals. In speaker recognition, all these differences can be used to discriminate between speakers.

Generic Speaker Verification

The general approach to ASV consists of five steps: digital speech data acquisition, feature extraction, pattern matching, making an accept/reject decision, and enrollment to generate speaker reference models. A block diagram of this procedure is shown in Figure 8.3. Feature extraction maps each interval of speech to a multidimensional feature space. (A speech interval typically spans 10 to 30 ms of the speech waveform and is referred to as a frame of speech.) This sequence of feature vectors $\mathbf{x}_i$ is then compared to speaker models by pattern matching. This results in a match score z_i for each vector or sequence of vectors. The match score measures the similarity of the computed input feature vectors to models of the claimed speaker or feature vector patterns for the claimed speaker. Last, a decision is made to either accept or reject the claimant according to the match score or sequence of match scores, which is a hypothesis-testing problem.

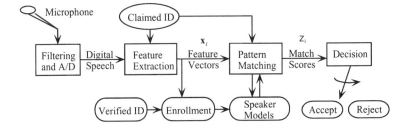

Figure 8.3 Generic speaker verification system.

For speaker recognition, features that exhibit high speaker discrimination power, high interspeaker variability, and low intraspeaker variability are desired. Many forms of pattern matching and corresponding models are possible. Pattern matching methods include dynamic time warping (DTW), hidden Markov modeling (HMM), artificial neural networks, and vector quantization (VQ). Template models are used in DTW, statistical models are used in HMM, and codebook models are used in VQ.

Previous Work

Table 8.2 shows a sampling of the chronological advancement in speaker verification. The following terms are used to define the columns in Table 8.2: "Source" refers to a citation in the references, "org" is the company or school where the work was done, "features" are the signal measurements such as linear prediction (LP) and log area ratio (LAR), "input" is the type of input speech (laboratory, office quality, or telephone), "text" indicates whether text-dependent or text-independent mode of operation is used, "method" is the heart of the pattern-matching process, "pop" is the population size of the test (number of people), and "error" is the equal error percentage for speaker verification systems "v" or the recognition error percentage for speaker identification systems "i" given the specified duration of test speech in seconds. This data is presented to give a simplified general view of past speaker-recognition research. The references should be consulted for important distinctions that are not included; e.g., differences in enrollment, differences in cross-gender impostor trials, differences in normalizing "cohort" speakers [40], differences in partitioning the impostor and cohort sets, and differences in known versus unknown impostors [5]. It should be noted that it is difficult to make meaningful comparisons between the text-dependent and the generally more difficult text-independent tasks. Text-independent approaches, such as Gish's segmental Gaussian model [15] and Reynolds' Gaussian Mixture Model (GMM) [36] need to deal with unique problems (e.g., sounds or articulations present in the test material, but not in training). It is also difficult to compare between the binary-choice verification task and the generally more difficult multiple-choice identification task [9,29].

There are over a dozen commercial ASV systems, including those from ITT, Lernout & Hauspie, T-NETIX, Veritel, and Voice Control Systems. Perhaps the largest scale deployment of any biometric to date is Sprint's Voice FONCARD®, which uses TI's voice-verification engine. Speaker verification applications include access control, telephone banking, and telephone credit cards. The accounting firm of Ernst and Young estimates that high-tech computer thieves in the U.S. steal $3 to $5 billion annually. Automatic speaker-recognition technology could substantially reduce this crime by reducing these fraudulent transactions. It takes a pair of subjects to make a false acceptance error: an impostor and a target. Because of this hunter and prey relationship, in this work, the impostor is referred to as a wolf and the target as a sheep. False acceptance errors are the ultimate concern of high-security speaker-verification applications; however, they can be traded off for false rejection errors.

The following section contains an overview of digital signal acquisition, speech production, speech signal processing, and speaker characterization based on linear prediction and mel cepstra modeling.

2.   Speech processing

Speech processing extracts the desired information from a speech signal. To process a signal by a digital computer, the signal must be represented in digital form so that it can be used by a digital computer.

Speech Signal Acquisition

Initially, the acoustic sound pressure wave is transformed into a digital signal suitable for voice processing. A microphone or telephone handset can be used to convert the acoustic wave into an analog signal. This analog signal is conditioned with antialiasing filtering (and possibly additional filtering to compensate for any channel impairments). The antialiasing filter limits the bandwidth of the signal to approximately the Nyquist rate (half the sampling rate) before sampling. The conditioned analog signal is then sampled to form a digital signal by an analog-to-digital (A/D) converter. Today's A/D converters for speech applications typically

Source	Org	Features	Method	Input	Text	Pop	Error
Atal [1]	AT&T	Cepstrum	Pattern Match	Lab	Dependent	10	i: 2%@0.5s v: 2%@1s
Markel and Davis [26]	STI	LP	Long Term Statistics	Lab	Independent	17	i: 2%@39s
Furui [12]	AT&T	Normalized Cepstrum	Pattern Match	Tele-phone	Dependent	10	v: 0.2%@3s
Schwartz, et al. [43]	BBN	LAR	Nonparametric pdf	Tele-phone	Independent	21	i: 2.5%@2s
Li and Wrench [23]	ITT	LP, Cepstrum	Pattern Match	Lab	Independent	11	i: 21%@3s i: 4%@10s
Doddington [9]	TI	Filter-bank	DTW	Lab	Dependent	200	v: 0.8%@6s
Soong, et al. [44]	AT&T	LP	VQ (size 64) Likelihood Ratio distortion	Tele-phone	10 isolated digits	100	i: 5%@1.5s i: 1.5%@3.5s
Higgins and Wohlford [19]	ITT	Cepstrum	DTW Likelihood Scoring	Lab	Independent	11	v: 10%@2.5s v: 4.5%@10s
Attili, et al. [3]	RPI	Cepstrum, LP, Autocorr	Projected Long Term Statistics	Lab	Dependent	90	v: 1%@3s
Higgins, et al. [18]	ITT	LAR, LP-Cepstrum	DTW Likelihood Scoring	Office	Dependent	186	v: 1.7%@10s
Tishby [47]	AT&T	LP	HMM (AR mix)	Tele-phone	10 isolated digits	100	v: 2.8%@1.5s v: 0.8%@3.5s
Reynolds [34]; Reynolds and Carlson [35]	MIT-LL	Mel-Cepstrum	HMM (GMM)	Office	Dependent	138	i: 0.8%@10s v: 0.12%@10s
Che and Lin [7]	Rutgers	Cepstrum	HMM	Office	Dependent	138	i: 0.56% @2.5s i: 0.14%@10s v: 0.62% @2.5s

Table 8.2 Selected chronology of speaker-recognition progress.

Source	Org	Features	Method	Input	Text	Pop	Error
Tishby [47]	AT&T	LP	HMM (AR mix)	Tele-phone	10 isolated digits	100	v: 2.8%@1.5s v: 0.8%@3.5s
Colombi, et al. [8]	AFIT	Cep, Eng dCep, ddCep	HMM monophone	Office	Dependent	138	i: 0.22%@10s v: 0.28%@10s
Reynolds [37]	MIT-LL	Mel-Cepstrum, Mel-dCepstrum	HMM (GMM)	Tele-phone	Independent	416	v: 11%/16% @3s v: 6%/8% @10s v: 3%/5% @30s matched/ mismatched handset

Table 8.2 Selected chronology of speaker-recognition progress (contd.).

sample with 12 to 16 bits of resolution at 8,000 to 20,000 samples per second. Oversampling is commonly used to allow a simpler analog antialiasing filter and to control the fidelity of the sampled signal precisely (e.g., sigma-delta converters).

In local speaker-verification applications, the analog channel is simply the microphone, its cable, and analog signal conditioning. Thus, the resulting digital signal can be very high quality, lacking distortions produced by transmission of analog signals over telephone lines.

YOHO Speaker-Verification Corpus

The work presented here is based on high-quality signals for benign-channel speaker verification applications. The primary database for this work is known as the YOHO Speaker Verification Corpus, which was collected by ITT under a U.S. Government contract. The YOHO database was the first large-scale, scientifically controlled and collected, high-quality speech database for speaker-verification testing at high confidence levels. Table 8.3 describes the YOHO database [17]. YOHO is available from the Linguistic Data Consortium (University of Pennsylvania) and test plans have been developed for its use [5]. This database already is in digital form, emulating the third generation Secure Terminal Unit's (STU-III) secure voice telephone input characteristics, so the first signal processing block of the verification system in Figure 8.3 (signal conditioning and acquisition) is taken care of.

In a text-dependent speaker-verification scenario, the phrases are known to the system (e.g., the claimant is prompted to say them). The syntax used in the YOHO database is "combination lock" phrases. For example, the prompt might read: "Say: twenty-six, eighty-one, fifty-seven."

YOHO was designed for U.S. Government evaluation of speaker-verification systems in "office" environments. In addition to office environments, there are enormous consumer markets that must contend with noisy speech (e.g., telephone services) and far-field microphones (e.g., computer access).

"Combination lock" phrases (e.g., "twenty-six, eighty-one, fifty-seven")
138 subjects: 106 males, 32 females
Collected with a STU-III electret-microphone telephone handset over 3 month period in a real-world office environment
4 enrollment sessions per subject with 24 phrases per session
10 verification sessions per subject at approximately 3-day intervals with 4 phrases per session
Total of 1380 validated test sessions
8 kHz sampling with 3.8 kHz analog bandwidth (STU-III like)
1.2 gigabytes of data

Table 8.3 The YOHO corpus [5].

Speech Production

There are two main sources of speaker-specific characteristics of speech: physical and learned. Vocal tract shape is an important physical distinguishing factor of speech. The vocal tract is generally considered as the speech production organ above the vocal folds. As shown in Figure 8.4 [11], this includes the following: laryngeal pharynx (beneath epiglottis), oral pharynx (behind the tongue, between the epiglottis and velum), oral cavity (forward of the velum and bounded by the lips, tongue, and palate), nasal pharynx (above the velum, rear end of nasal cavity), and the nasal cavity (above the palate and extending from the pharynx to the nostrils). An adult male vocal tract is approximately 17 cm long [11].

The vocal folds (formerly known as vocal cords) are shown in Figure 8.4. The larynx is composed of the vocal folds, the top of the cricoid cartilage, the arytenoid cartilages, and the thyroid cartilage (also known as "Adam's apple"). The vocal folds are stretched between the thyroid cartilage and the arytenoid cartilages. The area between the vocal folds is called the glottis.

As the acoustic wave passes through the vocal tract, its frequency content (spectrum) is altered by the resonances of the vocal tract. Vocal tract resonances are called *formants*. Thus, the vocal tract shape can be estimated from the spectral shape (e.g., formant location and spectral tilt) of the voice signal.

Voice verification systems typically use features derived only from the vocal tract. As seen in Figure 8.4, the human vocal mechanism is driven by an excitation source, which also contains speaker-dependent information. The excitation is generated by airflow from the lungs, carried by the trachea (also called the "wind pipe") through the vocal folds (or the arytenoid cartilages). The excitation can be characterized as phonation, whispering, frication, compression, vibration, or a combination of these.

For other aspects of speech production that could be useful for speaker recognition, please refer to [6].

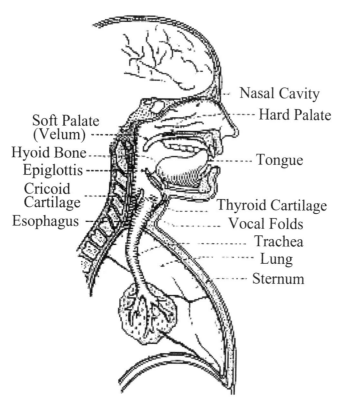

Figure 8.4 Human vocal system (reprinted with permission from J. Flanagan, *Speech Analysis Synthesis and Perception*, 2nd ed. New York and Berlin: Springer-Verlag, 1972, p. 10, Fig. 2.1 © Springer-Verlag).

Linear Prediction

The all-pole LP models a signal s_n by a linear combination of its past values and a scaled present input [24]

$$s_n = -\sum_{k=1}^{p} a_k \cdot s_{n-k} + G \cdot u_n \qquad (8.1)$$

where s_n is the present output, p is the prediction order, a_k are the model parameters called the predictor coefficients (PCs), s_{n-k} are past outputs, G is a gain scaling factor, and u_n is the present input. In speech applications, the input u_n is generally unknown, so it is ignored. Therefore, the LP approximation $\hat{s}_n$, depending only on past output samples, is

$$\hat{s}_n = -\sum_{k=1}^{p} a_k \cdot s_{n-k} \qquad (8.2)$$

The source u_n, which corresponds to the human vocal tract excitation, is not modeled by these PCs. It is certainly reasonable to expect that some speaker-dependent characteristics are present in this excitation signal (e.g., fundamental frequency). Therefore, if the excitation signal is ignored, valuable speaker-verification discrimination information could be lost.

Defining the prediction error e_n (also known as the residual) as the difference between the actual value s_n and the predicted value $\hat{s}_n$ yields

$$e_n = s_n - \hat{s}_n = s_n + \sum_{k=1}^{p} a_k \cdot s_{n-k} \qquad (8.3)$$

Using the a_k model parameters, Eq. (8.4) represents the fundamental basis of LP representation. It implies that *any* signal is defined by a linear predictor and the corresponding LP error. Obviously, the residual contains all the information not contained in the predictor coefficients (PCs).

$$s_n = -\sum_{k=1}^{p} a_k \cdot s_{n-k} + e_n \qquad (8.4)$$

From Eq. (8.1), the LP transfer function is defined as

$$H(z) \equiv \frac{S(z)}{U(z)} \equiv \frac{Z[s_n]}{Z[u_n]} \qquad (8.5)$$

which yields

$$H(z) = \frac{G}{1 + \sum_{k=1}^{p} a_k z^{-k}} \equiv \frac{G}{A(z)} \qquad (8.6)$$

where $A(z)$ is known as the pth-order inverse filter.

LP analysis determines the PCs of the inverse filter $A(z)$ that minimize the prediction error e_n in some sense. Typically, the mean square error (MSE) is minimized because it allows a simple, closed-form solution of the PCs. For example, an 8th-order 8 kHz LP analysis of the vowel /U/ (as in "foot") had the predictor coefficients shown in Table 8.4.

Power of z	0	−1	−2	−3	−4	−5	−6	−7	−8
Predictor Coefficient	1	−2.346	1.657	−0.006	0.323	−1.482	1.155	−0.190	−0.059

Table 8.4 Example of 8th-order linear predictor coefficients for the vowel /U/ as in "foot".

Evaluating the magnitude of the z transform of H(z) at equally spaced intervals on the unit circle yields the following power spectrum having formants (vocal tract resonances or spectral peaks) at 390, 870, and 3040 Hz (Figure 8.5). These resonance frequencies are in agreement with the Peterson and Barney formant frequency data for the vowel /U/ [33].

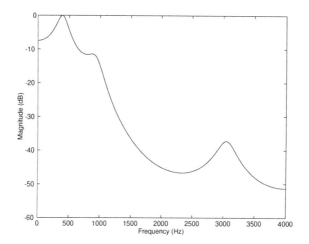

Figure 8.5 Frequency response for the vowel /U/.

Features are constructed from the speech model parameters; for example, the a_k shown in Eq. (8.6). These LP coefficients are typically nonlinearly transformed into perceptually meaningful domains suited to the application. Some feature domains useful for speech coding and recognition include reflection coefficients (RCs); log-area ratios (LARs) or arcsin of the RCs; line spectrum pair (LSP) frequencies [4,6,21,22,41]; and the LP cepstrum [33].

Reflection Coefficients and Log Area Ratios

The vocal tract can be modeled as an electrical transmission line, a waveguide, or an analogous series of cylindrical acoustic tubes. At each junction, there can be an impedance mismatch or an analogous difference in cross-sectional areas between tubes. At each boundary, a portion of the wave is transmitted and the remainder is reflected (assuming lossless tubes). The reflection coefficients k_i are the percentage of the reflection at these discontinuities. If the acoustic tubes are of equal length, the time required for sound to propagate through each tube is equal (assuming planar wave propagation). Equal propagation times allow simple z transformation for digital filter simulation. For example, a series of five acoustic tubes of equal lengths with cross-sectional areas A_1, ..., A_5 is shown in Figure 8.6. This series of five tubes represents a fourth-order system that might fit a vocal tract minus the nasal cavity. The reflection coefficients are determined by the ratios of the adjacent cross-sectional areas with appropriate boundary conditions [33]. For a pth-order system, the boundary conditions given in Eq. (8.7) correspond to a closed glottis (zero area) and a large area following the lips.

$$A_0 = 0$$

$$A_{p+1} \gg A_p \qquad (8.7)$$

$$k_i = \frac{A_{i+1} - A_i}{A_{i+1} + A_i} \quad for \quad i = 1,2,\ldots p$$

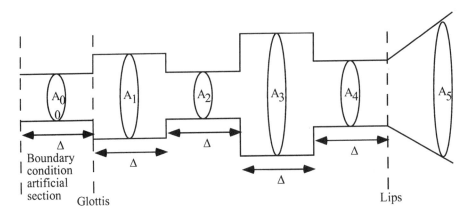

Figure 8.6 Acoustic tube model of speech production.

Narrow bandwidth poles result in $|k_i| \approx 1$. An inaccurate representation of these RCs can cause gross spectral distortion. Taking the log of the area ratios results in more uniform spectral sensitivity. The LARs are defined as the log of the ratio of adjacent cross-sectional areas

$$g_i = \log\left[\frac{A_{i+1}}{A_i}\right] = \log\left[\frac{1+k_i}{1-k_i}\right] = 2\tanh^{-1} k_i \quad for \quad i = 1,2,\ldots p \qquad (8.8)$$

Mel-Warped Cepstrum

The mel-warped cepstrum is a very popular feature domain that does not require LP analysis. It can be computed as follows: 1) window the signal, 2) take the fast Fourier transform (FFT), 3) take the magnitude, 4) take the log, 5) warp the frequencies according to the mel scale, and 6) take the inverse FFT. A variation on the cepstrum is the LP-cepstrum, where steps $1-3$ are replaced by the magnitude spectrum from LP analysis. The mel warping transforms the frequency scale to place less emphasis on high frequencies. It is based on the nonlinear human perception of the frequency of sounds [32]. The cepstrum can be considered as the spectrum of the log spectrum. Removing its mean reduces the effects of linear time-invariant filtering (e.g., channel distortion). Often, the time derivatives of the mel cepstra (also known as delta cepstra) are used as additional features to model trajectory information. The cepstrum's density has the benefit of being modeled well by a linear combination of Gaussian densities as used in the Gaussian Mixture Model [36]. Perhaps the most compelling reason for using the mel-warped cepstrum is that it has been demonstrated to work well in speaker-recognition systems [15] and, somewhat ironically, in speech-recognition systems [32], too. Furui addresses this irony and other issues plaguing speaker recognition in his set of open questions [14].

The next section presents feature selection, estimation of mean and covariance, divergence, and Bhattacharyya distance. It is highlighted by the development of the divergence shape measure and the Bhattacharyya distance shape.

3. Feature selection and measures

To apply mathematical tools without loss of generality, the speech signal can be represented by a sequence of feature vectors. The selection of appropriate features and methods to estimate (extract or measure) them are known as feature selection and feature extraction, respectively.

Traditionally, pattern-recognition paradigms are divided into three components: feature extraction and selection, pattern matching, and classification. Although this division is convenient from the perspective of designing system components, these components are not independent. The false demarcation among these components can lead to suboptimal designs because they all interact in real-world systems.

In speaker verification, the goal is to design a system that minimizes the probability of verification errors. Thus, the underlying objective is to discriminate between the given speaker and all others. A comprehensive review of discriminant analysis is given in [16]. For an overview of the feature selection and extraction methods, please refer to [6]. The next section introduces pattern matching.

4. Pattern matching

The pattern-matching task of speaker verification involves computing a match score, which is a measure of the similarity between the input feature vectors and some model. Speaker models are constructed from the features extracted from the speech signal. To enroll users into the system, a model of the voice, based on the extracted features, is generated and stored (possibly on an encrypted smart card). Then, to authenticate a user, the matching algorithm compares/scores the incoming speech signal with the model of the claimed user.

There are two types of models: stochastic models and template models. In stochastic models, the pattern matching is probabilistic and results in a measure of the likelihood, or conditional probability, of the observation given the model. For template models, the pattern matching is deterministic. The observation is assumed to be an imperfect replica of the template, and the alignment of observed frames to template frames is selected to minimize a distance measure d. The likelihood L can be approximated in template-based models by exponentiating the utterance match scores

$$L = \exp(-a\,\mathrm{d})$$

(8.9)

where a is a positive constant (equivalently, the scores are assumed to be proportional to log likelihoods). Likelihood ratios can then be formed using global speaker models or cohorts to normalize L.

The template model and its corresponding distance measure is perhaps the most intuitive method. The template method can be dependent or independent of time. An example of a time-independent template model is VQ modeling [45]. All temporal variation is ignored in this model and global averages (e.g., centroids) are all that is used. A time-dependent template model is more complicated because it must accommodate human speaking rate variability.

Template Models

The simplest template model consists of a single template $\overline{\mathbf{x}}$, which is the model for a frame of speech. The match score between the template $\overline{\mathbf{x}}$ for the claimed speaker and an input feature vector $\mathbf{x}_i$ from the unknown user is given by $d(\mathbf{x}_i, \overline{\mathbf{x}})$. The model for the claimed speaker could be the centroid (mean) of a set of N training vectors

$$\overline{\mathbf{x}} = \mu = \frac{1}{N} \sum_{i=1}^{N} \mathbf{x}_i \qquad (8.10)$$

Many different distance measures between the vectors $\mathbf{x}_i$ and $\overline{\mathbf{x}}$ can be expressed as

$$d(\mathbf{x}_i, \overline{\mathbf{x}}) = (\mathbf{x}_i - \overline{\mathbf{x}})^{\mathrm{T}} \mathbf{W}(\mathbf{x}_i - \overline{\mathbf{x}}) \qquad (8.11)$$

where $\mathbf{W}$ is a weighting matrix. If $\mathbf{W}$ is an identity matrix, the distance is *Euclidean;* if $\mathbf{W}$ is the inverse covariance matrix corresponding to mean $\overline{\mathbf{x}}$, then this is the *Mahalanobis distance*. The Mahalanobis distance gives less weight to the components having more variance and is equivalent to a Euclidean distance on principal components, which are the eigenvectors of the original space as determined from the covariance matrix [10].

Dynamic Time Warping

The most popular method to compensate for speaking-rate variability in template-based systems is known as DTW [42]. A text-dependent template model is a sequence of templates $(\overline{\mathbf{x}}_1, \ldots, \overline{\mathbf{x}}_N)$ that must be matched to an input sequence $(\mathbf{x}_1, \ldots, \mathbf{x}_M)$. In general, N is not equal to M because of timing inconsistencies in human speech. The asymmetric match score z is given by

$$z = \sum_{i=1}^{M} \mathrm{d}(\mathbf{x}_i, \overline{\mathbf{x}}_{j(i)}) \qquad (8.12)$$

where the template indices *j(i)* are typically given by a DTW algorithm. Given reference and input signals, the DTW algorithm does a constrained, piecewise linear mapping of one (or both) time axis(es) to align the two signals while minimizing z. At the end of the time warping, the accumulated distance is the basis of the match score. This method accounts for the variation over time (trajectories) of parameters corresponding to the dynamic configuration of the articulators and vocal tract. Figure 8.7 shows a warp path for two speech signals using their energies as warp features.

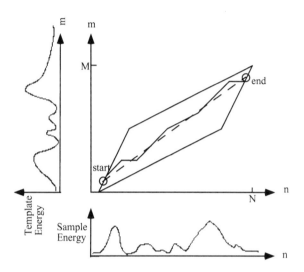

Figure 8.7 DTW path for two energy signals.

If the warp signals were identical, the warp path would be a diagonal line and the warping would have no effect. The Euclidean distance between the two signals in the energy domain is the accumulated deviation off the dashed diagonal warp path. The parallelogram surrounding the warp path represents the Sakoe slope constraints of the warp [42], which act as boundary conditions to prevent excessive warping over a given segment.

Vector Quantization Source Modeling

Another form of template model uses multiple templates to represent frames of speech and is referred to as VQ source modeling [45]. A VQ code book is a collection of codewords and it is typically designed by a clustering procedure. A code book is created for each enrolled speaker using his training data, usually based upon reading a specific text. A pattern match score can be formed as the distance between an input vector $\mathbf{x}_j$ and the minimum distance codeword $\overline{\mathbf{x}}$ in the claimant's VQ code book C. This match score for L frames of speech is

$$z = \sum_{j=1}^{L} \min_{\overline{\mathbf{x}} \in C} \left\{ d\left(\mathbf{x}_j, \overline{\mathbf{x}}\right) \right\} \tag{8.13}$$

The clustering procedure used to form the code book averages out temporal information from the codewords. Thus, there is no need to perform a time alignment. The lack of time warping greatly simplifies the system; however, it neglects speaker-dependent temporal information that may be present in the prompted phrases.

Nearest Neighbors

A technique combining the strengths of the DTW and VQ methods is called nearest neighbors (NN) [17,20]. Unlike the VQ method, the NN method does not cluster the enrollment training data to form a compact code book. Instead, it keeps all the training data and can, therefore, use temporal information.

As shown in Figure 8.8, the claimant's interframe distance matrix is computed by measuring the distance between test-session frames (the input) and the claimant's stored enrollment-session frames. The NN distance is the minimum distance between a test-session frame and the enrollment frames. The NN distances for all the test-session frames are then averaged to form a match score. Similarly, as shown in the rear planes of Figure 8.8, the test-session frames are also measured against a set of stored reference "cohort" speakers to form match scores. The match scores are then combined to form a likelihood ratio approximation [17].

The NN method is one of the most memory- and compute-intensive speaker-verification algorithms. It is also one of the most powerful methods, as illustrated later in Figure 8.10.

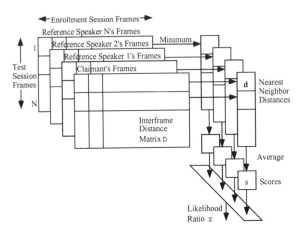

Figure 8.8 Nearest neighbor method.

Stochastic Models

Template models dominated early work in text-dependent speaker recognition. This deterministic approach is intuitively reasonable, but stochastic models recently have been developed that can offer more flexibility and result in a more theoretically meaningful probabilistic likelihood score.

Using a stochastic model, the pattern-matching problem can be formulated as measuring the likelihood of an observation (a feature vector of a collection of vectors from the unknown speaker) given the speaker model. The observation is a random vector with a conditional probability density function (pdf) that depends upon the speaker. The conditional pdf for the claimed speaker can be estimated from a set of training vectors and, given the estimated density, the probability that the observation was generated by the claimed speaker can be determined.

The estimated pdf can either be a parametric or a nonparametric model. From this model, for each frame of speech (or average of a sequence of frames), the probability that it was generated by the claimed speaker can be estimated. This probability is the match score. If the model is parametric, then a specific pdf is assumed and the appropriate parameters of the density can be estimated using the maximum likelihood estimate. For example, one useful parametric model is the multivariate normal model and it is parameterized by a mean vector μ and a covariance matrix $\mathbf{C}$. In this case, the probability that an observed feature vector $\mathbf{x}_i$ was generated by the model is

$$p(\mathbf{x}_i|\text{model}) = (2\pi)^{-k/2}|\mathbf{C}|^{-1/2}\exp\left\{-\tfrac{1}{2}(\mathbf{x}_i-\mu)^{\mathrm{T}}\mathbf{C}^{-1}(\mathbf{x}_i-\mu)\right\} \quad (8.14)$$

Hence, $p(\mathbf{x}_i|\text{model})$ is the match score. If nothing is known about the true densities, the unknown densities can be approximated by a GMM or nonparametric statistics can be used to find the match score.

The match scores for text-dependent models are given by the probability of a sequence of frames without assuming independence of speech frames. Although a correlation of speech frames is implied by the text-dependent model, deviations of the speech from the model are usually assumed to be independent. This independence assumption enables estimation of utterance likelihoods by multiplying frame likelihoods. The model represents a specific sequence of spoken words.

A stochastic model that is very popular for modeling sequences is the HMM. In conventional Markov models, each state corresponds to a deterministically observable event; thus, the output of such sources in any given state is not random and lacks the flexibility needed here. In an HMM, the observations are a probabilistic function of the state; i.e., the model is a doubly embedded stochastic process where the underlying stochastic process is not directly observable (it is hidden). The HMM can only be viewed through another set of stochastic processes that produce the sequence of observations [32]. The HMM is a finite-state machine, where a pdf (or feature vector stochastic model) $p(\mathbf{x}|s_i)$ is associated with each state s_i (the main underlying model). The states are connected by a transition network, where the state transition probabilities are $a_{ij} = p(s_i|s_j)$. For example, a hypothetical three-state HMM is illustrated in Figure 8.9.

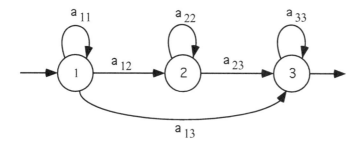

Figure 8.9 An example of a three-state HMM.

The probability that a sequence of speech frames was generated by this model can be found by Baum-Welch decoding [30,31]. This likelihood is the score for L frames of input speech given the model

$$p(\mathbf{x}(1; L)|\text{model}) = \sum_{\substack{\text{all state} \\ \text{sequences}}} \prod_{i=1}^{L} p(\mathbf{x}_i|s_i)p(s_i|s_{i-1}) \qquad (8.15)$$

This is a theoretically meaningful score. HMM-based methods have been shown to be comparable in performance to conventional VQ methods in text-independent testing [47] and more recently to outperform conventional methods in text-dependent testing (e.g., [35]).

5. Classification and Decision Theory

Having computed a match score between the input speech-feature vector and a model of the claimed speaker's voice, a verification decision is made whether to accept or reject the speaker or request another utterance (or, without a claimed identity, an identification decision is made). If a verification system accepts an impostor, it makes a false acceptance (FA) error. If the system rejects a valid user, it makes a false rejection (FR) error. The FA and FR errors can be traded off by adjusting the decision threshold, as shown by a Receiver Operating Characteristic (ROC) curve. The operating point where the FA and FR are equal corresponds to the equal error rate.

The accept or reject decision process can be an accept, continue, time-out, or reject hypothesis-testing problem. In this case, the decision making, or classification, procedure is a sequential hypothesis-testing problem [48]. For a brief overview of the decision theory involved, please refer to [6].

6. Performance

Using the YOHO prerecorded speaker-verification database, the following results on wolves and sheep were measured. The impostor testing was simulated by randomly selecting a valid user (a potential wolf) and altering his/her identity claim to match that of a randomly selected target user (a potential sheep). Because the potential wolf is not intentionally attempting to masquerade as the potential sheep, this is referred to as the "casual impostor" paradigm. Testing the system to a certain confidence level implies a minimum requirement for the number of trials. In this testing, there were 9,300 simulated impostor trials to test to the desired confidence [5,17].

DTW System

The DTW ASV system tested here was created by Higgins, *et al.* [18]. This system is a variation on a DTW approach that introduced likelihood ratio scoring via cohort normalization in which the input utterance is compared with the claimant's voice model and with an alternate model composed of models of other users with similar voices. Likelihood ratio scoring allows for a fixed, speaker-independent, phrase-independent acceptance criterion. Pseudorandomized phrase prompting, consistent with the YOHO corpus, is used in combination with speech recognition to reduce the threat of playback (e.g., tape recorder) attacks. The enrollment algorithm creates users' voice models based upon subword models (e.g., "twen," "ti," and "six"). Enrollment begins with a generic male or female template for each subword and results in a speaker-specific template model for each subword. These models and their estimated word endpoints are successively refined by including more examples collected from the enrollment speech material [18].

Cross-speaker testing (casual impostors) was performed, confusion matrices for each system were generated, wolves and sheep of DTW and NN systems were identified, and errors were analyzed.

Table 8.5 shows two measures of wolves and sheep for the DTW system: those who were wolves or sheep at least once and those who were wolves or sheep at least twice. Thus, FA errors occur in a very narrow portion of the 186-person population, especially if two errors are required to designate a person as a wolf or sheep. The difficulty in acquiring enough data to adequately represent the wolf and sheep populations makes it challenging to study these errors.

186 Subjects of the YOHO Database	
At least one FA Error	At least two FA Errors
17 Wolves (9%)	2 Wolves (1%)
11 Sheep (6%)	5 Sheep (3%)

Table 8.5 Known wolves and sheep of the DTW system.

The DTW system made 19 FA errors over the 9,300 impostor trials. Table 8.6 shows that these 19 pairs of wolves and sheep have interesting characteristics. The database contains four times as many males as it does females, but the 18:1 ratio of male wolves to female wolves is disproportionate. It is also interesting to note that one male wolf successfully preyed upon three different female sheep. The YOHO corpus provides at least 19 pairs of wolves and sheep under the DTW ASV system for further investigation.

19 FA errors across 9300 impostor trials		
Number of FA errors	Wolf sex	Sheep sex
15	Males	Males
1	Female	Female
3	1 Male	3 Females

Table 8.6 Wolf and sheep distribution by sex.

ROC of DTW and NN Systems

Figure 8.10 shows the NN system's ROC curve and a point on the ROC for the DTW system (ROCs of better systems are closer to the origin). The NN system was the first one known to meet the 0.1% FA and 1% FR performance level at the 80% confidence level and it outperforms the DTW system by about half an order of magnitude.

These overall error rates do not show the individual wolf and sheep populations of the two systems. As shown in Figures 8.11-8.14, the two systems commit different errors.

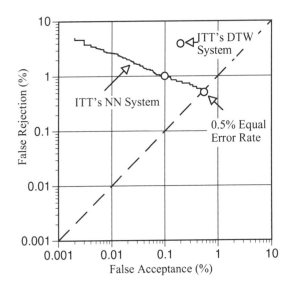

Figure 8.10 Receiver operating characteristics.

Wolves and Sheep

FA errors due to individual wolves and sheep are shown in the 3-D histogram plots of Figures 8.11 through 8.14. Figure 8.11 shows the individual speakers who were falsely accepted as other speakers by the DTW system. For example, the person with an identification number of 97328 is never a wolf and is a sheep once under the DTW system.

The DTW system rarely has the same speaker as both a wolf and a sheep (there are only two exceptions in this data). These exceptions, called *wolf-sheep*, probably have poor models because they match a sheep's model more closely than their own and a wolf's model also matches their model more closely than their own. These *wolf-sheep* would likely benefit from retraining to improve their models.

Now let us look at the NN system. Figure 8.12 shows the FA errors committed by the NN system. Two speakers, who are sheep, are seen to dominate the NN system's FA errors. A dramatic performance improvement would result if these two speakers were recognized correctly by the system.

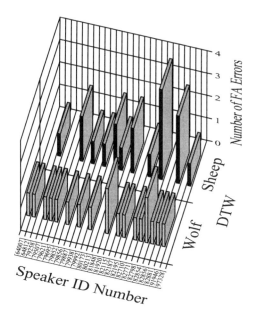

Figure 8.11 Speaker versus FA errors for the DTW system's wolves and sheep.

Now we will investigate the relations between the NN and DTW systems. Figure 8.13 shows the sheep of the NN and DTW systems. The two sheep that dominate the FA errors of the NN system are shown not to be sheep in the DTW system. This suggests the potential for making a significant performance improvement by combining the systems.

Figure 8.14 shows that the wolves of the NN system are dominated by a few individuals who do not cause errors in the DTW system. Again, this suggests the potential for realizing a performance improvement by combining elements of the NN and DTW systems. Along these lines, a high-performance speaker detection system consisting of eight combined systems has been demonstrated recently [27].

7.   Conclusions

Automatic speaker recognition is the use of a machine to recognize a person from a spoken phrase. Speaker-recognition systems can be used to *identify* a particular person or to *verify* a person's claimed identity. Speech processing, speech production, and features and pattern matching for speaker recognition were introduced. Recognition accuracy was shown by coarse-grain ROC curves and fine-grain histograms revealed the wolves and sheep of two example systems. Speaker recognition systems can achieve 0.5% equal error rates at the 80% confidence level in the benign real-world conditions considered here.

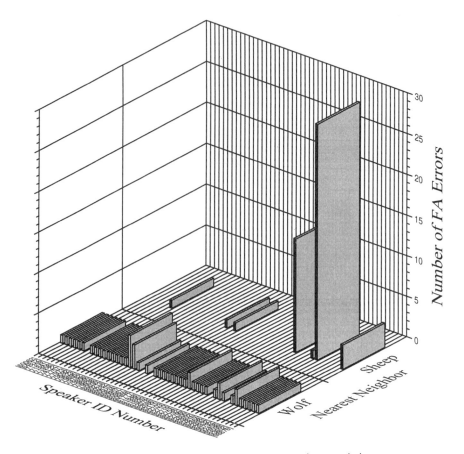

Figure 8.12 Speaker versus FA errors for NN system's wolves and sheep.

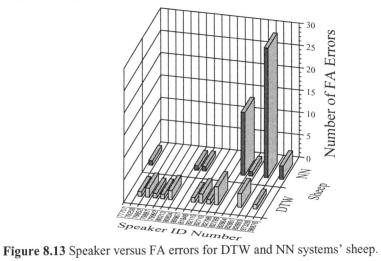

Figure 8.13 Speaker versus FA errors for DTW and NN systems' sheep.

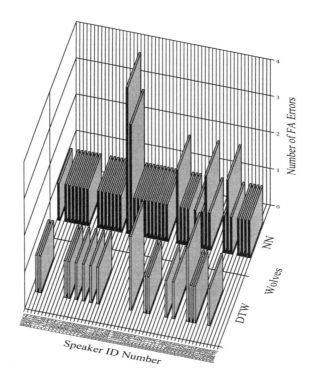

Figure 8.14 Speaker versus FA errors for DTW and NN systems' wolves.

References

[1] B. S. Atal, "Effectiveness of Linear Prediction Characteristics of the Speech Wave for Automatic Speaker Identification and Verification," *Journal of the Acoustical Society of America,* Vol. 55, No. 6, pp. 1304-1312, 1974.

[2] B. S. Atal, "Automatic Recognition of Speakers from Their Voices," *Proceedings of the IEEE* Vol. 64, pp. 460-475, 1976.

[3] J. Attili, M. Savic, and J. Campbell. "A TMS32020-Based Real Time, Text-Independent, Automatic Speaker Verification System," In *International Conference on Acoustics, Speech, and Signal Processing in New York,* IEEE, pp. 599-602, 1988.

[4] J. P. Campbell, T. E. Tremain, and V. C. Welch. "The Federal Standard 1016 4800 bps CELP Voice Coder," *Digital Signal Processing,* Vol. 1, No. 3, pp. 145 -155, 1991.

[5] J. P. Campbell, "Testing with The YOHO CD-ROM Voice Verification Corpus," In *International Conference on Acoustics, Speech, and Signal Processing in Detroit,* IEEE, pp. 341-344, 1995. Available: http://www.biometrics.org/.

[6] J. P. Campbell, "Speaker Recognition: A Tutorial." *Proceedings of the IEEE,* Vol. 85, No. 9, pp. 1437-1462, 1997.

[7] C. Che and Q. Lin, "Speaker recognition using HMM with experiments on the YOHO database," In *EUROSPEECH in Madrid*, ESCA, pp. 625- 628, 1995.

[8] J. Colombi, D. Ruck, S. Rogers, M. Oxley, and T. Anderson, "Cohort Selection and Word Grammer Effects for Speaker Recognition," In *International Conference on Acoustics, Speech, and Signal Processing in Atlanta,* IEEE, pp. 85- 88, 1996.

[9] G. R. Doddington, "Speaker Recognition—Identifying People by their Voices," *Proceedings of the IEEE,* Vol. 73, No. 11, pp. 1651-1664, 1985.

[10] R. Duda and P. Hart. *Pattern Classification and Scene Analysis.* New York: Wiley, 1973.

[11] J. Flanagan, *Speech Analysis Synthesis and Perception.* 2nd ed., Berlin: Springer-Verlag, 1972.

[12] S. Furui, "Cepstral Analysis Technique for Automatic Speaker Verification." *IEEE Transactions on Acoustics, Speech, and Signal Processing,* Vol. ASSP-29, No. 2 , pp. 254-272, 1981.

[13] S. Furui, "Speaker-Dependent-Feature Extraction, Recognition and Processing Techniques." *Speech Communication,* Vol. 10, pp. 505-520, 1991.

[14] S. Furui, "Recent Advances in Speaker Recognition," *Pattern Recognition Letters,* Vol. 18, pp. 859-872, 1997.

[15] H. Gish and M. Schmidt, "Text-Independent Speaker Identification," *IEEE Signal Processing Magazine,* Vol. 11, No. 4, pp. 18-32, 1994.

[16] R. Gnanadesikan and J. R. Kettenring. "Discriminant Analysis and Clustering," *Statistical Science,* Vol. 4, No. 1, pp. 34-69, 1989.

[17] A. Higgins, "YOHO Speaker Verification," Speech Research Symposium, Baltimore, 1990.

[18] A. Higgins, L. Bahler, and J. Porter, "Speaker Verification Using Randomized Phrase Prompting," *Digital Signal Processing,* Vol. 1, No. 2 , pp. 89-106, 1991.

[19] A. L. Higgins and R. E. Wohlford, "A New Method of Text-Independent Speaker Recognition," In *International Conference on Acoustics, Speech, and Signal Processing in Tokyo,* IEEE, pp. 869-872, 1986.

[20] A. Higgins, L. Bahler, and J. Porter, "Voice Identification Using Nearest Neighbor Distance Measure," In *International Conference on Acoustics, Speech, and Signal Processing in Minneapolis,* IEEE, pp. 375-378, 1993.

[21] F. Itakura, "Line Spectrum Representation of Linear Predictive Coefficients," *Transactions of the Committee on Speech Research, Acoustical Society of Japan,* Vol. S75, No. 34, 1975.

[22] G. Kang and L. Fransen, *Low Bit Rate Speech Encoder Based on Line-Spectrum-Frequency,* NRL, NRL Report 8857, 1985.

[23] K. P. Li and E. H. Wrench, "Text-Independent Speaker Recognition with Short Utterances," In *International Conference on Acoustics, Speech, and Signal Processing in Boston,* IEEE, pp. 555-558, 1983.

[24] J. Makhoul, "Linear Prediction: A Tutorial Review," *Proceedings of the IEEE,* Vol. 63, pp. 561-580, 1975.

[25] R. Mammone, X. Zhang, and R. Ramachandran, "Robust Speaker Recognition-A Feature-based Approach," *IEEE Signal Processing Magazine,* Vol. 13, No. 5, pp. 58-71, 1996.

[26] J. D. Markel and S. B. Davis, "Text-Independent Speaker Recognition from a Large Linguistically Unconstrained Time-Spaced Data Base," *IEEE Transactions on Acoustics, Speech, and Signal Processing* ASSP-27, No. 1, pp. 74-82, 1979.

[27] A. Martin and M. Przybocki, "1997 Speaker Recognition Evaluation," In *Speaker Recognition Workshop,* editor A. Martin (NIST). Section 2. Maritime Institute of Technology, Linthicum Heights, Maryland, June, pp. 25-26, 1997. Available: ftp://jaguar.ncsl.nist.gov/speaker/ and http://www.nist.gov/itl/div894/894.01/.

[28] D. O'Shaughnessy, *Speech Communication, Human and Machine.* Digital Signal Processing, Reading: Addison-Wesley, 1987.

[29] G. Papcun, "Commensurability Among Biometric Systems: How to Know When Three Apples Probably Equals Seven Oranges," In *Proceedings of the Biometric Consortium, 9th*

Meeting, editor J. Campbell (NSA). Holiday Inn, Crystal City, Virginia, April, 8-9, 1997. Available: http://www.biometrics.org/.

[30] L. R. Rabiner, "A Tutorial on Hidden Markov Models and Selected Applications in Speech Recognition," *Proceedings of the IEEE,* Vol. 77, No. 2, pp. 257-286, 1989.

[31] L. Rabiner and B. H. Juang, "An Introduction to Hidden Markov Models," *IEEE ASSP Magazine,* Vol. 3, pp. 4-16, January 1986.

[32] L. Rabiner and B. H. Juang, *Fundamentals of Speech Recognition,* Signal Processing, editor A. Oppenheim. Englewood Cliffs: Prentice-Hall, 1993.

[33] L. Rabiner and R. Schafer. *Digital Processing of Speech Signals,* Signal Processing, editor A. Oppenheim. Englewood Cliffs: Prentice-Hall, 1978.

[34] D. Reynolds, "Speaker Identification and Verification using Gaussian Mixture Speaker Models," *Speech Communication,* Vol. 17, pp. 91-108, 1995.

[35] D. Reynolds and B. Carlson, "Text-Dependent Speaker Verification Using Decoupled and Integrated Speaker and Speech Recognizers," In *EUROSPEECH in Madrid,* ESCA, pp. 647-650, 1995.

[36] D. Reynolds and R. Rose, "Robust Text-Independent Speaker Identification Using Gaussian Mixture Speaker Models," *IEEE Transactions on Speech and Audio Processing,* Vol. 3, No. 1, pp. 72-83, 1995.

[37] D. Reynolds, "M.I.T. Lincoln Laboratory Site Presentation," In *Speaker Recognition Workshop,* editor A. Martin (NIST). Section 5. Maritime Institute of Technology, Linthicum Heights, Maryland, March, pp. 27-28, 1996. Available: ftp://jaguar.ncsl.nist.gov/speaker/ and http://www.nist.gov/itl/div894/894.01/.

[38] A. Rosenberg, "Automatic Speaker Verification: A Review," *Proceedings of the IEEE,* Vol. 64, No. 4, pp. 475-487, 1976.

[39] E. Rosenberg and F. K. Soong, "Recent Research in Automatic Speaker Recognition," In *Advances in Speech Signal Processing,* ed. S. Furui and M. M. Sondhi. pp. 701-738. New York: Marcel Dekker, 1992.

[40] E. Rosenberg, J. DeLong, C-H. Lee, B-H. Juang, and F. K. Soong, "The Use of Cohort Normalized Scores for Speaker Verification," In *International Conference on Spoken Language Processing in Banff,* University of Alberta, pp. 599-602, 1992.

[41] S. Saito and K. Nakata. *Fundamentals of Speech Signal Processing.* Tokyo: Academic Press, 1985.

[42] H. Sakoe and S. Chiba. "Dynamic Programming Algorithm Optimization for Spoken Word Recognition." *IEEE Transactions on Acoustics, Speech, and Signal Processing* ASSP-26, No. 1, pp. 43-49, 1978.

[43] R. Schwartz, S. Roucos, and M. Berouti, "The Application of Probability Density Estimation to Text Independent Speaker Identification," In *International Conference on Acoustics, Speech, and Signal Processing in Paris,* IEEE, pp. 1649-1652, 1982.

[44] F. Soong, A. Rosenberg, L. Rabiner, and B-H. Juang, "A Vector Quantization Approach to Speaker Recognition," In *International Conference on Acoustics, Speech, and Signal Processing in Florida,* IEEE, pp. 387-390, 1985.

[45] F. K. Soong, A. E. Rosenberg, L. R. Rabiner, and B. H. Juang, "A Vector Quantization Approach to Speaker Recognition." *AT&T Technical Journal,* Vol. 66, No. 2 , pp. 14-26, 1987.

[46] A. Sutherland and M. Jack, "Speaker Verification." In *Aspects of Speech Technology,* editors M. Jack and J. Laver, Edinburgh: Edinburgh University Press, pp. 185-215, 1988.

[47] N. Z. Tishby, "On the Application of Mixture AR Hidden Markov Models to Text Independent Speaker Recognition," *IEEE Transactions on Acoustics, Speech, and Signal Processing,* Vol. 39, No. 3, pp. 563 – 570, 1991.

[48] A. Wald, *Sequential Analysis.* New York: Wiley, 1947.

9 INFRARED IDENTIFICATION OF FACES AND BODY PARTS

Francine J. Prokoski
Mikos Ltd, Fairfax Station, VA
mikos@gte.net

Robert B. Riedel
Mikos Ltd, Fairfax Station, VA
riedel@his.com

Abstract *Infrared imaging offers a robust technique for identification of faces and body parts. The anatomical information which is utilized by the infrared identification (IRID) technology involves subsurface features unique to each person. Those features may be imaged at a distance, using passive infrared sensor technology, with or without the cooperation of the subject. IRID therefore provides an unique capability for rapid, on-the-fly identification, under all lighting conditions including total darkness. A comparison between minutiae from facial thermograms and from fingerprints, based upon the anatomical structures underlying each, supports the experimental findings that facial thermograms are as unique as fingerprints. For many biometric applications, identification based upon facial thermograms is preferable over fingerprints, since the former requires no physical contact with the subject and can be collected on-the-fly. Also, a significant percentage of the population does not produce good enough fingerprints for identification; in contrast with the fact that every living person presents a thermal pattern. Thermal images of part of the face may be identified if a sufficient area is seen, just as with latent fingerprints. Systems for analyzing fingerprint minutiae may be utilized to analyze facial thermal minutiae with good results.*

Keywords: *IRID, infrared, disguise, face, thermograms, minutiae, non-contact, biometric, passive sensing, security, intrusion.*

1. Introduction

Thermography, which is the use of cameras sensitive in the infrared spectrum, provides highly secure, rapid, noncontact positive identification of human faces or other parts of the body, even with no cooperation from the subject being identified. When the need is for "on the fly" identification such as to maintain accountability during emergency evacuations, or to spot targeted *Faces in a Crowd*, only imaging technologies in either the visible or infrared spectra are suitable. Infrared identification (IRID) is similar to visual identification in that both are completely passive and able to be performed from a distance using either manual or automatic comparison with previously collected images.   However, identification from infrared images provides several significant benefits.

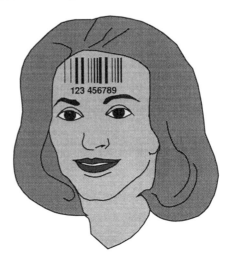

Figure 9.1 The ideal biometric.

Human thermograms are affected by changes in ambient temperature, by ingestion of substances which are vasodilators or vasoconstrictors, by sinus problems, inflammation, arterial blockages, incipient stroke, soft tissue injuries, and other physiological conditions. Radiometric IR camera systems can even produce a non-contact EKG by exhibiting local temperature fluctuations associated with the cycle of heart beats. Medical providers can utilize these time-varying temperature maps for triage, diagnosis, and treatment monitoring.

When the objective is identification only, the temperature data itself is not directly used. Rather, the thermal data is analyzed to yield anatomical information which is invariant to such changes. The IR image provides such detailed anatomical information that each person's information is unique, and is constant regardless of the medical condition variables. Consider this analogous to analyzing fingerprints; whether the print is formed by coating the finger with oil, blood, or ink, and whether a full rolled or partial latent print is seen, the same fingerprint pattern emerges. Similarly, facial thermograms yield the same blood vessel pathways and minutiae regardless of apparent temperatures.

Infrared video cameras are passive, emitting no energy or other radiation on the subject, but merely collecting and focusing the thermal radiation spontaneously and continuously emitted from the surface of the human body. IR cameras operating in the mid (3-5 micron) or long (8-12 micron) infrared bands produce images of patterns caused by superficial blood vessels which lay up to 4 cm below the skin surface. The human body is bilaterally symmetrical. Any significant asymmetry indicates a potential health-related abnormality. The assumption of symmetry facilitates assigning face axes. The reality of minor local asymmetries in each person's face facilitates alignment of images for comparison, classification, and identification of unique thermograms for each person.

2. Comparison of IRID to Other Biometrics

All biometric techniques deal with the physiology of the human body, which involves aging effects, seasonal variations, biorhythm cycles, variations due to medical conditions, and changes associated with metabolic effects.

Uniqueness and Repeatability

Biometric technologies intended to be used in automated identification systems are routinely advertised as using data which is unique to each person, and which is repeatable over time and under varying conditions. In fact, there is no proof that any biometric signature, including fingerprints, is unique for each person. At best, statistically significant sampling can be performed, over time and under a set of controlled conditions, to support the hypothesis of uniqueness. Improvements in sensor technology generally aid in the demonstration of uniqueness by detecting finer inherent details, which can lead to lower rates of false positive errors. However, those improvements may exaggerate variations due to aging, physiological or ambient conditions, or system equipment modifications, which can lead to higher false negative errors. Therefore, a proper balance must be established between tuning a biometric system to yield a unique signature for each enrollee, and accommodating expected variations in those signatures over time. Facial thermography is a robust biometric, meeting the dual requirements of uniqueness and repeatability, as can be demonstrated through comparison with fingerprints.

The thermal patterns seen by an infrared camera derive primarily from the pattern of blood vessels under the skin, which transport warm blood throughout the body. Figure 9.6 illustrates the major superficial blood vessels of the face. Figure 9.7 shows the added details discernible from hotter areas near the heart. Current commercially-available infrared cameras provide sufficient thermal sensitivity and spatial resolution to produce images such as shown in Figure 9.2 wherein the shape of the thermal contours is anatomically determined by the structure of the head and face, and the position of the blood vessels. Aside from growth, accidental injury, and surgical intervention, a person's anatomy does not change, and the complexity of the vast network of blood vessels assures that each person's vascular patterns are unique. Even identical twins have different thermograms, as shown in Figure 9.3. In contrast, visual ID systems cannot differentiate between identical twins. In fact, many people look very similar in visual images, and the use of disguises can often enable one person to look like any other, without detection.

While there is no way to prove that facial thermograms are unique, it is possible to show they contain inherently more information than fingerprints. Similarly, it has never been proven that fingerprints are unique. However, that working hypothesis has been supported by many years of experience. Facial thermogram minutiae is analogous to fingerprint minutiae in that two sets of minutiae may be considered to identify the same person if a significant number of the minutiae in the two sets have corresponding positions and characteristics. In the United States, 16 minutiae points must correspond in order for an unknown fingerprint to be considered a match to a known print for evidentiary purposes. Efforts are underway to establish guidelines for similar evidentiary rules regarding the use of infrared facial imagery for positive identification in legal proceedings.

Immunity from Forgery

Infrared Identification systems can detect attempted disguise. The IRID technique utilizes hidden micro parameters which lie below the skin surface, and which cannot be forged. Although that IR signature can be blocked, it cannot be changed by appliqué or surgical intervention, without those processes being detectable. The temperature distribution across artificial facial hair or other appliqués is readily distinguished from normal hair and skin. Plumpers in cheeks, dental reconstruction, and external skin tightening appliances distort the skin surface but do not add minutiae. The distortion they cause can be modeled by using finite element analysis of the local rubber sheeting effects applied to groups of minutiae.

Plastic surgery done for reconstruction or intentional disguise may add or subtract skin tissue, redistribute fat, add silicone or other inert materials, create or remove scars, resurface the skin via laser or chemicals, apply permanent color by tattooing eyelids and lips, remove tattoos and birthmarks by laser, implant or dipilate hair. Any one or combination of such procedures would probably defeat a visual identification system, but would generally not affect infrared identification unless the blood vessels were repositioned across most of the face. Procedures which are sufficiently invasive to reroute the patterns of superficial blood flow would necessarily cause incisions detectable in infrared, and would risk damaging facial nerves. It is, therefore, considered possible that a person could surgically distort his facial thermogram to avoid recognition, but the thermogram would contain evidence that he had done so.

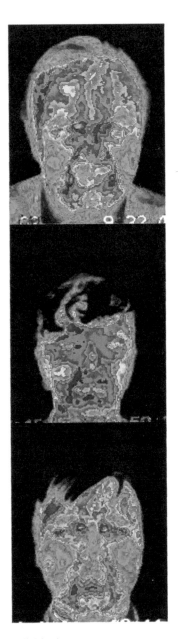

Figure 9.2 Visual and infrared images of three individuals.

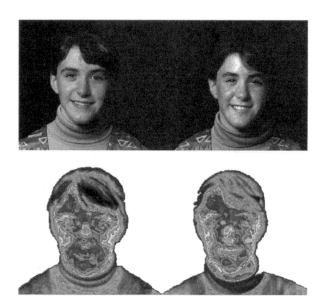

Figure 9.3 Visible and infrared imagery of identical twins.

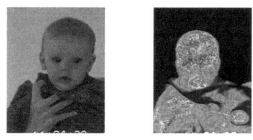

Figure 9.4 Visible and infrared imagery of an infant.

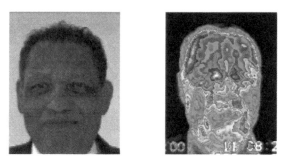

Figure 9.5 Visible and infrared imagery of an individual having dark complexion.

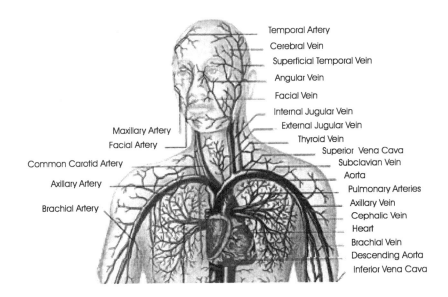

Figure 9.6 Facial and thoracic arteries and veins.

In general, only a single frame of infrared imagery is required to make a positive identification, given that it includes sufficient area of the face. It has been hypothesized but never demonstrated, that a person could paint his face with different emissivity materials in order to mimic another's facial thermogram. If that were ever determined to be a credible threat, then two or more frames would be captured and compared, to detect the presence of the minute thermal variations associated with heart rate and respiration cycles. The painted face could not evince such variations. Slight air flow past the face can be used to enhance the detectability of such materials.

Operation under Uncontrolled or Dim Light, or Total Darkness

IRID systems work accurately even in dim light or total darkness, whereas visible light systems are poor in dim light and useless in the dark. Using IR or UV strobes to illuminate the face for a visible identification system produces poor quality images not suitable for accurate identification. When the requirement is for identification under poor lighting conditions, only IRID technology is effective.

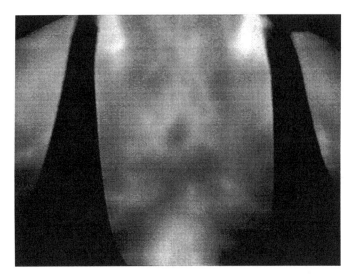

Figure 9.7 Infrared image of chest and neck showing vasculature.

Tolerance for Aging and Changes in Appearance

The body's pattern of blood vessels is "hardwired" into the face at birth (Figure 9.4) and remains relatively unaffected by aging, except for predictable growth as with fingerprints. This provides for less inherent variability in the identifying features than is provided by visual images. Visible changes to hair color and style, colored contact lenses, suntan and makeup have no effect on the accuracy of thermal analysis. Other changes, such as weight loss and gain, cause rubber sheeting distortions only, which are dealt with in the matching algorithm.

Non-Intrusiveness

Both IRID and visual ID share the advantageous features of non-contact and hands-free on-the-fly operation, causing no interference with or slowing down of the entrants. All persons provide usable images, whereas significant percentages of persons do not provide usable fingerprints, retinal/iris scans, or voiceprints. Therefore, no secondary identification technique is needed to accommodate special cases. Most people are accustomed to the presence of surveillance cameras, and little if any training is required for cooperative ID system use. Facial recognition systems can provide accurate real-time automated identification even without the entrant's cooperation or awareness, which is not true of other biometrics.

Identification of Dark-Skinned Subjects

Reported results on visual recognition systems include low incidences of dark skinned subjects in the databases. It is expected that inclusion of a representative distribution of skin tones might significantly degrade the performance results cited unless the lighting were adjusted for proper illumination of all skin tones. By contrast, skin tone does not affect the accuracy of thermal identification (Figure 9.5). This is an important factor in commercialization of any system which will be used for high security applications, or in relation to the criminal justice system. Wide implementation and acceptance of automated facial recognition will require that the accuracy of the system is irrespective of race and gender, as well as other characteristics such as age.

Scaling up to Large Populations

The underlying features behind IRID are the locations of specific junctions among blood vessels in the face, each of which is about 0.1" in diameter. Minutiae points may be either derived from thermal contours created by less sensitive thermal cameras, or may be extracted from absolute locations in imagery from more sensitive cameras. In either case, 175 minutiae can be extracted from a full facial image, each of which is associated with a relative location on the face, one or more vectors, and relative thermal band. The totality of possible minutiae configurations is significantly greater than the foreseeable maximum human population.

Use of a FaceCode technique as illustrated in Figure 9.1 would in theory eliminate the need for image matching. Given the coding were sufficiently robust to compensate for expected variations in appearance and expression, a person's FaceCode would be read similarly to scanning a barcoded product.

Maximum Throughput

The use of infrared identification provides the fastest throughput of any biometric technique. Since it is non-contact, it can be done at a distance limited only by the selection of optics. It does not require processing to correct lighting effects, and can recognize multiple faces in a single frame simultaneously. The non-contact nature of IRID helps to minimize maintenance costs and reduce the chance of accidental or intentional damage or vandalism. The limiting factor in IRID throughput is the matching engine decision time, which is less than 1 second for verification on a Pentium II/333 with current software. Any contact technique (such as fingerprints or hand geometry) or any action technique (such as voice or handwriting signature) requires more entrant time, and also requires training of the user and maintenance of the user interface. Other non-contact techniques include: iris scan, retinal scan, and visual images. Iris and retinal scans require more precise positioning, and the cooperation of the entrant in looking at a particular point. Additionally, these techniques can identify only one person at a time. Visual image identification throughput speed can match IRID under ideal conditions, but typically requires additional processing to analyze the visual images under varying lighting conditions, resulting in a processing rate slower than for IRID.

Minimum False Rejects (False Negatives)

IRID works automatically with essentially all of the population; more than other biometrics. IRID systems are expected to systematically produce false rejects only for persons who have a new gross thermal condition such as severe sunburn or extensive facial surgery. Less than 1% of the travel population is affected by such conditions. This contrasts with a larger percentage of persons who do not produce good quality fingerprints on livescan systems due to the ridge structure of their fingers and/or dry skin. Contact lens wearers cause problems with the iris scan if they wear colored lenses or bifocal lenses, or wear their lenses only sometimes. Also, persons with glaucoma or cataracts may not be reliably identified by iris scanning. The situation is similar for retinal scans. Voice recognition is affected by colds, allergies, tiredness, dryness, and other variables. As speech processing technology improves, these effects will be less of a problem. However, current technology seems to be inadequate in dealing with the effects of airline travel on the vocal tract, especially lengthy foreign travel on speaker recognition. Visual imaging may produce false rejects for changes to facial hair, hair style, makeup, headgear, facial expression, and lighting. Even under laboratory conditions, false rejects occur due to normal daily variations in appearance. Hand geometry should have a low false reject rate since there is little variation in the basic features being measured, but variations in the mechanical readers and in the training and behavior of the users commonly produce false rejects.

Minimum False Accepts (False Positives)

Using thresholding for match/no match decisions to compare facial thermograms, the probability of false positives can be set as tightly as desired. Depending on the quality of the imagery and the repeatability of the face position, a single correlation value can be produced for each match, or local correlation values can be produced for each area of the face or cluster of minutiae. Prior to correlation, relatively cold pixels representing background, hair, and eyeglasses are converted to random noise so as to not influence the correlation process. The autocorrelation values for whole frame or local areas are used to tailor the match decision thresholds for each entrant. These steps avoid the occurrences of the matching engine recognizing hair styles, glasses, or head outline rather than the true biometric of facial thermal patterns, and thereby minimize the rate of false accepts.

Other biometrics often rely on more sophisticated analysis to reduce false accepts when allowing for normal variations in biometric features. For example, detecting papercuts on fingerprints, varied contact lens patterns on iris scanning, effects of a sore throat on voice recognition, and facial hair changes in visual face recognition requires more computationally intense processing than does converting cold pixels to noise in IRID.

Correlation between Visual and Infrared Images

The two frequently cited disadvantages of IRID relative to visual recognition are that infrared cameras cost more, and there do not exist large databases of thermal images comparable to mug shot files. Both those citations deserve comment. The camera cost disparity will remain, but will continue to be reduced during the next three years as IR camera production increases. When the cost of providing controlled lighting for a visible ID system is included, the current cost discrepancy is about $10,000 per system. That figure has been decreasing by about 30% per year for the past 10 years, which reduction is expected to continue. The strong advantages of infrared identification are expected to justify the higher camera price for many applications.

IRID matching techniques can utilize existing visible image databases. There are several correspondences between an IR and a visible image of the same person as can be seen in the figures. In particular, the head shape and size and the position of features such as the eyes, mouth, nose and ears are the same in both. The infrared image lacks the color of the skin, hair, and eyes; however, it provides detailed anatomical information missing from the visible image. Infrared images are unique to each person; visual images of corresponding resolution are not unique because many people look similar and can disguise themselves to look enough like one another that an automated system cannot distinguish them. Therefore, in a large database, it is not possible to automatically perform a one-to-one linkage between infrared and visual images because the visual images are not sufficiently unique and so a number of people may have visual images which could match the IR image.

However, for each infrared image an automated system can eliminate all visual images which cannot be a match because the corresponding features are not similar enough. In general, more than 95% of the persons in a visual database can be eliminated as a match to a given infrared image. This has application to the use of infrared surveillance imagery to identify wanted persons for whom we may have only a visual image database. The IR-Visual correlation system compares each person seen in infrared to a visual image database, and determines all the possible matches. Therefore, IRID can be used to correlate between IR surveillance images and mug shot, DMV, and other available visual databases. Large scale use of infrared imagery can therefore proceed without waiting until large databases of infrared images are established.

3. Principles of Infrared Identification

Human Identification Accuracy is the Standard

Psychological research on how humans recognize one another's face indicates the importance of the location and shape of eyes, nose, eyebrows, face shape, chin, lips, and mouth, in order of decreasing utility. Humans can in general achieve accuracy on the order of 97% in identification tasks involving determining the identity of a person whom they have previously seen. The 3% error is primarily due to the fact that people are very similar to one another in visual imaging, and other cues are used in daily life to compensate for that similarity. For example, height, voice, hair style, location, style of movement, body language, etc. all contribute to visual recognition. Automated identification systems for recognition of persons based upon visual facial images in general seek to duplicate the behavior of humans performing that same task. The use of neural nets, adaptive clustering networks, retinal sampling, template matching, eigenfaces, etc. all seek to replicate human performance. The best systems currently achieve results comparable to human identification.

Earlier research by the authors found that humans were error-free in recognizing various thermal images as belonging to the same person when the imagery was taken under a wide variety of temperature, pose, hair style, lighting, and clothing changes. When asked to describe the features used in making their determinations, it is the connectedness of the thermal contours which are most considered, along with the thermal shapes of specific areas such as the canthi, and the degree of asymmetry in the forehead, cheek, and nose area. Whether the nose is hot or cold, and the amount of definition within the nose area are significant sorting features for quickly considering possible matches. Therefore, attempts to automate identification based upon thermal images can be expected to be optimized when different features are used compared to visual image recognition. However, the fact that humans can, with little experience, learn to accurately recognize thermal facial images demonstrates that sufficient information exists within the thermal face to allow very accurate automated identification.

Preprocessing Requirements

Certain standard preprocessing requirements exist for any pattern recognition task, regardless of the category: (i) the target must be found in the data; (ii) the quality of the target data must be analyzed and found to be adequate; (ii) the target data must be normalized for amplitude, distribution, and orientation; and (iv) background, clutter, and noise must be subtracted.

For facial identification tasks in either visible or IR, the recognition engine must contend with appearance variations that include: (i) hair falling on forehead, (ii) side and top hair style changes - on and off the face, (iii) facial hair added or removed, (iv) facial expression changes, (v) rotation, tilt, and tip of the head, (vi) sunglasses and eyeglasses, (vii) sunburn, (viii) hats, (ix) distinguishing chin edge from neck, (x) focus and motion blur, and (xi) makeup.

Other variables pose different problems between visual and thermal imagery. In particular, illumination variations in intensity and direction are the greatest source of error in visual identification tasks. The resulting shadows and glare and apparent changes in facial relief outlines cause problems for template matching and feature-based automated systems. Thermal identification is relatively immune to illumination changes, unless the illumination produces measurable thermal changes in the face temperature patterns. Eyeglasses are a potential problem in both modes of identification, although visual identification of persons who always wear the same eyeglasses can improve the overall accuracy of the system, since the resulting image is probably more distinguishable from other faces in the database. Glare from the glass can mask the area behind it from visual detection, and that area is always masked from infrared view due to the inability of the thermal energy to transmit through most glasses and plastics. The effect of any eyeglasses on thermal images can be considered analogous to the effect of sunglasses or photosensitive lenses on visual images, in terms of impact on identification.

Edge Effects and Rotation

The accuracy of matching images where one is rotated with respect to the other varies with the amount of overlap between the two images. If analysis is performed using template matching, the size and position of the templates limits the amount of rotation which can be tolerated. If feature analysis is used for matching, then the edge effect impact on features affected by the rotation will limit the potential accuracy. To accommodate greater repositioning of the head, small templates and small features should be selected. This also produces less accumulated error at the edges of templates due to tilt or rotation or tip, and also reduces the error introduced by assumptions of flatness for local shape areas. If minutiae matching is used, each minutia is either seen or not; its positions and characteristics are not affected otherwise by head position.

4. Facial Identification using Templates and Matching of Shapes

IRID systems built and tested to date have employed either template or shape matching engines.

Common Processing Approaches

The two generic classical approaches to facial recognition, namely template matching and feature matching, apply both to visual and thermal images. It would be expected, therefore, that the same general rules governing use of each approach for visual faces would apply to thermal images, although the specific implementations would differ. For template matching, if the templates represent the entire database, the system may lose the utility of fine differences among subjects. If the templates are individually established for each subject, that may drastically lengthen the processing time. However, when the task to be considered is verification of purported identity, that implies a cooperative situation where the verification time is typically expected to be on the order of 3 seconds. Therefore, there would seem to be enough time to apply two sets of templates: an individual one representing the purported identity, and a global one representing the database. Persons who are highly differentiated from the norm would be more reliably identified using their own templates, whereas persons closer to the average would generate fewer false negatives (or correspondingly allow a tighter acceptance threshold) if they were compared to a standard template set.

The template areas which seem to be most effective for visual and thermal recognition have some similarity. In both modalities, the nose is considered a poor choice for discrimination of frontal images, and so is not included in the templates. For both modalities, the upper face offers more significance for recognition of identity, where the lower face presents more clues to gender and expression recognition. Eye areas are important to both modalities, however for visual images the total eye area is used and for thermal the canthi or areas between the nose and inner edge of the eyes are most significant. Visual recognition often utilizes the mouth, chin, head shape, and eyebrows for template areas, with the eyes and mouth the primary ones selected. For thermal imaging, however, the mouth is not used, since the thermal signature varies greatly depending on whether the mouth is open or shut. Face edges can be used with the same effectiveness in either modality. In addition, however, the inner thermal curve produced by the main facial artery of each cheek is found to be highly discriminating.

Feature extraction approaches follow the same mathematical derivations, but result in consideration of different sets of features in the visible and IR. For example, simple metric graph matching in visual images commonly selects the corners of the mouth, centers of the eyes, tip of the nose, and edges of the face as feature points. However, the nose tip and mouth corners experience rubber sheeting effects with changes in expression, which adds fuzziness to their positions and distances. For thermal images, there are many more feature points which can be selected which are in the upper portion of the face and less prone to rubber sheeting effects. In particular, centroids and maxima on the canthi, the angle and position of the major arteries through the forehead, and the uppermost thermal cheek contour projections on each side seem to be reliable markers whether used as kernels for Gabor wavelet transforms or as anchor points for grid matching. The use of a standard rectangular overlaid grid is expected to be equally useful in both modalities. However, for IR there are many more feature points having greater variability.

Different Processing Approaches

Other differences between visual and thermal processing result from the more consistent and sharper edge effects found in thermal faces. Visual face gray values are produced from a combination of changes in apparent skin tone (with perhaps overlaid cosmetics) resulting from shadows and depth of facial features --- strongly influenced by the intensity and directionality of the ambient light. There simply is not a lot of consistent variation in gray values across a face other than at the boundaries and within the facial features (eyes, nostrils, mouth). In contrast, thermal images have a wealth of variation in gray values, regardless of ambient conditions. Therefore, whereas contour matching has not been found to provide very good results for visual identification, it does provide reasonable results for thermal imagery.

Current Results for Thermal Images

Several IRID tests have been performed using various IR cameras and matching engines. No training of the system was performed and no adaptive algorithms were used in any of the exercises. The databases included a broad variety of individuals of varying height, skin tone, and hair style, with and without glasses. Images were obtained over a period of weeks. A single try was allowed for each attempted entry. The best results generated a crossover point for Type I and Type II errors of 1%.

Computational Intensity. The current IRID matching engine program is not computationally intensive, relative to some of the visual recognition systems. It utilizes only spatial information rather than feature extraction to transform space. Transformation of the image can often reduce the computation time, enhance the separability of images in the database over that obtained in the original spatial presentations, and reduce the impact of minor distortions such as linear translations and rotations. With the use of a dedicated processor or faster PC, significantly more computation could be performed on the thermal image without exceeding the 3 second desired decision time. A wavelet technique for alignment and scaling has been evaluated and shown to significantly improve accuracy in matching images taken over long intervals. That technique currently requires a workstation and 15-30 seconds per match decision.

Enrolling and Classifying Infrared Images. Only a single frame of infrared video (1/30 second), full frontal view, is required to uniquely identify a person. It can be taken at the same time and with the same care as a standard photograph used for passports. There are strong correlations between the infrared and visual facial images: head shape and size, location and shape and size of features. The database of images can be segmented into classes using those values, and the same classification system will work for visual or infrared images.

Template Matching. If thermograms are first standardized as to size, and then normalized as to gray scale, the areas about the canthi and the inner cheeks can be compared using template matching with rather good results. On databases of 250 people, a crossover error of 7.5% was obtained by matching the four template areas, using older cooled IR cameras having NETD (noise equivalent temperature difference) of approximately $0.1°$ C. At that level of sensitivity, detailed thermal contours are not seen and so shape analysis does not apply. Inexpensive uncooled IR cameras can now provide that level of verification accuracy with a simple processor chip installed in the camera, for a system cost below $5000.

Elemental Shape Real-Time Matching. With IR cameras having NETD of $0.7°$ C or less, 100 or more different closed thermal contours are seen in each face. The sets of shapes are unique for each individual, even in the case of identical twins, because they result from the underlying complex network of veins and arteries. Variation in defining the thermal slices from one image to another has the effect of shrinking or enlarging the resulting shapes, while keeping the centroid location and other features of the shapes constant. Each nesting of thermal closed contours is called an "elemental shape". Pre-production systems based on elemental shape analysis have been built and have undergone extensive testing with more then 250 persons, and databases of more than 10,000 images over the past three years. This has proven the persistence of those features of the facial thermograms which are used for identification and the uniqueness of each person's thermal image. Facial thermograms for a limited number of subjects have been obtained for periods of 8 to 23 years, demonstrating the required persistence over those durations. Automated IRID using elemental shapes in real time has achieved 96% accuracy. Studies on pairs of identical twins have demonstrated that they can be separately identified through elemental shape analysis.

Elemental Shape Post-Processed Matching. The totality of shapes in a library of facial thermal images were analyzed. Eigenshape analysis was used to compare eleven characteristics of each shape including: perimeter, area, centroid (x, y) locations, minimum and maximum chord length through the centroid, standard deviation of that length, minimum and maximum chord length between perimeter points, standard deviation of that length, and area/perimeter. Shapes whose edges are interrupted may need to be ignored when compared against images with different edge effects. As examples, shapes along the face edge will change when the face is turned; shapes around the eyes will be affected when glasses are put on.

Each person's image was then characterized by a set of 11-coefficient vectors. The difference in eigenspace between any two images was calculated to yield a measurement to which a threshold was applied to make a "match/no match" decision. The resulting system accuracy was 97% for non-cooperative imagery; and 98.5% for cooperative imagery, when tuned for 0% false positive identification. While these studies proved that IRID can produce high accuracy identification for cooperative access control applications, the calculation techniques are computationally intensive and would need to be hardware-based for real-time use. The shape analysis approach is adversely impacted by edge effects due to head rotation, tip and tilt especially in the noncooperative mode. Future systems using facial minutiae matching will reduce the impact of edge effects on IRID system accuracy.

5. Current IRID System Design, Cost, Features, and Performance

Prototype IRID systems for access control were developed by Unisys Corporation. The configured systems were designed for unattended cooperative access control. Components include IR camera, Pentium I PC-class computer, face acquisition assembly, keypad, and the enrollment subsystem of a monitor and keyboard. The face acquisition assembly includes the mechanisms which move the camera and focus it so that the face is within the field of view. The primary cost component of the system is the infrared camera. The camera in the current system was designed and built by Lockheed Martin. It is uncooled and produces images of 320 x 240 pixels in the 8-12 micron band. The stand-alone access control system version of the IRID system was designed to sell for $25,000 initially with reduction to $5000 in quantity, and to offer a false positive rate of less than 1% and false negative rate of less than 2% for single entry attempts.

A non-cooperative, non-real-time faces-in-the-crowd version of the IRID system was also built and demonstrated, with more than 100 persons represented in a database of 500 images. Twelve targeted persons were selected, and the database was searched for all appearances by the targeted persons. An accuracy of 98% was achieved, with no false positives. All the database images were manually selected from IR videotapes obtained at trade shows. Only in-focus, essentially full face images were selected and manually scaled and centered. Processing was not done in real time. The 11-coefficient eigenanalysis required 20 hours on a 486/66 PC. Future faces-in-the-crowd use of IRID will utilize minutiae analysis with the goals of permitting the system to run in real time and utilizing a larger percentage of non-full frontal images.

Operational use of IRID systems requires setting the decision threshold to minimize false positive errors. This requires quality imagery and selection of unique and persistent features. False negative errors are reduced by providing feedback to the entrants and/or performing additional analysis of the imagery. This accomplishes maximum overall throughput while providing additional processing for more difficult identifications in those cases where a particular entrant has a wider day-to-day variation than the average entrant; or where multiple individuals in the database have some common or similar features in their thermograms that require additional analysis to separate them.

6. Future IRID Systems using Thermal Minutiae

IRID systems now under development will utilize lower cost cooled and uncooled infrared cameras, and will perform identification based upon thermal minutiae.

Infrared Camera Sensitivity Determines Minutiae Analysis

Images produced by current commercial IR cameras with NETD of approximately $0.07°$ C exhibit thermal contours from which minutiae may be derived. On the order of 100 sizable closed contour shapes are produced for each face. One or more minutiae may be associated with each shape, such as by using the centroid, or using all inflection points. Since each shape has 11 or more measurable characteristics, which can be associated with that minutia point, with wide ranges of possible values, the total set of possible facial thermograms would appear to be far more numerous than the set of possible fingerprints. However, matching of eigenshape-derived minutiae is computationally intensive, requiring hardware embodiment for a PC-based system. Also, those points will vary in location and characteristics if their shape encounters edge effects due to the face being turned, or its appearance changed by glasses or facial hair, causing a major impact on the accuracy of systems using the derived minutiae. More sensitive infrared cameras offer the potential for absolute minutiae extraction, which is more accurate and more comparable to fingerprint analysis.

Cameras with an NETD of 0.04°C or better are able to directly image the superficial blood vessels, displaying them as a network of hot pathways superimposed on the thermal facial image. Such cameras permit the direct extraction of minutiae, and do not require production or consideration of elemental shapes. Comparison of these "absolute" thermal minutiae is more akin to fingerprint minutiae matching, rather than the computationally intensive approaches required for eigenshape comparisons. Images have been obtained from prototype improved infrared cameras and used to develop the techniques summarized here.

Fingerprints are characterized by a limited range of intensity values corresponding to three dimensional ridges which are essentially concentric rings about a single center, plus anomalous arches, line endings, and bifurcations. High sensitivity facial thermograms are characterized by a continuously varying wide distribution of temperatures overlaid with the pattern of underlying blood vessels, which appear as relatively hot pathways superimposed on the varying background.

Fingerprint matching techniques commonly extract minutiae points from the prints, and then compare the sets of minutiae rather than compare the entire prints. Generally, line endings, branch points, and islands are considered to be minutiae. Features of each minutiae point are considered. They may includes: the type, orientation, location, and the count of the numbers of skin ridges lying on each of the lines between that minutia and each and every other minutia. Using such an approach, on the order of 80 to 150 minutiae points are identified in each fingerprint. Other fingerprint minutiae extraction and matching approaches produce essentially the same number of minutiae, with differences in what features are considered in attempted matching and in how the matching is performed.

For facial thermograms, the analogous features which are most easily extracted are branch points of the superficial blood vessels. A face includes on the order of 175 such points. Once the minutiae have been extracted, prior techniques associated with fingerprint minutiae matching can then be applied.

Minutiae Variations

Matches between different prints taken from the same finger are never perfect, since the fingers are deformable three-dimensional connected and jointed structures which leave two-dimensional prints on surfaces they encounter through pressure. The exact angles between the fingers and the surfaces, the amount and direction of pressure, and the effect of movement between the fingers and the surfaces all cause variations in the exact prints produced. Even when prints are produced by a live scan technique, variations in the scanner optics, hand position, oil or dust on the fingers, use of lotions, and scratches or paper cuts will produce variations in the prints produced. Therefore, the exact number, position, and characteristics of minutiae extracted from two prints may be different even though they are produced by the same finger.

Analogous techniques and systems may be applied to the extraction and matching of minutiae points from human faces for identification of individuals. Persons who have previously been identified and logged-into a facial recognition system can later be automatically identified from live or recorded images, by comparing the facial minutiae of the unknown image with that of all known images in the database. The database can be partitioned by first classifying the faces either based on minutiae-related characteristics or on other characteristics such as the degree of bilateral asymmetry. Such classification reduces the search requirements for identification against a reference database.

The underlying anatomical features which produce facial thermograms aid in locating and orienting the face for analysis. The face is basically bilaterally symmetrical. There is variation of temperatures from the hot areas on either side of the nose to the relatively cool areas of the ears and cheeks. The eyes appear to be cooler than the rest of the face. The nostrils and mouth, and surrounding areas, will look warm or cool depending upon whether the subject is inhaling or exhaling though them. Since the face surface can be distorted through changes in expression, through activities such as eating and talking, and as a result of weight gain and loss, analysis of the minutiae points must consider those changes, as with analogous fingerprint deformations.

Also, since it is of interest to identify faces seen in crowds, or faces turned at any angle, a significant number of minutiae points must be extractable for those applications so that even a partial face can be used for ID. This is again analogous to the situation with fingerprints, where a partial latent print may be matched against a rolled print obtained at a booking station and entered into the FBI database, if enough minutiae are found in the latent. The particular technique used to extract thermal minutiae from facial images, and the number extracted, depend on the sensitivity of the IR camera used; just as the number of fingerprint minutiae found depends on the resolution of the image scanner.

Minutiae Matching Approaches

Various minutiae extraction algorithms are used by fingerprint identification systems, some of which merely utilize the location of the minutiae and others which also utilize additional information about the type of minutia each point represents. For example, simple graph matching techniques can be used to compare two clusters of minutiae locations. Or, the ridge angle at each minutia point can be considered and matched along with the coordinates. Or the type of minutia can also be considered and matched along with coordinates and angles. A measure of goodness of fit can then be computed and used to rank order possible matches. The distances between ridges of a fingerprint average 0.4 millimeters but can vary by a factor of two for any individual finger depending on skin displacement when the finger contacts the hard surface normally encountered in establishing a print. The matching algorithm should consider such possible deformation in the print and accommodate local warping to the grid of minutiae.

Thermal face identification systems based on absolute minutiae can similarly consider only the location of the minutiae, or can also consider the vectors of the branching blood vessels. In addition, however, the thermogram provides a third dimension, apparent temperature, which is significantly more varied than is ridge depth for fingerprint analysis. Thermal minutiae can be characterized by associated temperatures, as well as by location and vector. Deformations in the facial minutiae grid due to varying facial expressions can be modeled similarly as with deformation in prints.

There are other direct comparisons which can be made between fingerprint and facial thermogram minutiae analysis, such as techniques for locating the center and axes of the face or finger. The inherent bilateral symmetry of the facial thermogram, as well as the universality of the underlying vascular structure, produces a more consistent and unambiguous center and axes for the face, even when only a partial face, analogous to a partial latent, is seen. Consideration of ridge crossings in fingerprints is analogous to consideration of thermal contour crossings in facial thermograms.

Minutiae may be extracted manually or automatically. Automatic systems generally require better quality imagery. The matcher engine must allow for some degree of inaccuracy or variability with respect to each of the encoded coordinates and their characteristics, to accommodate distortions caused by the subject's movements and by the limits to precision of the scanning and analysis subsystems.

The same thermal minutiae are repeatedly extracted from a given individual. They may be overlaid and annotated on the infrared image, or in fact on a visual image or on any image obtained from another medical sensor having the same orientation to the subject. The thermal minutiae, therefore, provide reference or fiducial points for manual or automated comparison, merging, or registration among a set of images taken at different times, with different orientations, and with different cameras. The merging of infrared and visual identification can effectively use superimposed thermal minutiae on visual mug shots.

7. FaceCode and the Future of IRID

Once a large number of subjects' minutiae patterns are collected, it may be possible to derive a unique FaceCode repeatably from each person, eliminating the need for comparing against a database. As an example, by restricting attention to a particular area of faces (such as a square whose top edge is determined by a line through inner corners of the eyes, with the edge length equal to twice the inter-eye spacing), a standard grid can be superimposed on that square area and a binary or other code produced corresponding to the location of specific vascular branches seen in the thermal image. If the grid is fine enough, that code is expected to be unique for each person, and be persistent for the person's life and under all imaging conditions. If that technique is successful, then a FaceCode can be read directly from a person's face without the need to match facial thermogram contours or minutiae patterns. It would be as if the person's personal ID number were barcoded across his face as suggested in Figure 9.1.

It is expected that an effective thermal face encoding technique will be developed in the future. The coding scheme will take account of head position, and allow for degraded accuracy of identification when only a partial face is seen. The resulting systems would not require use of a keypad, ID card, or other technique by which the entrant asserts his identity for verification by the system. Depending on the computational complexity of the processing required, such a capability could allow for rapid throughput with minimal cooperation by the entrant. This technique would also support real time faces-in-the-crowd and digital signature applications.

Identification of Body Parts

Infrared images can be processed to yield repeatable minutiae points corresponding to specific vascular and other physiological locations under the skin from any extended area of the body, generating a corresponding set of minutiae which is unique to each individual. Expanding the facial minutiae approach to whole body applications produces the technique called SIMCOS (Standardized Infrared Minutiae CO-ordinate System). SIMCOS-derived minutiae use standardized anatomical references to obtain a standard set of minutiae for each body part for each person. Knowing the wiring chart between blood vessels at the different locations, evidence of vasoactivity, which is seen as thermal activity, can provide information about otherwise inaccessible portions of the body hidden by hair, clothing, or other coverings. For example, SIMCOS points provide consistent and meaningful nodes for a wireframe or finite element analysis of the thermodynamic behavior of the face in response to specific protocols of anesthesia or drug use.

The MIKOS SIMCOS technique provides a built-in set of registration points on the body's surface, which can be annotated onto images produced by visible or infrared cameras, or by any medical imager used in conjunction with a thermal sensor. The registration points then can be used to compare and combine images taken with different equipment at different times and under different conditions, facilitating comparison of those images. Also, for medical uses the minutiae provide reference points for continuous re-alignment of surgical instruments, radiation sources, and other diagnostic equipment. Since the infrared camera is totally passive, it can be used continuously during other surveillance or medical procedures to overlay precise registration points on all images while also monitoring for overheating, shock, hypothermia, renal failure, and other conditions observable from the thermal data.

8.   Future IRID System Design, Cost, Features, and Performance

Uncooled IRID systems within the next three years are expected to incorporate better thermal sensitivity, autofocus, expanded depth of focus, and prices below $1000 in large quantities. Cooperative IRID access control systems will then match the initial cost of current live scan fingerprint readers, but will offer non-contact passive operation, will apply to all persons without exception, and will be more secure and accurate. Applications to computer security, ATM and point of sales terminals, and electronic passports will offer the ease of photographs with the security of fingerprints at a cost competitive with other biometric techniques.

9. Conclusions

Infrared imaging analysis offers a robust technique for classification, recognition, and identification of faces and body parts. Depending upon the sensitivity and resolution of the thermal camera used, accuracies of 85% to 98% are currently obtained from pre-production systems in Beta testing. Further improvements in the cameras, and in image centering and scaling techniques, are expected to offer further accuracy gains. The inherent anatomical information which is utilized by the infrared identification (IRID) technology involves subsurface features unique to each person. Those features may be imaged at a distance, using passive infrared sensor technology, with or without the cooperation of the subject.

IRID, therefore, provides an unique capability for rapid, on-the-fly identification, under all lighting conditions including total darkness. As the cost of uncooled infrared cameras continues to decline, and medical applications for thermal imaging become commonplace, it is anticipated that IRID will increasingly provide cost effective security of physical spaces, computer systems, distribution of goods and services, and evidentiary proof of identity for law enforcement.

References

[1]   F. J. Prokoski, J. S. Coffin, and R. B. Riedel., "Method and Apparatus for Identification of Individuals and their Conditions from Analysis of Elemental Shapes in Biosensor Data Represented as N-dimensional Images, " *U.S. patent 5,163,094,* November 1992.

[2]   F. J. Prokoski, "Security and Non-Diagnostic Medical Uses for Thermal Imaging," *American Academy of Thermology, 1997 Annual Meeting, Pittsburgh,* April 1997.

[3]   F. J. Prokoski, "Patient ID and Fusion of Medical Images Using SIMCOS," *MEDTEC Medical Technology Conference,* Orlando, Florida, July 1996.

[4]   P. J. B. Hancock, A. M. Burton, and V. Bruce, "Face Processing: Human Perception and Principal Components Analysis," *University of Stirling,* Scotland, UK.

[5]   F. J. Prokoski, "Accuracy of Facial Identification from Infrared Imagery – IRID," *SBIR Final Report,* Hanscom Air Force Base, MA, U.S. Air Force, February 1994.

[6]   R. Chellappa, S. Sirohey, C.I.Wilson, and C.S. Barnes, "Human and Machine Recognition of Faces: A Survey," *University of MD,* College Park, MD, 1994.

[7]   M. M. Menon and E. R. Boudreau, "An Automatic Face Recognition System Using the Adaptive Clustering Network," *in The Lincoln Laboratory Journal, Vol. 7, No. 1, 1994.*

10 KEYSTROKE DYNAMICS BASED AUTHENTICATION

M. S. Obaidat
Monmouth University
W. Long Branch, NJ
obaidat@monmouth.edu

B. Sadoun
Applied Science University
Amman, Jordan
bsadoun@aol.com

Abstract *This chapter deals with the applications of keystroke dynamics to authenticate/verify access to computer systems and networks. It presents our novel contribution to this area along with other related works. The use of computer systems and networks has spread at a rate completely unexpected a decade ago. Computer systems and network are being used in almost every aspect of our daily life. As a result, the security threats to computers and networks have also increased significantly. We give a background information including the goals of any security system for computers and networks, followed by types of security attacks on computers and networks. We present the applications of keystroke dynamics using interkey times and hold times as features to authenticate access to computer systems and networks.*

Keywords: *Keystroke dynamics, computer security, computer verification/authentication, interkey times, hold times, neural networks, pattern recognition, system identification.*

1. Introduction

Computer systems and networks are now used in almost all technical, industrial, and business applications. The dependence of people on computers has increased tremendously in recent years and many businesses rely heavily on the effective operations of their computer systems and networks. The total number of computer systems installed in most organizations has been increasing at a phenomenal rate.

Corporations store sensitive information on manufacturing process, marketing, credit records, driving records, income tax, classified military data, and the like. There are many other examples of sensitive information that if accessed by unauthorized users, may entail loss of money or releasing confidential information to unwanted parties [1-9].

Many incidents of computer security problems have been reported in the popular media [1]. Among these is the recent incident at Rice University where intruders were able to gain high level of access to the university computer systems which forced the administration to shut down the campus computer network and cut its link with the Internet for one week in order to resolve the problem. Other institutions such as Bard College of the University of Texas Health Science center reported similar breaches. Parker [10] reported that one basic problem with computer security is that the pace of the technology of data processing equipment has outstripped capability to protect the data and information from intentional misdeeds.

Attacks on computer systems and networks can be divided into active and passive attacks [11-12].

1. Active attacks: These attacks involve altering of data stream or the creation of a fraudulent stream. They can be divided into four subclasses: masquerade, replay, modification of messages, and denial of service. A masquerade occurs when one entity fakes to be a different entity. For example, authentication sequence can be collected and replayed after a valid authentication sequence has taken place. Replay involves the passive capture of data unit and its subsequent retransmission to construct an unapproved access. Modification of messages simply means that some portion of a genuine message is changed, or that messages are delayed or recorded, to produce an unauthorized result.

2. Passive attacks: These are inherently eavesdropping on, or snooping on, transmission. The goal of the attacker is to access information that is being transmitted. Here, there are two subclasses: release of message contents, and traffic analysis. In the first subclass, the attack occurs, for example, on an e-mail message, or a transferred file that may contain sensitive information. In traffic analysis, which is more sophisticated, the attacker could discover the location and identity of communicating hosts and could observe the frequency and length of encrypted messages being exchanged. Such information could be useful in guessing the nature of information/data.

Passive attacks are difficult to detect, however, measures are available to prevent them. On the other hand, it is difficult to prevent the occurrence of active attacks.

Computer security goals consist of maintaining three main characteristics: integrity, confidentiality, and availability [12]. These goals can overlap, and they can even be mutually exclusive. For example, strong protection of confidentiality can severely restrict availability to authorized parties.

1. Integrity: This characteristic means that the assets can be modified (e.g., substitution, deletion, or insertion) only by authorized parties or only in authorized ways. Integrity means different things in different contexts [12]. Among the meanings of integrity are precise, accurate, unmodified, consistent,

and correct result. Three aspects of integrity are commonly recognized: (i) authorized actions, (ii) separation and protection of resources, and (iii) error detection and correction.

2. Confidentiality: This is also called privacy or secrecy. It means that the computer and network systems are accessible only to authorized parties. The type of access can be read-only access; the privileges include viewing, printing, or even just knowing the existence of an object.

3. Availability: This term is also known by its opposite, denial of service. Here, the term means that assets are accessible to authorized parties. An authorized individual should not be prevented from accessing objects to which he/she has legitimate access. Availability applies both to data and service.

One major aspect of a multiuser computer system that can be a significant threat to security arises from access to remote terminals. Denning [13] states that the effectiveness of access control is based on two ideas: (1) user identification and (2) protection of the access rights of users. Protecting the access rights of users is generally done at the system level, by not allowing access permissions to be altered except by authorized "super-users". Denning [13] presents several cryptographic types of user authentication, in addition to password schemes. To properly identify a valid user, one or more of the following techniques are commonly used [1-8,14]:

• What the user knows or has memorized (password).

• What the user carries or possesses (e.g., a physical key).

User passwords are the most common means of identification, but they are subject to compromise, either by interception as the user types it, or by a direct attack. Hardware locks are secure, but there is no way for a computer system to know that the users who have logged on are really who they say they are.

A third method, using biometric characteristic such as the user's typing technique, was discounted by Walker as impractical [14]. However, more recent work by Obaidat et al. [1-8], Gaines et al. [22], Umphress and Williams [15], Leggett and Williams [25], Yong and Hammon, and Joyce and Gupta [21] has shown that a user can be identified based on his/her typing technique using traditional pattern recognition and neural network techniques. These research efforts in keystroke dynamics have focused on attributes like stream of interkey times (latency periods between keystrokes) and hold times (durations between the hit and release moments of key hold) to provide a unique feature/identifier/signature for authenticating an individual's identity.

2. Types of Security Attacks

The attacks on the security of a computer system or network can be characterized by viewing the function of the computer system/network as a provider of information. The possible attacks that may occur on a computer and networking system are as follows [11]:

1. Interruption: In this case, an asset of a system becomes unavailable, lost, or unusable due to alteration. Clearly, this is an attack on availability. Examples include vicious destruction of hardware devices, deletion of a program/data file, cutting of a communication link, disabling of a file management system, failure of an operating system function, etc.

2. Interception: This means that an unauthorized individual has gained access to an asset. Clearly, this is considered an attack on confidentiality. The consequences range from inconvenience to catastrophe. Examples include copying of data files/programs, and wiretapping to obtain data in a network. The unauthorized party could be a person, a program, or a computer system.

3. Modification: Here the unauthorized party not only accesses but also tampers with an asset. Clearly, this is an attack on integrity. Examples include changing values in a record or data file, altering a program so that it performs differently, and modifying the contents of messages being sent over a network. The modification can be done on the hardware configuration as well. Some cases of modification can be detected with simple schemes, but others may be more difficult if not impossible to detect.

4. Fabrication: Here an unauthorized party inserts counterfeit objects into the system. This is considered an attack on the authenticity of the computer system or network. The intruder may insert spurious transactions into the system, or add records to an existing data base. In some cases, these additions can be detected as forgeries, but if done skillfully, they are virtually indistinguishable from the real thing.

Computer networks, in particular, have security problems due to the following reasons [11-12]:

1. Sharing: Since resources and work are shared, more users have the potential to access networked systems than a single computer node.

2. Anonymity: An intruder can attack from thousands of kilometers away and thus, never have to touch the system attacked or come into contact with any of its managers or users.

3. Complexity of system: Operating systems tend to be very complex. Reliable security is not easy to implement on a large operating system, especially one not designed specifically for security. Designing a secure computer network is even more difficult since it combines two or more computer systems with possibly dissimilar operating systems.

4. Multiple points of attack: When a file physically exists on a remote host, the file may pass via many nodes in order to reach to the user over the computer network.

5. Unknown path: network users seldom have control on the routing paths of their own packets. Routes taken depend on many factors including load conditions and traffic patterns.

3. Predicting Human Characteristics

As early as the beginning of the 20th century, psychologists, and mathematicians have experimented with human actions. Psychologists have demonstrated that human actions are predictable in the performance of repetitive, and routine tasks [15]. In 1895, observation of telegraph operators showed that each operator had a distinctive pattern of keying messages over telegraph lines [16]. Furthermore, an operator often recognized who is typing on the keyboard and sending information simply by listening to the characteristic pattern of dots and dashes. Since the beginning of civilization, humans are able to recognize the person coming into a room from the sound of steps of the individual. Clearly, each person has a unique way of walking. Similarly, telegraph operators were able to find out who was sending message by just listening to the characteristics of dots and dashes.

Today, the telegraph keys have been replaced by other input/output devices such as keyboard and mouse. It has been established that keyboard characteristics are rich in cognitive qualities and hold promise as an individual identifier. Anyone sitting close to a typist or has an office next to a typist is usually able to recognize the typist by keystroke patterns.

Over many centuries, humans have relied on written signatures to verify the identity of an individual. It has been proven that human hand and its environment make written signatures difficult to forge. It has been shown [21] that the same neurophysiological factors that make written signature unique are also exhibited in an individual typing pattern. Once a computer user types on the keyboard of a computer, he/she leaves a digital signature in the form of keystroke latencies (elapsed time between keystrokes and hold times).

Human nature dictates that a person does not just sit before a computer and deluge the keyboard with a furious and continuous stream of non-stop data entry. Instead, the person types for a while, pauses to collect thoughts and ideas, pauses again to take a rest, continues typing, and so forth. In developing a scheme for identity verification, a common baseline must be established for determining which keystrokes characterize the individual's key pattern and which do not. Physiologists have studied human interface with computer systems and developed several models describing the interface to computers. One of the popular models is the keystroke-level model developed by Card et al. [17]. Their model describes the human-machine interaction during a session at a computer terminal. It was intended as a vehicle for the evaluation and comparison of competing designs for highly interactive programs. The keystroke level model summarizes the terminal session as follows:

$$T_t = T_a + T_e,$$

where T_t represents the duration of the terminal session; T_a represents the time required to assess the task, build mental representation of the functions to be performed, and choose a method for solving the problem; and T_e represents the time needed to execute all functions constituting the task.

Note that T_a varies according to the extent of the considered task, experience of the user, and understanding of the functions to be performed. Clearly, this term is not quantifiable. Thus T_a cannot be used to characterize a person. On the other hand, T_e describes mechanical actions which itself can be expressed as:

$$T_e = T_k + T_m,$$

where T_k is the time to key in information and T_m is the time needed for mental preparation. Note that when interacting with a program, the user does not divide his actions into mental time followed by keystroke time. Instead, the two are intermixed.

Shaffer [18] has shown that when a typist is keying data, the brain acts as a buffer, which then outputs the text onto the keys of the keyboard. Average capacity of the buffer is about 6-8 characters in length [19]. Because of the limited size of the buffer, typists group symbols into smaller cognitive units and pause between each unit. Cooper [19] established that the typical pause points are between words as well as within words that are longer than 6-8 characters.

4. Applications of Keystroke Dynamics Using Interkey Times as Features

Although handwriting and typing are distinct manual skills, they both have measurable characteristics that are unique to those who perform the task [5,6]. Umphress and Williams [15] have conducted an experiment for keystroke characterization. They used two sets of inputs for user identification, namely, a reference profile and a test profile. Each keystroke was time-tagged to the nearest hundredth of a second and stored on a floppy disk. Another program was used to analyze the keystrokes and produced a database of reference profiles for each individual participating in the experiment. A third program was used to compare test profile keystrokes to reference profiles. Seventeen persons participated in that experiment. Each person was asked to take two typing tests. These tests were separated over several days. In the first test, the participants were asked to type about 1400 characters of prose. The second typing test, the test profile, consisted of 300 characters of prose. It was found that a high degree of correlation could be obtained if the same person typed both the reference and test profiles. Several medium confidence levels were assigned in cases where the typists of the profile differed. However, in most cases test profiles had low scores when the typists was not the same person who typed the reference profile.

Obaidat and his colleagues [3,5,6] described a method of identifying a user based on the typing technique of the user. The inter-character time intervals measured as the user types a known sequence of characters was used with traditional pattern recognition techniques to classify the users, with good verification results. By requiring the character sequence to be typed two times, and by using the shortest measurements of each trial, better results were obtained than if the user typed the sequence only once. The minimum-distance classifier provided the best classification accuracy. In order to obtain a better classification accuracy, their analysis considered the effect of the dimensionality reduction, and the number of classes in the identification system. The measurement vector is obtained by computing the real-time durations between the characters entered in the password. Figure 10.1 shows the flow chart of the overall steps general recognition system.

Obaidat and Macchiarolo [4, 7-8] used some traditional neural network paradigms along with classical pattern recognition techniques for the classification/identification of computer users using as feature the interkey times of the keystroke dynamics. They considered six users in their work. The dataset used for the recognition of computer users is made up of the time intervals between successive keystrokes by users while typing a known sequence of characters (phrase). The participants in the experiment were asked to enter the same phrase, which was not visible during the process of typing; therefore, it was important to display the message on the monitor after entering it. The phrase was retyped by the participant if it was entered incorrectly. The time duration between keystrokes was then collected by using an IBM compatible PC-based data acquisition system which used Fortran and assembly language programming. The assembly language procedures make use of the software keyboard interrupt facility and provide the main program with the time duration between keystrokes. For example, if the password "OBAIDAT" were entered, then the assembly language program would compute the time duration between the letter pairs (O, B), (B, A), (A, I), (I, D), (D, A), and (A, T). An open period of time was given to the participants to conduct the experiment. This helped in averaging out the effect of uncorrelated sources of noise that could be introduced by instruments and participants. Furthermore, it helped to gather data that represent the different modes of the participants. A phrase that consists of 30 vector components was used first; however, only the first 15 vector components were used later since using the remaining vectors did not change the results. The data were collected from six different users over a six-week period. The total number of measurement vectors per user was 40. The raw data were arranged as follows:

- each pattern consisted of 15 values, which were the time durations in milliseconds between successive keystrokes of a known character sequence;

- there were 40 trials per user (class) (600 values per class), and

- there were six classes that were defined (3600 values total).

For training purposes, the raw data were separated into two parts: all of the odd-numbered patterns of each class, and all of the even-numbered patterns. In any given simulation run, only half of the data were used to form the training set. After each network was trained, the entire pattern set (24 patterns) was presented to the network for classification.

Several versions of the training data were created to investigate the network's ability to generalize, rather than to memorize the training set. The difference in the training pattern sets are: (a) whether the patterns are from the odd or even half of the raw data, and (b) the granularity of the training set which is defined by the number of raw patterns averaged to compose each training pattern. For example, if all raw patterns in a class are averaged together to form a single training pattern, the granularity is low. On the other hand, if no averages are used, i.e., all of the patterns are used in training, the granularity is high. Intuitively, when a higher granularity is used for training then better classification performance should be obtained. Table 1 shows one example of a training set used [4].

During the investigation phase, various combinations of these patterns were created to test the learning abilities of the three different neural network paradigms. After experimentation determined the best neural network architecture for this application, the network was incorporated into an "on-line" system that would collect the character time intervals from users in real-time and perform a classification immediately. The simulators used to simulate the neural network paradigms were written using C programming language. Some critical timing functions were written in assembly language. The on-line computer security system consists of the following major tasks:

Data Input

The timing functions used the 8253 timer that is located in all IBM-PC compatible computers to measure the interkey time intervals. In the case of a PC-AT computer, a BIOS microsecond timing function [4] is used instead, as the 8253 timer outputs are not accessible. Similar schemes can be used for other computer platforms. In all platforms, a calibration subroutine is called before any timings are measured. The calibration routine first determines which timing method to use based on the computer type and then calibrates the timer using the time-of-the day clock. During actual timing of keystrokes, the routine gets each keystroke, stores it, and then begins timing, while waiting for the next keystroke. When the next keystroke occurs, it stores the time intervals and the key hit. This process is then repeated, and a second set of measurements is recorded. It has been shown that taking the lowest value of each interval, based on two sets of values, improves the classification accuracy.

Training

To train the neural network, a set of measurement vectors from each user class was required. These vectors are collected from each user and stored. When a sufficient number of vectors have been collected, they may be averaged and normalized to form a set of patterns that will be used to train the network. The number of pattern vectors is defined by the user of the program. The user can describe the network configuration to the program, and memory is allocated for the processing units (neurons), training pattern storage, and weight vector storage. Training consists of applying a pattern vector to the input, comparing the current output with the target output, and adjusting the weight values according to the training algorithm. When the error of the training vector set is reduced to a pre-defined threshold which is the total summed squared (TSS) error less than or equal to 0.01 in our work, training is stopped, and the entire network is saved to a disk file.

Classification

To run the program as an on-line classifier/identifier, the network is recalled from the file saved after training. Memory is allocated as needed, and the weight vectors are read from the file. The user is prompted to type the keyword phrase. The inter-character intervals are stored, normalized using the user-selected normalization function (either the percentage of the largest value, unit-length vector, or none) and presented to the network inputs. The input values are propagated through the network, which has the same number of output units as there are defined user classes. The output unit which is strongly activated (above a user-defined threshold) represents the classification of the input measurement vector.

Normalization, Performance, and Incorporation

In an operational test, six user typed a 15-character phrase 20 times, each over a period of 6 weeks. The raw data were used to create pattern sets to train the network. Two types of normalization of data were investigated: unit length vector, and fraction of the largest element. The unit length vector normalization is obtained by dividing each element of the measurement vector by the total magnitude of the vector (square root of the sum of the square of each element's value). This proved to be unsatisfactory in that the vectors of different users were made more similar, and the network could not distinguish the difference between them during training. By dividing each element's value by the largest element value, the elements were simply rescaled into a range from 0 to 1. This is the range needed for the inputs of the neural network, while preserving the relative differences in the elements. To create the training patterns, two normalized vectors were averaged together to create each training pattern.

The training time of the network can be varied by adjusting the learning rate and momentum parameters. The learning rate is the fraction of the error value that is used to compute the weight adjustments. The momentum value is the fraction of the previous adjustment that is added into the current adjustment. After training is finished, each user tested the network. The overall accuracy was 97.8%.

The system can be easily incorporated into a computer security system. Initially, each user that is to have privileged access would be required to submit samples of his inter-character typing for a known phrase. These samples are acquired through the use of the data input module, and are kept by an administrator. The administrator then generates the training set, and configures a network using the training module. The network will have a number of inputs equal to the number of measurements in each vector, a number of hidden units, and a number of outputs equal to the number of users. The weight values are then determined through training, which could take place off line or as a background process. After training, the weights are stored and can be quickly recalled for on-line classification. When a user needed to be removed or added to the authorized list, the training set would have to be regenerated, however, the training module can automatically regenerate a training set from the existing and new sample data. Adding a user would require adding another output unit to the network, and the additional weights adjusted through training.

In practice, a user would identify himself by using the number assigned to his sample classification. He would then be asked to type the keyword phrase. His inter-character typing intervals would be collected and classified. If the user's number matches the class assigned by the classification system, then the user is granted access. If the classification does not match, several things could happen:

1. The user is denied access. This is the highest security level.

2. The user is granted access, after providing a higher-level password.

3. The user is granted only limited access.

4. The user is granted access, but a "warning" is signaled to the administrator, and the user's actions are intercepted for later analysis. This is considered the lowest security level.

There is a tradeoff to consider with any security system; the risk of security breach balanced with the user inconvenience.

Bleha and Obaidat [6] experimented with the Percepton algorithm as a classifier to verify the identity of computer users. By performing the real-time measurements of the time durations between keystroke entered in the user's password, data was collected from 10 valid users and 14 invalid users over a period of 8 weeks. The password used was the user's name. Decision functions were derived using half of the data (training data) to compute the weight vectors. The decision functions were applied to the remaining half of the data (testing data) to verify the users. An error of 9% in rejecting valid users, and an error of 8% in accepting invalid users were achieved. The percepton algorithm was found to be robust with respect to the choice of the initial weight vector.

Obaidat [13] evaluated the performance of five pattern recognition algorithms as applied to the identification of computer users using the time intervals between successive keystrokes created by users while typing a known sequence of characters. These algorithms are potential function, Bayes classifier, minimum distance and the cosine measure. A 100% accuracy was achieved when the potential function algorithm was used. The least successful algorithm was the cosine measure. Obaidat and Sadoun [2] evaluated the performance of a newly devised neural network scheme, called Hybrid-Sum-Of-Products (HSOP) [27] for computer users verification and other classification problems. They compared the performance of HSOP to the Sum-Of -Products and Backpropagation neural network paradigms. They found that HSOP performs better than the other two paradigms. In their work they used interkey time intervals between keystrokes while typing a known phrase.

5. Applications of Keystroke Dynamics using Hold Times as Features

Obaidat and Sadoun [1] verified computer users using hold times of keystroke dynamics as features to authenticate computer users. The participants in the experiment were asked to enter their login user ID during an eight-week period. The program collected key hit and key release times on an IBM compatible PC to the nearest 0.1 ms. The program was implemented as a *terminate and stay resident program* in an MS-DOS based environment. The standard keyboard interrupt handler was replaced by one that could sense the incoming keyboard scan codes and record them along with a time stamp. The program measures the time durations between the moment every key button is hit to the moment it is released. This procedure was performed for each letter of the user ID and for each participant. A scan code is generated for both the hit and release of any key.

The login routine was modified so that each time a login attempt was made, the timing vector of the assart (hit and release time) was stored for analysis. This procedure increases the dimensionality of an N character string to (2N-1). Such a high dimensionality can provide better discrimination even if the number of characters is not large. The login monitoring results were collected from 15 users who were given open period of time to conduct the experiment. Such approach averaged out the effects of fatigue and stress as well as the uncorrelated sources of noise. The forgery attempts of the 15 ID's used were collected from each of the 15 invalid users who attempted each of the 15 ID's 15 times. All attempted forgeries were collected in one session for each invalid user. Participants used the system interactively and the results were recorded. The interkey times were collected using the key interrupt facility. The average user ID length was seven characters. The data set was divided into two parts: the training part and testing part. Pattern recognition and neural network [28] techniques were used for the classification process. It was found that hold times are more effective than interkey times and the best identification performance was achieved by using both time measurements. An identification accuracy of 100% (zero false accept and zero false reject) was obtained when the combined hold times and interkey times-based approaches were considered as features using the fuzzy ARTMAP, radial basis function network (RBFN), and learning vector quantization (LVQ) neural network. Other neural network and classical pattern recognition algorithms such as backpropagation with sigmoid transfer function (BP, Sig), hybrid sum-of-products (HSOP), sum-of-products (SOP), potential function, and Bayes' decision rule also gave good accuracy.

The success of this approach was measured mainly in terms of false rejection rate (type I error) and false acceptance rate (type II error), cost of recognition system, and time to access identity verification. The two important measures considered in our work are type I error rate and type II error rate. The false rejection rate (type I error rate) of a verification system gives an indication of how often an authorized individual will not be properly recognized. Type II error describes how often an unauthorized individual will be mistakenly recognized and accepted by the system. It is generally more indicative of the level of a mechanism. This is due to the fact that it describes the degree to which the security measure may be breached by intruders. Type I error is important since it describes the amount of user frustration in using the security system. Our research results have shown that the most successful pattern recognition technique was the potential function followed by the Bayes' rule. The least successful algorithm was the cosine measure. The hold time-based verification/authentication scheme gave better accuracy than the interkey time-based scheme. When neural network paradigms were used for the classification process, it was found that the hold time-based verification/authentication scheme is superior to the interkey time-based scheme. Furthermore, the combined hold and interkey time-based approach gave the least misclassification error. The most successful neural network paradigms for the verification/authentication task are the LVQ, RBFN, and Fuzzy ARTMAP. They basically gave a zero misclassification error for both false acceptance rate and false rejection rate. Figures 10.2-10.7 illustrate these findings.

The average string length used in this recent work was just seven characters. In our previous work [3-8], we obtained lower classification accuracy with a password of 15 characters long. In all the experiments we conducted, it was observed that when considering hold times alone we obtained better accuracy as compared when interkey times are considered as the only characterizing features. Clearly, hold times are more effective for identification than interkey times. Such results suggest that hold times may in general provide better characterization of the typing skills than the interkey times. Also, we found that the most successful neural network paradigm provides better authentication/verification accuracy than the best classical pattern recognition schemes.

One recent related work was conducted by Robinson et al. [20] in which the authors used key hold times to characterize typing style more effectively. They applied some traditional pattern recognition schemes for the classification procedure. They used hold times and interkey times as features and the best performance was obtained when the inductive learning classifier was used.

6. Conclusions

To conclude, keystroke dynamics are rich with individual mannerism and traits and they can be used to extract features that can be used to authenticate/verify access to computer systems and networks. The keystroke dynamics of a computer user's login string provide a characteristic pattern that can be used for verification of the user's identity. Keystroke patterns combined with other security schemes can provide a very powerful and effective means of authentication and verification of computer users. Neither our work nor any other work we are aware of has dealt with typographical errors. Further research into reliable methods for handling typographical errors is needed in order to make keystroke-based authentication systems non-irritating and widely accepted by the computing and network security community. Finally, it is found that artificial neural network paradigms are more successful than classical pattern recognition algorithms in the classification of users.

References

[1] M. S. Obaidat and B. Sadoun, "Verification of Computer users using Keystroke Dynamics," *IEEE Trans. on Systems, Man and Cybernetics*, Vol. 27, No. 2, pp. 261-269, April 1997.

[2] M. S. Obaidat and B. Sadoun, "An Evaluation Simulation Study of Neural Network Paradigm for Computer Users Identification," *Information Sciences Journal-Applications, Elsevier*, Vol. 102, No. 1-4, pp. 239-258, November 1997.

[3] M. S. Obaidat, "A Methodology for Improving Computer Access Security," *Computers & Security*, Vol. 12, pp. 657-662, 1993.

[4] M. S. Obaidat and D. T. Macchairolo, "An On-line Neural Network System for Computer Access Security," *IEEE Trans. Industrial Electronics*, Vol. 40, No. 2, pp. 235-241, April 1993.

[5] S. Bleha and M. S. Obaidat, "Dimensionality Reduction and Feature Extraction Applications in Identifying Computer Users," *IEEE Trans. Systems, Man, and Cybernetics*, Vol. 21, No. 2, March/April 1991.

[6] S. Bleha and M.S. Obaidat, "Computer User Verification Using the Perceptron," *IEEE Trans. Systems, Man, and Cybernetics*, Vol. 23, N0. 3, pp. 900-902, May/June, 1993.

[7] M. S. Obaidat et al., "An Intelligent Neural Network System for Identifying Computer Users," In *Intelligent Engineering Systems through Artificial Neural Networks* (C. Dagli et al. editors), pp. 953-959, ASME Press, New York, 1991.

[8] M. S. Obaidat and D. T. Macchairolo, "A Multilayer Neural Network System for Computer Access Security," *IEEE Trans. on Systems, Man and Cybernetics*, Vol. 24, No. 5, pp. 806-813, May 1994.

[9] J. A. Adam, "Threats and Countermeasures," *IEEE Spectrum*, Vol. 29, No. 8, 21-28, August 1992.

[10] D. B. Parker, *Computer Security Management*, Reston Publishing Co., Reston, VA, 1981.

[11] W. Satllings, *Network and Internetwork Security*, Prentice Hall, Upper Saddle River, NJ, 1995.

[12] C. P. Pfleeger, *Security in Computing*, Prentice Hall, Upper Saddle River, NJ, 1997.

[13] D. Denning, *Cryptography and Data Security*, Addison-Wesley, Reading, MA, 1983.

[14] B. Walker, *Computer Security and Protection Structures*, Dowden, Hutchinson and Ross Inc., 1977.

[15] D. Umphress and G. Williams, "Identity Verification Through keyboard Characteristics," *International Journal Man-Machine Studies*, Vol. 23, pp. 263-273, Academic Press, 1985.

[16] W. L. Bryan and N. Halter, "Studies in the Physiology and Psychology of the Telegraphic Language," *The Psychology of Skill: Three Studies*, (E. H. Gardener and J. K. Gardner, editors), pp. 35-44, NY Time Co., NY 1973.

[17] S. Card, T. Moran, and A. Newell, "The Keystroke Level Model for User Performance Time with Interactive Systems," *Communications of ACM*, Vol. 23, pp. 396-410, 1980.

[18] L. H. Shaffer, "Latency Mechanisms in Transcription," *Attention and Performance*, Vol. IV, (S. Kornblum, editor), Academic Press, 1973.

[19] W. E. Cooper, *Cognitive Aspects of Skilled Typewriting*, pp. 29-32, Springer-Verlag, 1983.

[20] J. Robenson, V. Liang, J. Chambers, and C. MacKenzie, "Computer User Verification Using Login String Keystroke Dynamics," *IEEE Transactions on Systems, Man and Cybernetics*, Vol. 28, No. 2, pp. 236-244, March 1998.

[21] R. Joyce and G. Gupta, "Identity Authentication Based on keystroke Latencies," *Communications of ACM*, Vol. 33, No. 2, pp. 168-176, February 1990.

[22] R. Gaines, W. Lisowski, S. Press, and N. Shapiro, "Authentication by Keystroke Timing: Some Preliminary Results," *Rand Report, R-256*, NSF, Rand Corp., Santa Monica, CA, 1980.

[23] J. Garcia, "Personal Identification Apparatus," *Patent No. 4,621,334*, U.S. Patent and Trademark Office, Washington, DC, 1986.

[24] L. Leggett, G. Williams, and D. Umphress, "Verification of User Identity via Keyboard Characteristics," In *Human Factors in Management Information Systems*, (J. Carey, editor), Ablex, Norwood, NJ, 1988.

[25] L. Leggett and G. Williams, "Identity Verification Through Keyboard Characteristics," *International Journal Man-Machine Studies*, Vol. 23, No. 3, pp. 263-273, September 1985.

[26] J. R. Young and R. W. Hammon, "Method and Apparatus for Verifying and Individual's Identity," *Patent No. 4, 805,222*, U.S. Patent and Trademark Office, Washington, DC, 1989.

[27] M. S. Obaidat and B. Sadoun," HSOP: A Neural Network Paradigm and Its Applications," *Neural Computing & Applications Journal*, Springer, Vol. 2, No. 2, pp. 89-96, 1994.

[28] *NeuralWorks Reference Guide*, NeuralWare, Inc., Pittsburgh, PA, 1993.

Training Set			
File Name	*Patterns per Class*	*Raw Patterns per Average Pattern*	*Odd/Even Half*
all.pat	40	1	All
Odd20	1	20	Odd
Even20	1	20	Even
Odd5	4	5	Odd
Even5	4	5	Even
Odd2	10	2	Odd
Even2	10	2	Even

Table 10.1 An example of a training set used.

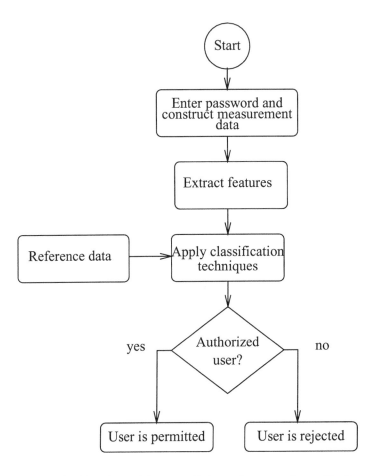

Figure 10.1 Flowchart of the overall steps of a computer verification system.

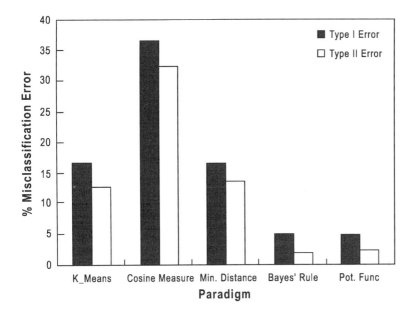

Figure 10.2 Interkey time-based results using pattern recognition techniques.

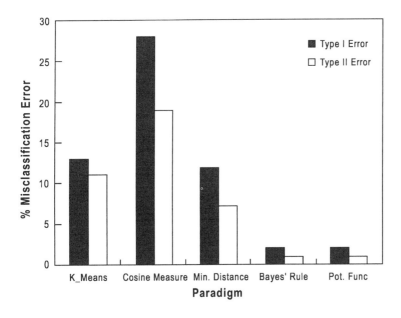

Figure 10.3 Hold time-based classification results using pattern recognition techniques.

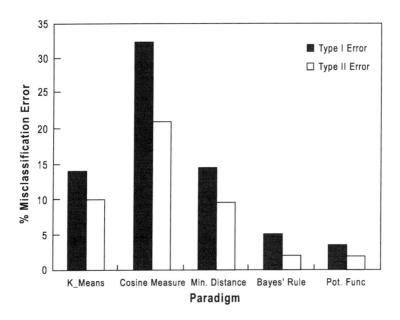

Figure 10.4 Combined interkey and hold time-based classification results using pattern recognition techniques.

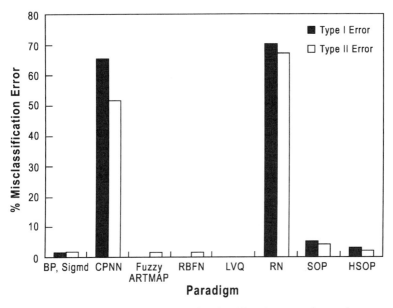

Figure 10.5 Interkey time based classification results using neural network techniques.

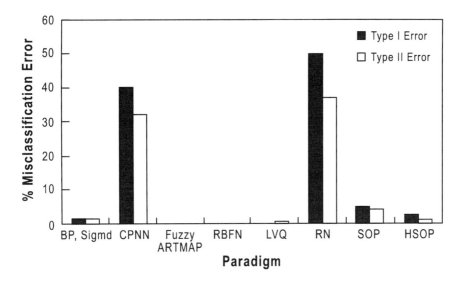

Figure 10.6 Hold time-based classification results using neural network techniques.

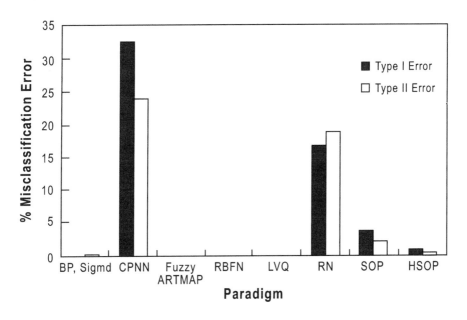

Figure 10.7 Combined interkey and hold time-based classification results using neural network techniques.

11 AUTOMATIC GAIT RECOGNITION

M.S. Nixon, J.N. Carter,
D. Cunado, P.S. Huang, S.V. Stevenage
University of Southampton
Southampton, UK
{msn,jnc,dc95r,psh95r}@ecs.soton.ac.uk
{S.V.Stevenage}@soton.ac.uk

Abstract *Gait is an emergent biometric aimed essentially to recognise people by the way they walk. Gait's advantages are that it requires no contact, like automatic face recognition, and that it is less likely to be obscured than other biometrics. Gait has allied subjects including medical studies, psychology, human body modelling and motion tracking. These lend support to the view that gait has clear potential as a biometric. Essentially, we use computer vision techniques to derive a gait signature from a sequence of images. The majority of current approaches analyse an image sequence to derive motion characteristics that are then used for recognition; only one approach is feature based. Early results by these studies confirm that there is a rich potential in gait for recognition. Only continued development will confirm whether its performance can equal that of other biometrics and whether its application advantages will indeed make it a pragmatist's choice.*

Key words: *Gait, walking, biometrics.*

1. Introduction

In many applications of person identification, many established biometrics can be obscured. The face may be hidden or at low resolution; the palm is obscured; the ears cannot be seen. However, people need to walk, so their gait is usually apparent. This motivates using gait as a biometric and it has recently attracted interest. Gait is attractive since it requires no subject contact, in common with automatic face recognition and other biometrics. The Oxford Dictionary definition of gait is "*manner of walking, bearing or carriage as one walks*" suggesting that studies can concentrate on different facets of a person's walk. Apart from perceptibility, another attraction of using gait is that motion can be hard to disguise. Consider for example a robbery: the robber will need to make access either quickly, to minimize likelihood of capture, or without being obvious in order not to provoke attention. On escape, again the robber will either exit at speed, or in (apparent) leisure. The motion in both cases is natural, for the subject will either not want to attract attention or to move quickly.

Clearly, there are limits to the use of gait as a biometric, a detailed study of the limitations awaits development of technique. However, it is not unlikely that footwear can affect gait, as can clothing. Equally, physical condition can affect gait such as pregnancy, affliction of the legs or feet, or even drunkenness. These factors are not new to biometrics: a face can be made up or have spectacles, ears can be obscured by hair, hands can even be cut off, as acknowledged in other chapters. As usual, a major question concerns whether these are part of human perception whereas a biometric system can perceive the underlying characteristics of the biometric - in the case of gait, the individual's musculature which essentially limits the variation of motion. As such, these factors await investigation.

The view that gait can be used to recognize individual is not new: Shakespeare used a rich lexicon of adjectives to describe gait, including princely, lion's, heavy, humble, weary, forced, gentle, swimming, and majestic. Further, in The Tempest [Act 4 Scene 1], Ceres observes

"High'st Queen of state, Great Juno comes; I know her by her gait".
Even more, in Troilius and Cressida [Act 4 Scene 5], Ulysses states

"Tis he, I ken the manner of his gait; He rises on the toe: that spirit of his in aspiration lifts him from the earth".
The former is one of Shakespeare's many observations on recognizing people by their gait; the latter includes a concise description of Diomedes' demeanour.

Accordingly, there appears much potential for using gait as a biometric. There have been allied studies, particularly those in medical studies for therapy, but there have also been psychological studies, and approaches aimed to model and track human targets through an image sequence, though not usually for recognition, as discussed in Section 2. Current approaches to automatic gait recognition are surveyed in Section 3, together with a more detailed examination of two extant approaches to automatic recognition. Possibilities for further work are discussed in Section 4 prior to the conclusions concerning the potential for gait as a biometric.

2.   Allied Research

Medical Studies

The aim of medical research has been to classify the components of gait for the treatment of pathologically abnormal patients. Murray et al. [34] produced standard movement patterns for pathologically normal people which were used to compare the gait patterns for pathologically abnormal patients [35]. The data collection system used required markers to be attached to the subject. This is typical of most of the data collection systems used in the medical field, and although practical in that domain, they are not suitable for identification purposes. Gait was considered by Murray as "a total walking cycle" - the action of walking can be thought of as a periodic signal. The following terms are used to describe the gait cycle, as given in [34], and are used throughout the report. Fig. 11.1 illustrates the terms described. A gait cycle is the time interval between successive instances of initial foot-to-floor contact 'heel strike' for the same foot. Each leg has two distinct periods; a stance phase, when the foot is in contact with the floor, and a swing phase, when the foot is off the floor moving forward to the next step. The cycle begins with the heel strike of one foot which marks the start of the stance phase. The ankle flexes to bring the foot flat on the floor and the body weight is transferred onto it. The other leg swings through in front as the heel lifts of the ground. As the body weight moves onto the other foot, the supporting knee flexes. The remainder of the foot, which is now behind lifts off the ground ending the stance phase.

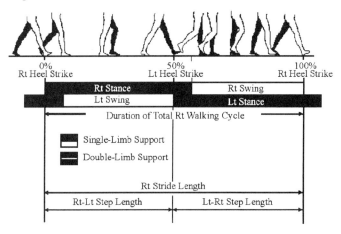

Figure 11.5 Relationship between temporal components of the walking cycle and the step and stride lengths during the cycle.

The start of the swing phase is when the toes of the foot leave the ground. The weight is transferred onto the other leg and the leg swings forward to meet the ground in front of the other foot. The gait cycle ends with the heel strike of the foot. Stride length is the linear distance in the plane of progression between successive points of contact of the same foot. Step length is the distance between successive contact points of opposite feet. A step is the motion between successive heel strikes of opposite feet; a complete gait cycle is comprised of two steps.

Murray *et al.*'s work [34,35] suggests that if all gait movements were considered, gait is unique. In all there appear to be twenty distinct gait components, some of which can only be measured from an overhead view of the subject. Murray found "the pelvic and thorax rotations to be highly variable from one subject to another" [35]. These patterns would be difficult to measure even from an overhead view of the subject, which would not be suited to application in many practical situations. Murray also suggested that these rotation patterns were not found to be consistent for a given individual in repeated trials. In [34,35] ankle rotation, pelvic tipping and spatial displacements were shown to possess individual consistency in repeated trials. Unfortunately, these components would be difficult to extract from real images.

The normal hip rotation pattern of the angle of the thigh (the angle between the thigh and horizontal) is characterized by one period of extension and one period of flexion in every gait cycle, as described by Murray. Fig. 11.2 gives the average rotation pattern: the upper and lower lines indicate the standard deviation from the mean. In the first half of the gait cycle, the hip is in continuous extension as the trunk moves forward over the supporting limb. In the second phase of the cycle, once the weight has been passed onto the other limb, the hip begins to flex in preparation for the swing phase. This flexing action accelerates the hip, directing the swinging limb forward for the next step. Later, we will see how these angles have featured in a model-based recognition system.

There is an extensive literature on studies of gait for medical use, none of which is concerned primarily with biometrics. Intuitively, measurements by gait researchers could prove to be of benefit in biometrics, though there is natural concern that the markers used do not realistically capture individual characteristics. Using gait as a biometric concerns its derivation by computer vision, for this is the only way it can satisfy its purpose. Some insight into gait as a biometric can however be drawn from psychology.

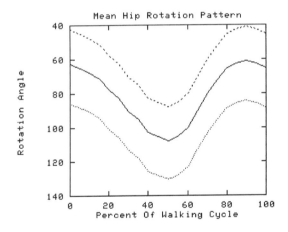

Figure 11.6 Mean hip rotation pattern [35].

Psychology of Gait

In the earliest studies of gait perception [21] participants were presented with images produced from points of light attached to body joints. When the points were viewed in static images, they were not perceived to be in human form, rather that they formed a picture - of a Christmas tree even. When the points were animated, they were immediately perceived as representing a human in motion. Later work showed how by point light displays a human could be rapidly extracted and that different types of motion could be discriminated, including jumping and dancing [15]. More recently Bingham [5] has shown that point light displays are sufficient for the discrimination of different types of object motion and that discrete movements of parts of the body can be perceived. As such, human vision appears adept at perceiving human motion, even when viewing a display of light points. Indeed, the redundancy involved in the light point display might provide an advantage for motion perception [39] and could perhaps offer improved performance over video images.

Naturally, studies in perception have also addressed gender as well as pure motion, again using point light displays. One early study [25] showed how gender could be perceived, and how accuracy was improved by inclusion of height information [44]. The ability to perceive gender has been attributed to anatomical differences which result in greater shoulder swing for men, and more hip swing for women [30]. Indeed, a torso index (the hip:shoulder ratio) has been shown to discriminate gender [14], and the identification of gender by motion of the center of moment was also suggested.

Gender identification would appear to be less demanding than person identification. However, it has been shown that subjects could recognize themselves and their friends [12], and this has been explained by considering gait as a synchronous, symmetric pattern of movement from which identity can be perceived [13]. Like Shakespeare's observations, these studies encourage the view that gait can indeed be used as a biometric. Surprisingly, research into the psychology of gait has not received much attention, especially using video, in contrast with the enormous attention paid to face recognition. One recent study [45], using video rather than point light displays, has shown that humans can indeed recognize people by their gait, and can learn their gait for purposes of recognition. The study concentrated on determining whether illumination or length of exposure could impair the ability of gait perception. The study confirmed that, even under adverse conditions, gait could still be used as a cue to identity.

Clearly, psychological studies confirm Shakespeare's earlier observations, and support the view that gait can indeed be used for recognition. Prior to study of automatic recognition, we shall consider some of the (many) approaches to human body and motion modeling, for these are of potential benefit in recognition. Indeed, some of the approaches have found deployment in automatic gait recognition.

Modeling the Human Body and its Motion

Many studies have considered human motion extraction and tracking, though not for recognition purposes. The selection of good body models is important to efficiently recognize human shapes from images and analyze human motion properly. Stick figure models and volumetric models are commonly used for three-dimensional tracking, and the ribbon model and blob model are also used but are not so popular. Stick figure models connect sticks at joints to represent the human body. Akita [1] proposed a model consisting of six segments: two arms, two legs, the torso and the head. Lee and Chen's model [27] uses 14 joints and 17 segments. Guo *et al.* [18] represent the human body structure in the silhouette by a stick figure model which has ten sticks articulated with six joints.

On the other hand, volumetric models are used for a better representation of the human body. One model [38] consists of 24 segments and 25 joints and those segments and joints are linked together into a tree-structured skeleton. The "flesh" of each segment is defined by a collection of spheres located at fixed positions within the segment's co-ordinate system. Concurrently, angle limits and collision detection are incorporated in the motion restrictions of the human model. Among the different volumetric models, generalized cones are the most commonly used. A generalized cone [29] is the surface swept out by moving a cross-section of constant shape but smoothly varying size along an axis. Generalized cylinders are the simplified case of generalized cones that have a cross-section of constant shape and size. Fig. 11.3 shows examples for a stick figure model, a cylinder model and a blob model.

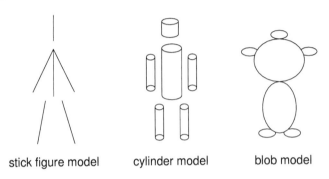

stick figure model   cylinder model   blob model

Figure 11.7 Human body models.

Later work developed Marr's approach [19, 42] to a set of 14 elliptical cylinders representing the feet, legs, thighs, hands, arms, upper-arms, head and torso. Kurakake and Nevatia [26] treat the human body as an articulated object having parts that can be considered as almost rigid and connected through articulations. They use the ribbon which is the two-dimensional version of the generalized cylinder to represent the parts. The blob model was developed by Kauth *et al.* [24] for application to multi-spectral satellite (MSS) imagery and used in human motion tracking [3]. The person is modeled as a connected set of blobs, each of which serves as one class. Each blob has a spatial and color Gaussian distribution, and a support map that indicates which pixels are members of the blobs.

However, these structural models need to be modified according to different applications and are mainly used in human motion tracking. The alternative is to consider the property of the spatio-temporal pattern as a whole. Among the current research, human motion can be defined by the different gestures of body motion, different athletic sports (tennis, ballet) or human walking or running. The analysis varies according to different motions. There are two main methods to model human motion. The first is model-based: after the human body model is selected, the 3-D structure of the model is recovered from image sequences with [27,41] or without moving light displays [1,18,19,42]. The second emphasizes determining features of motion fields without structural reconstruction [28,33,39].

Ideas from human motion studies [34] can be used for modeling the movement of human walking. Hogg [19] and Rohr [43] use flexion/extension curves for the hip, knee, shoulder and elbow joints in their walking models. Guo *et al.* [18] use joint angles between different sticks as features of different walking persons. A different approach for the modeling of motion was taken by Akita [1], who used a sequence of stick figures, called the key frame sequence, to model rough movements of the body.

In his key frame sequence of stick figures, each figure represents a different phase of body posture from the point view of occlusion. The key frame sequence is determined in advance and referred to in the prediction process. In order to find out the interpretation tree of human body and reduce its computational complexity, Chen and Lee [10] applied general walking-model constraints from walking motion knowledge to eliminate the number of unfeasible solutions. Campbell and Bobick [9] proposed techniques for representing movements based on space curves in subspaces of a "phase space", a symbolic description that translates the continuous domain of human motion into a discrete sequence of symbols. The phase space has axes of joint angles and torso location and attitude, and the axes of the subspaces are subsets of the axes of the phase space.

Other approaches that are different from above consider the properties of the spatio-temporal pattern as a whole. Polana and Nelson [40] define temporal textures to be the motion patterns of indeterminate spatial and temporal extent, activities to be motion patterns which are temporally periodic but are limited in spatial extent, and motion events to be isolated simple motions that do not exhibit any temporal or spatial repetition. Little and Boyd's approach [28] is similar to Polana and Nelson's, but they derive dense 2-D optical flow of the person and derive a series of measures of the position of the person and the distribution of the flow. The frequency and phase of these periodic signals are determined and used as features of the motions.

Tracking People

There have been a number of approaches to tracking humans in scenes, more for security applications rather than for recognition. A model-based approach was used in one of the earliest tracking studies [19]. The WALKER model mapped images into a description in which a person was represented using a series of hierarchical levels. The performance of the system was illustrated by superimposing the machine-generated picture over the original photographic images.

There has been much progress since, and we shall review only some of the more recent work. Gavrila and Davis [17] presented a vision system for the 3-D model-based tracking of unconstrained human movement. Multiple view images were employed, to avoid using markers, to recover 3D body pose. Initial results were presented for performance on a (large) Humans-In-Action database and shown to track some demanding postures. In an extension to learned dynamical models for robust curve tracking [4] (for describing non-rigid motions of articulated and deformable objects) an improved model is derived from a set of examples in which the object deforms resulting in a shape description of low dimensionality. This is applied to automatically describe the shape of a moving pedestrian. A new method for the 3D model-based tracking of human body uses multiple views to avoid occlusion of body parts [23]. Available parts are then tracked between frames of a video sequence in a model aimed to minimize the difference between the human model and the imaged views. Initial results were presented showing how humans could be tracked in the presence of severe occlusion.

Parameterised optical flow has actually been used to track articulated motion in an image sequence [22]. Limbs were represented as a set of connected cardboard patches where analyzed motion was constrained to enforce articulated motion. The approach was demonstrated to track humans walking, over long image sequences. The possibility of recognition from these data was noted, but not explored greatly. One system used 3D planar projections to achieve better tracking than contemporaneous 2D trajectory-based systems [7]. The system was based on detecting and segmenting optical flow from within a central region. Then 3D planar geometry was used with an active camera system to ensure focus on the central region. Extended Kalman filters were used to analyze the trajectories and the system was shown to successfully track moving objects, including people, and pursuit performance was shown to improve on 2D performance.

Another recent approach, the "person finder" Pfinder system [46], has been aimed to solve the problem of tracking a single person given a fixed-camera. The system is based on the blob description of human motion [24] which uses coherent connected regions. The statistics of the blobs are then recursively updated to combine present information with prior knowledge. The system then learns the scene of the fixed camera and then detects a person as a large deviation from that scene. Then, the person can be tracked through an image sequence. The system is not aimed at recognition and applications include real-time interface devices and video games.

Most tracking approaches naturally lack the accuracy required for recognition since that was not their original purpose. However, it would seem reasonable to assume that tracking procedures could be deployed to develop a gait signature. The result of tracking a subject's progress through a sequence of frames is shown in Fig. 11.4 showing the estimates, provided by a Kalman filter, of the horizontal position of the waist/crotch and of the ankle/foot. These estimates are derived by tracking the position of regions of high curvature in successive frames. Clearly, the position of the waist gives a better estimate of velocity than a measurement which can be used for recognition. However, the position of the ankle alters according to Murray's earlier model, as such potentially leading to a biometric.

Figure 11.8 Tracking human motion for recognition.

Each of the allied subjects continues to support the notion that gait can be used as a biometric. The physical characteristics of gait are established and viewed to be unique, humans can perceive gait and gait can be modeled and extracted by computer vision techniques. Much of this work has been of benefit to the approaches to automatic gait recognition.

3.   Automatic Gait Recognition

Current Approaches

In what was perhaps the earliest approach to automatic recognition by gait, the gait signature was derived from the spatio-temporal pattern of a walking person [37]. Here, in the XT dimensions (translation and time), the motions of the head and of the legs have different patterns. These patterns were processed to determine the body motion's bounding contours and then a five stick model was fitted. The gait signature was derived by normalizing the fitted model for velocity and then by using linear interpolation to derive normalized gait vectors. This was then applied to a database of 26 sequences of five different subjects, taken at different times during the day. Depending on the values used for the weighting factors in a Euclidean distance metric, the correct classification rate varied from nearly 60% to just over 80%, a promising start indeed.

Later, optical flow was used to derive a gait signature [28]. This did not aim to use a model of a human walking, but to describe features of an optical flow distribution. The optical flow was filtered to produce a set of moving points together with their flow values. The geometry of the set of points was then measured using a set of basic measures and further information was derived from the flow information. Then, the periodic structure of the sequence was analyzed to show several irregularities in the phase differences; measures including the difference in phase between the centroid's vertical component and the phase of the weighted points were used to derive a gait signature. Experimentation on a limited database showed how people could be discriminated with these measures, appearing to classify all subjects correctly.

Another approach was aimed more at generic object-motion characterization [33], using gait as an exemplar of their approach. The approach was similar in function to spatio-temporal image correlation, but used the parametric eigenspace approach to reduce computational requirement and to increase robustness. The approach first derived body silhouettes by subtracting adjacent images, with further processing to reduce noise. Then, the images were projected into eigenspace, a well established approach in automatic face recognition. Eigenvalue decomposition was then performed on the sequence of silhouettes where the order of the eigenvectors corresponds to frequency content. Recognition from a database of 10 sequences of seven subjects showed classification rates of 100% for 16 eigenvectors and 88% for eight, compared with 100% for the (more computationally demanding) spatio-temporal correlation approach. Further, the approach appears robust to noise in the input images. Recently, this has been extended to include Canonical Analysis (CA) [20] for better discrimination, as described in the next section.

In the only model-based approach, the gait signature is derived from the spectra of measurements of the thigh's orientation [11]. This was demonstrated to achieve a recognition rate of 90% on a database of 10 subjects. Contemporaneously, the nature of gait has been recognized by "probabilistic decomposition of human dynamics at multiple abstractions" [8] where the dynamics of gait in video sequences were recognized. Also, different types of gait have been recognised from the trajectories of tracked body parts [32]. The feature vector was extracted from optic flow and from trajectory information and then classified by use of Hidden Markov models, showing good gait discrimination.

Recognition by Statistical Measurement

Eigenspace transformation (EST) based on Principal Component Analysis (PCA) or the Karhunen-Loeve Transform has been demonstrated to be a potent metric in automatic face recognition and gait analysis, but without using data analysis to increase classification capability. A new approach combines canonical space transformation based on CA or Linear Discriminant Analysis (LDA), with the eigenspace transformation, for gait analysis [20]. This gives a 'statistical' approach to automatic gait recognition where the image sequence is described as a whole, and neither by a model- nor by a motion-based approach, but one which describes the motion content.

Face image representations based on PCA have been used successfully for various face recognition applications. However, PCA based on the global covariance matrix of the full set of image data is not sensitive to class structure in the data. In order to increase the discriminatory power of various facial features, LDA has been used to optimize the class separability of different face classes and improve the classification performance [16]. Unfortunately, this has high computational cost. Moreover, the within-class covariance matrix obtained via CA alone may be singular. Combining EST with canonical space transformation (CST) reduces data dimensionality and optimizes class separability of different gait sequences simultaneously.

Given c training classes to be learnt, where each class represents a walking sequence of a single subject, $\mathbf{x}'_{i,j}$ is the j-th image (of n pixels) in class i and N_i is the number of images in i-th class. The total number of training images is

$$N_T = N_1 + N_2 + \ldots + N_C \tag{11.1}$$

and the training set is represented by $\left[\mathbf{x}'_{1,1} \ldots \mathbf{x}'_{1,N_1}, \mathbf{x}'_{2,1}, \ldots, \mathbf{x}'_{c,N_c} \right]$. First, the brightness of each sample image is normalized by

$$\mathbf{x}_{i,j} = \mathbf{x}'_{i,j} / \left\| \mathbf{x}'_{i,j} \right\|. \tag{11.2}$$

After normalization, the mean pixel value for the full image set is:

$$m_x = \frac{1}{N_T} \sum_{i=1}^{c} \sum_{j=1}^{N_i} \mathbf{x}_{i,j}. \tag{11.3}$$

Then we form an $n \times N_T$ matrix $\mathbf{X}$, where each column is formed from each of $\mathbf{x}_{i,j}$ less the mean as:

$$\mathbf{X} = \left[\mathbf{x}_{1,1} - m_x, \ldots, \mathbf{x}_{1,N_1} - m_x, \ldots, \mathbf{x}_{c,N_c} - m_x \right]. \tag{11.4}$$

EST uses the eigenvalues and eigenvectors, generated by the data covariance matrix derived from the product $\mathbf{XX}^T$, to rotate the original data co-ordinates along the direction of maximum variance. Calculating the eigenvalues and eigenvectors of the $n \times n$ matrix $\mathbf{XX}^T$ is computationally intractable for typical image sizes. Based on singular value decomposition, we can compute the eigenvalues of $\mathbf{X}^T\mathbf{X}$, where the matrix size is $N_T \times N_T$ and is much smaller than $n \times n$. The eigenvectors of $\mathbf{X}^T\mathbf{X}$ are used as an orthogonal basis to span a new vector space. Each image can be projected to a single point in this space. According to the theory of PCA, the image data can be approximated by taking only the largest eigenvalues and their associated eigenvectors. This partial set of k eigenvectors spans an eigenspace in which the points $\mathbf{y}_{i,j}$ are the projections of the original images $\mathbf{x}_{i,j}$ by the eigenspace transformation matrix, $[\mathbf{e}_1,...,\mathbf{e}_k]$, as

$$\mathbf{y}_{i,j} = [\mathbf{e}_1,...,\mathbf{e}_k]^T \mathbf{x}_{i,j}. \tag{11.5}$$

After this transformation, each original image can be approximated by the linear combination of these eigenvectors.

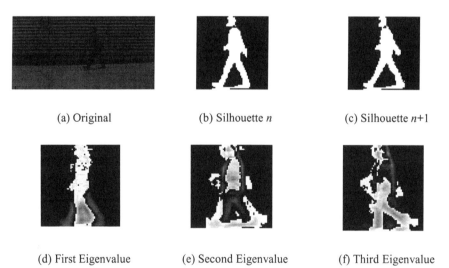

(a) Original	(b) Silhouette n	(c) Silhouette $n+1$
(d) First Eigenvalue	(e) Second Eigenvalue	(f) Third Eigenvalue

Figure 11.9 Original image and derived silhouettes and eigenvalues.

In CST, the classes of the transformed vectors resulting from eigenspace calculation are used to calculate a scatter matrix $\mathbf{S}_t$, a within-class matrix $\mathbf{S}_w$ and a between class matrix $\mathbf{S}_b$ which reflect the dispersion, the variance and the variance of the difference, respectively. The objective of CST is to minimize $\mathbf{S}_w$ and to maximize $\mathbf{S}_b$, simultaneously. This is achieved by minimizing the generalized Fisher linear discriminant function $\mathbf{J}$, where

$$\mathbf{J}(\mathbf{W}) = \mathbf{W}^T\mathbf{S}_b\mathbf{W} / \mathbf{W}^T\mathbf{S}_w\mathbf{W}. \tag{11.6}$$

This ratio is maximised by the selection of the feature $\mathbf{W}$ if

$$\partial \mathbf{J} / \partial \mathbf{W} = 0 \tag{11.7}$$

Supposing $\mathbf{W}^*$ to be the optimal solution and that $\mathbf{w}_i^*$ be its column vector which is a generalised eigenvector and corresponds to the i-th largest eigenvector λ_i, then

$$\mathbf{S}_b \mathbf{w}_i^* = \lambda_i \mathbf{S}_w \mathbf{w}_i^* . \tag{11.8}$$

After the generalised eigenvalue equation is solved, we obtain a set of eigenvalues and eigenvectors that span the canonical space where the classes are much better separated and the clusters are much smaller.

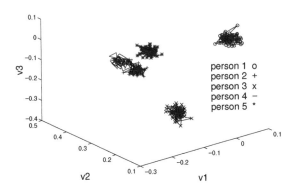

Figure 11.10 Canonical space trajectories for five subjects.

In application, this analysis is applied to human silhouettes derived by subtracting the background from the image and then thresholding the result. Fig. 11.5 shows in (a) one image from the original sequence and in (b) and (c) two of the extracted silhouettes. The eigenvalues are then extracted from the sequence of silhouettes, the first three of which for the subject are shown in Fig. 11.5(d-f). The trajectories in eigenspace overlap and their centroids are very close together. After CST, the trajectories are much better separated and with lower individual variance as in Fig. 11.6 (though for three dimensions only, for visualization purposes). Recognition from the canonical space is accomplished using the distance between the accumulated center to each centroid. On five sequences of five people from the Visual Computing Group, University of California, San Diego [28], an 85% classification rate was achieved by CST alone whereas 100% was achieved with combined EST and CST (as evidenced by the cluster size and separation in Fig. 11.6).

Recognition by Feature-Based Measurement

An alternative approach to collecting the motion information in an image sequence is to find feature(s) and collect their motion information. Here we show how gait signatures can be derived from spectra of the variation in inclination of the thigh, as extracted by computer vision techniques [11]. The bi-pendular model is used as the leg motion is periodic and each part of the leg (upper and lower) appear to have pendulum-like motion, consistent with earlier studies [34,35]. Fourier theory allows periodic signals to be represented as a fundamental and harmonics - the gait motion of the lower limbs can be described in such a way. The model of legs for gait motion allows these rotation patterns to be treated as periodic signals, so Fourier Transform (FT) techniques can be used to obtain a frequency spectrum. The spectra of different subjects can then be compared for distinctive, or unique, characteristics.

(a) Original Image   (b) After Canny Edge Detection and the
                                                           Hough Transform

Figure 11.11 Example image of walking subject (a) original and (b) extracted lines.

To collect data, the camera was situated with a plane normal to the subject's path, and with a static background, an example is given in Fig. 11.7(a). The environment was controlled to improve the data collection with a simple, plain background, with controlled lighting. To resolve difficulty in occlusion, subjects wore trousers which had a stripe painted on the outside. As such, some useable data could still be collected when the legs crossed. The video sequences were averaged to reduce high frequency noise and edge images were produced by applying the Canny operator with hysteresis thresholding.

The Hough Transform (HT) was then applied to the edge image resulting in an accumulator space that has several maxima, each corresponding to a line in the edge image. A peak detection algorithm is applied to extract the parameters of each of these lines (in x and y co-ordinates) using the standard foot-of-normal form

$$s = x\cos\phi + y\sin\phi,$$ (11.9)

where s and ϕ are the distance and angle to the foot of normal. There are several methods for peak detection. In back-mapping, the peak in the accumulator at $(spk, \phi pk)$ is found. For each edge point in the image which lies on the line represented by $(spk, \phi pk)$, the points in the accumulator associated with that edge point are decremented. This effectively removes the votes cast by the line $(spk, \phi pk)$, and so the peak is reduced. This process is repeated until the parameters for all the lines have been found, the result of processing Fig. 11.7(a) is shown in Fig. 11.7(b). To infill for missing data, and to smooth noisy components, the thigh angles given by the lines' inclinations were fitted to a high order polynomial by least squares. For variation in the thigh angle θ with time t, we have an eighth-order polynomial:

$$\theta(t) = a_0 + a_1 t + a_2 t^2 + \ldots + a_8 t^8. \qquad (11.10)$$

An example of the least squares fit for four sequences of single cycles of a particular subject are shown in Fig. 11.8. These fit nicely within the range of Murray's data, Fig. 11.2.

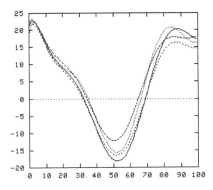

Figure 11.12 Least squares fitting.

These data were then analyzed using the DFT to provide phase and magnitude spectra. The magnitude spectrum dropped to near-zero above the fifth harmonic, again agreeing with earlier work [2]. The magnitude spectra for the two subjects can be used to distinguish between them but the phase spectra were much more different though some components carry little information since their respective magnitude component is very small.

The k-nearest neighbor rule was then used to classify the transform data using the 'leave one out' rule. Four video sequences were acquired for each of ten subjects. Note that the magnitude component of the FT is time shift invariant; it will retain its spectral envelope regardless of where in time the FT is performed. The phase component does not share this characteristic, and a time shift in the signal will change the shape of the phase envelope. Accordingly, the rotation patterns were aligned to start at the heel-strike, to allow phase comparison. This is because the magnitude plots do not confer discriminatory ability whereas the phase plots do. The multiplication appears reasonable, since gait is not characterized by extent of flexion alone, but is controlled by musculature which in turn controls the way the limbs move. Accordingly, there is physical constraint on the way we move our limbs. However, we cannot use phase alone, since some of the phase components occur at frequencies for which the magnitude component is too low to be of consequence. By multiplication of the spectra, we retain the phase for significant magnitude components. Clearly, in this analysis, using phase-weighted magnitude spectra gives a much better classification rate (90%) than use of magnitude spectra alone (40%), for $k = 3$. Selecting the nearest neighbor, as opposed to the 3-nearest neighbor, reduced the classification capability, to 40% for the magnitude and 50% for the phase weighted magnitude, as expected.

As such, feature based metrics can be used to provide gait signatures in a way which agrees with human insight, allowing the results to be validated visually.

4.   Further Work

Clearly, gait research is still in an exploratory phase, rather than at an established one. Accordingly, gait extraction and recognition offers a rich avenue of research opportunity. These opportunities exist not only in development and extension of basic technique, but also in application and as a potential contributor to multi-biometric systems.

In terms of technique, analysis of image sequences is at a very rudimentary stage, especially for the purposes of moving-feature extraction. As yet there are few techniques aimed primarily to integrate the whole image sequence, only one of which uses evidence gathering [36]. These could be extended further, not only to incorporate image context, perhaps including ego-motion or perspective mappings, but also to focus on description and extraction of complex objects, such as a walking human. For gait, we require techniques to isolate moving articulated objects. Naturally, these will focus more on the legs but there is also potential for extension to the thorax. As noted earlier, upper torso movement might offer greater potential for recognition. Equally, these techniques could be aimed at extracting the generic shape of the human body. As such, this requires extraction of arbitrary moving articulated shapes, as required for human motion analysis and recognition.

In terms of gait recognition, we require to derive a signature from an image sequence. So far, this has been derived by approaches that integrate motion across the sequence, and by one which is feature based. A common paradigm in many statistical approaches has been to use binary images for recognition. There are many other approaches which offer similar capabilities, such as Fourier descriptors of an object's shape. Extension of feature based measurement (or a model based approach) would naturally focus on development of a technique, that was mentioned previously.

As with all biometrics, gait research will benefit from an established database for purposes of development, preferably with a separate database for test purposes and hopefully with the stringency of the FERET test. Clearly any database will need to include variation in factors which can affect the perception of gait. These include variety in clothing (especially skirts), and in footwear and with subjects carrying common articles such as handbags or shopping. Also, we will require subjects walking with a wide variety of trajectories relative to the camera together with normal views as used in preliminary studies. Such a database will allow establishment of the properties and limits of signatures derived from gait. As such, they will provide an estimate of the confidence that can be associated with the use of gait to buttress other biometric measures. As stated earlier, gait may be evident where other biometrics can be assessed with limited precision, or are obscured. Clearly, a good database can only serve to evidence the uniqueness or otherwise of automatic gait measurement.

5.   Conclusions

Gait represents a potential biometric as humans can perceive it. It is attractive because it requires no contact and less likely to be concealed. Allied studies in physiology suggest that it can be modeled and is unique, as supported by psychological studies. A number of approaches have modeled the body and tracked it through image sequences, though not for recognition. All these allied subjects either lend support to the potential for gait as a biometric, or suggest that its analysis can be achieved by computer vision.

Indeed, a number of approaches have already shown that it is possible to recognize people by their gait. Naturally, this work is more exploratory than established system development, but results suggest that further development is warranted. The majority of current approaches are motion-based, combining the image sequence by its motion or by statistical analysis. Only one technique is feature-based and its results can clearly be identified with the data from which they were derived.

There is great scope for future research effort, both in application and development. Clearly, gait would benefit from an established database on which to assess new developments. These developments could be improvements in recognition procedure or in automated technique. As such, future work will establish more precisely the results that can be achieved by this new biometric. If its performance can equal that of other biometrics, then by its practical advantages it could indeed become a pragmatist's choice.

Acknowledgements

We gratefully acknowledge financial and research support from the UK Home Office, via Tony Kitson and Bob Nicholls. The input of James Shutler and Cranch Lamble is gratefully appreciated.

References

[1]   K. Akita, Image sequence analysis of real world human motion, *Pattern Recognition*, Vol. 17, No. 1, pp. 73-83, 1984.

[2]   C. Angeloni, P. O. Riley, and D. E. Krebs, "Frequency content of whole body gait kinematic data," *IEEE Trans. Rehabilitation Engineering*, Vol. 2, No. 1, pp. 40-46, 1994.

[3]   A. Azarbayejani, C. Wren, and A. Pentland, "Real-time 3-D tracking of the human body," *Proceedings IMAGE'COM*, May 1996. (Media Lab. Rept. #374, available at http://vismod.www.media.mit.edu/cgi-bin/tr_pagemaker)

[4] A. Baumberg and D. Hogg, "Generating spatiotemporal models from examples," *Image and Vision Computing*, Vol. 14, No. 8, pp. 525-532, 1996.

[5] G. P. Bingham, R. C. Schmidt, and L. D. Rosenblum, "Dynamics and the orientation of kinematic forms in visual event recognition," *Journal of Experimental Psychology: Human Perception and Performance*, Vol. 21, No. 6, pp. 1473-1493, 1995.

[6] M. J. Black, Y. Yacoob, A. D. Jepson, and D. J. Fleet, "Learning parameterized models of image motion," *Proceedings International Conference Computer Vision and Pattern Recognition,* pp. 561-567, 1997.

[7] K. J. Bradshaw, I. D. Reid, and D. M. Murray, "The active recovery of 3D motion trajectories and their use in prediction," *IEEE Trans. Pattern Analysis and Machine Intelligence. Vol.* 19, No. 3, pp. 219-233, 1997.

[8] C. Bregler, "Learning and recognizing human dynamics in video sequences," *Proceedings International Conference on Computer Vision and Pattern Recog*nition, pp. 568-574, June 1997.

[9] L. Campbell and A. Bobick, "Recognition of human body motion using phase space constraints," MIT Media Lab Perceptual Computing Report 309, 1995.

[10] Z. Chen and H-J Lee, "Knowledge-guided perception of 3-D human gait from a single image sequence," *IEEE Trans. Systems, Man, and Cybernetics,* Vol. 22, No. 2, pp. 336-342, 1992.

[11] D. Cunado, M. S. Nixon, and J. N. Carter, "Using gait as a biometric, via phase-weighted magnitude spectra," *In* J Bigun, G. Chollet, and G. Borgefors (editors): *Lecture Notes in Computer Science,* 1206 (Proceedings of *1st International Conference on Audio- and Video-Based Biometric Person Authentication*), pp. 95-102, 1997.

[12] J. E. Cutting and L. T. Kozlowski, "Recognizing friends by their walk," *Bulletin of the Psychonomic Society,*Vol. 9, No. 5, pp. 353-356, 1977.

[13] J. E. Cutting, D. R. Proffitt, and L. T. Kozlowski, "A biochemical invariant for gait perception," *Journal of Experimental Psychology: Human Perception and Performance*, Vol. 4, pp. 357-372, 1978.

[14] J. E. Cutting and D. R. Proffitt, "Gait perception as an example of how we perceive events," In R. D. Walk and H. L. Pick (editors), *Intersensory Perception and Sensory Integration*, Plenum Press, London UK, pp. 249-273, 1981.

[15] W. H. Dittrich, "Action categories and the perception of biological motion," *Perception*, 22, pp. 15-22, 1993.

[16] K. Etemad and R. Chellappa, "Discriminant analysis for recognition of human face images," *J. Opt. Sci. Am.*, Vol. 14, No. 8, pp. 1724-1733, 1997

[17] D. M. Gavrila and L. S. Davis, "3-D model-based tracking of humans in action: a multi-view approach," *Proceedings International Conference Computer Vision and Pattern Recognition*, pp. 73-80, June 1996.

[18] Y. Guo, G. Xu, and S. Tsuji, "Understanding human motion patterns," *Proceedings 12th International Conference on Pattern Recognition*, 2, pp. 325-329, 1994.

[19] D. Hogg, "Model-based vision - a program to see a walking person," *Image and Vision Computing*, Vol. 1, No. 1, pp. 5-20, 1983.

[20] P. S. Huang, C. J. Harris, and M. S. Nixon, "Canonical space representation for recognizing humans by gait or face," *Proceedings IEEE Southwest Symposium on Image Analysis and Interpretation*, Arizona, pp. 180-185, April 1998.

[21] G. Johansson, "Visual perception of biological motion and a model for its analysis," *Perception and Psychophysics*, 14, pp. 201-211, 1973.

[22] S. X. Ju, M. J. Black, and Y. Yacoob, "Cardboard people: a parameterized model of articulated image motion," *Proceedings International Conference on Automatic Face and Gesture Recog*nition, pp. 38-44, 1996.

[23] A. Kakadiaris and D. Metaxas, "Model-based estimation of 3D human motion with occlusion based on active multi-viewpoint selection," *Proceedings International Conference on Computer Vision and Pattern Recognition*, pp. 81-87, June 1996.

[24] R. J. Kauth, A. P. Pentland, and G. S. Thomas, "Blob: an unsupervised clustering approach to spatial pre-processing of MSS imagery," *Proceedings 11th International Symposium on Remote Sensing of the Environment*, April, Ann Arbor, MI, USA, pp.1309-1317, 1977.

[25] L. T. Kozlowski and J. E. Cutting, "Recognizing the sex of a walker from a dynamic point light display," *Perception and Psychophysics*, Vol. 21, pp. 575-580, 1977.

[26] S. Kurakake and R. Nevatia, "Description and tracking of moving articulated objects," *Systems and Computers in Japan*, Vol. 25, No. 8, pp. 16-26, 1994.

[27] H-J Lee and Z. Chen, "Determination of 3D human body postures from a single view," *Computer Vision, Graphics and Image Processing*, Vol. 30, pp. 148-168, 1985.

[28] J. Little and J. Boyd, Describing motion for recognition, *Proceedings International Symposium on Computer Vision*, Coral Gables, FL, USA, pp. 235-240, Nov. 1995.

[29] D. Marr and H. K. Nishihara, "Representation and recognition of the spatial organization of three-dimensional shapes," *Proceedings Royal Society London,* Vol. B:200, pp. 269-294, 1978.

[30] G. Mather and L. Murdock, "Gender discrimination in biological motion displays based on dynamic cues," *Proceedings Royal Society London,* Vol. B:258, pp. 273-279, 1994.

[31] D. Meyer, J. Denzler, and H. Niemann, "Model based extraction of articulated objects in image sequences for gait analysis," *Proceedings IEEE International Conference on Image Processing ICIP98,* Vol. III, pp. 78-81, 1998.

[32] D. Meyer, Human Gait Classification Based on Hidden Markov Models. In H.-P. Seidel, B. Girod, and H. Niemann Ets. *3D Image Analysis and Synthesis '97*, pp.39-146, Erlangen, November 1997.

[33] H. Murase and R. Sakai, "Moving object recognition in eigenspace representation: gait analysis and lip reading," *Pattern Recognition Letters*, Vol. 17, pp. 155-162, 1996.

[34] M. P. Murray, A. B. Drought, and R. C. Kory, "Walking patterns of normal men," *Journal of Bone and Joint Surgery*, Vol. 46-A, No. 2, pp. 335-360, 1964

[35] M. P. Murray, "Gait as a total pattern of movement," *American Journal of Physical Medicine*, Vol. 46, No. 1, pp. 290-332, 1967.

[36] J. M. Nash, J. N. Carter, and M. S. Nixon, "Dynamic Feature Extraction via the Velocity Hough Transform," *Pattern Recognition Letters*, Vol. 18, No. 10, pp. 1035-1047, 1997.

[37] S. A. Niyogi and E. H. Adelson, "Analyzing and recognizing walking figures in XYT," *Proceedings Conference Computer Vision and Pattern Recognition 1994*, pp. 469-474, 1994.

[38] J. O'Rourke and N. Badler, "Model-based image analysis of human motion using constraint propagation," *IEEE Trans. on Pattern Analysis Machine Intelligence*, Vol. 2, No. 6, pp. 522-536, 1980.

[39] G. L. Pellechia and G. E. Garrett, "Assessing lumbar stabilistation from point light and normal video displays of lumbar lifting", *Perceptual and Motor Skills,,* Vol. 85, No. 3, pp. 931-937, 1997

[40] R. Polana and R. Nelson, "Detecting activities," *Proceedings Conference on Computer Vision and Pattern Recognition*, pp. 2-7, June 1993.

[41] R. F. Rashid, "Towards a system for the interpretation of moving light displays," *IEEE Trans. Pattern Analysis Machine Intelligence*, Vol. 2, No. 6, pp. 574-581, 1980.

[42] K. Rohr, "Incremental recognition of pedestrians from image sequences," *Proceedings Conference Computer Vision and Pattern Recognition*, New York, USA, pp. 8-13, June 1993.

[43] K. Rohr, "Towards model-based recognition of human movements in image sequences," *Computer Vision, Graphics and Image Processing*, Vol. 59, No. 1, pp. 94-115, 1994.

[44] S. Runenson and G. Frykholm, "Kinematic specification of dynamics as an informational basis for person-and-action perception: expectation, gender recognition and deceptive intention," *Journal of Experimental Psychology: General*, Vol. 112, pp. 585-615, 1983.

[45] S. V. Stevenage, M. S. Nixon, and K. Vince, "Visual Analysis of Gait as a Cue to Identity," *At Press*, 1998.

[46] C. R. Wren, A. Azarbayejani, T. Darrell, and A. P. Pentland, "Pfinder: real-time tracking of the human body," *IEEE Trans. Pattern Analysis Machine Intelligence.*, Vol. 19, No. 7, pp. 780-785, 1997.

12 OBJECTIVE ODOUR MEASUREMENTS

K. C. Persaud, D-H. Lee, H-G. Byun
University of Manchester Institute of Science &
Technology, Manchester, UK
kcpersaud@umist.ac.uk
dhlee4@samsung.co.kr
byun@mail.samchok.ac.kr

Abstract *The biological chemoreception mechanisms are complex, and the sense of smell is extremely powerful in terms of discrimination between complex mixtures of chemicals, sensitivity to certain classes of chemicals and the range of concentrations that are detectable. This chapter introduces the biological concepts of chemoreception and information processing and goes on to describe approaches in producing biomimetic devices that may ultimately be used for biometric applications. Promising sensor technologies applicable for use in sensor arrays are introduced, and information processing strategies applicable to the pattern recognition problems are presented.*

Keywords: *Odour descriptors, perceptrons, electronic nose, chemoreception, taste, olfactometer, odour classification, chemosensory systems.*

1. Introduction

There has been a large growth of interest in building systems that imitate the five senses of the mammal. An instrument that could perform simple odour discrimination and provide measurement of odour intensity, without the influences mentioned above, would be very useful in modern industry. This chapter investigates possible approaches to biomimetic chemosensory systems that may someday approach the capacity and ability of the olfactory system in mammals. Using instruments for smell recognition of humans is currently under investigation by a number of research groups and companies, but this is a very complex task, and no practical devices are yet available.

The Mammalian Chemoreception System

When an animal or a person sniffs an odorant, molecules carrying the scent are captured by receptor neurons located in the roof of the nose. The receptors are loosely specialised in the kinds of odorants to which they respond. Cells that become excited fire action potentials, or pulses, that propagate through neural projections called axons to a part of the brain known as the olfactory bulb. The number of activated receptors indicates the intensity of the stimulus, and the pattern of activity in the nose conveys the nature of the scent. The olfactory receptor neurons are special nerve cells that interact with molecules contained in the inhaled air. Since the chemical receptors interact only with specific molecules, olfactory sensitivity is restricted to a limited repertoire of odours. The molecules detected are of low molecular weight, typically less than 300 Daltons. They are diverse in structure and size, containing hydrophobic regions as well as functional groups that may determine the classification of the odour [1]. The chemical sensors in the olfactory system act as receptors and the response forms a pattern (a set of signals) that is processed in the brain to produce an appropriate response due to a stimulus. Figure 12.1 shows a block diagram of the olfactory system. In a human, there is an array of 100 million odour sensors. Within this array, there are a number of different types of sensors, which display differing odour specificity to particular classes of odours. Individual elements in the array show broad and overlapping selectivity to chemical species. Importantly, olfactory receptors are not highly selective, but selectivity is achieved by the unique patterns of responses from numbers of such receptors [3].

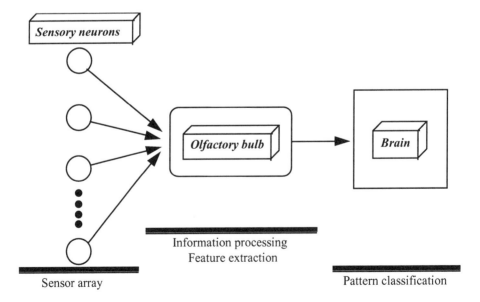

Figure 12.1 The functional components of the olfactory system.

The olfactory bulb carries out a great deal of pre-processing and feature extraction of the signals arriving from the olfactory receptors, analysing each input pattern and then producing specific messages, which it transmits via axons to another part of the olfactory system, the olfactory cortex. From there, new signals are sent to many parts of the brain, including an area called the entorhinal cortex, where the signals are combined with those from other sensory systems. The result is a meaning-laden perception, that is unique to each person, where some scents may produce a sense of well-being, and others a sense of nausea, and others may be linked to specific memories or events [2].

Odour classification studied using psychophysical methods would suggest that there are several primary odour groups. These include camphor-like, musky, floral, pepperminty, ether-like, pungent, and putrid. Unique mixtures of aromatic compounds including phenolics, ketones, and terpenes [4] determine the characteristic odours of fruits and vegetables. Smell sensitivity is genetically determined and varies from person to person, hence different people perceive or may react to different odours, or react differently to the taste of foods.

The nose is an important regulator of social life. Pheromones are a class of long-distance chemical messenger hormones that regulate social relations, behavioral and physiological responses in insects and many mammals. Observations of human behavior have linked smell messages to menstrual cycle timing, and women living in a dormitory over an extended period of time tend to menstruate at about the same time [5].

Taste and smell are highly associated with each other, and flavours are composite sensations, derived from primary taste and smell sensations by processes within the limbic brain areas that participate in emotion. The human emotional system emerged in evolutionary terms from the olfactory brain of earlier animals and remains closely linked to taste, smell, and eating behaviours. Taste and smell evaluation may be an important diagnostic screen for early brain changes destined to result in serious dementia [6]. The odours of coffee, chocolate, almond (benzaldehyde), and oil of lemon are often used for casual testing of olfactory function.

Figure 12.2 shows a block diagram of the major pathways of olfactory information in the brain. The brain actively seeks information about the outside world, mainly by directing an individual to look, listen and sniff. The search results from self-organising activity in the limbic system (a part of the brain that includes the entorhinal cortex and is thought to be involved in emotion and memory), that sends commands to the motor systems. As the motor command is transmitted, the limbic system issues what is called a reafference message, alerting all the sensory systems to prepare to respond to new information, and this cycle of activity is repeated [2].

The human nose is a very interesting chemosensory system that is full of contrasts. On the one hand, there is exquisite sensitivity to some chemicals such as mercaptans while on the other hand there is very poor sensitivity to other classes such as hydrocarbons. The estimation of odour intensity is on a very poor and compressed scale. A trained human nose is able to discriminate between subtle differences in the odours between almost identical mixtures of chemicals, but the perception of odour quality is not constant and is dependent on the concentration of the mixture of chemicals being assessed. The sensors in the nose are capable of regeneration and reconnection to the correct places in the olfactory bulb, and as yet these processes are poorly understood.

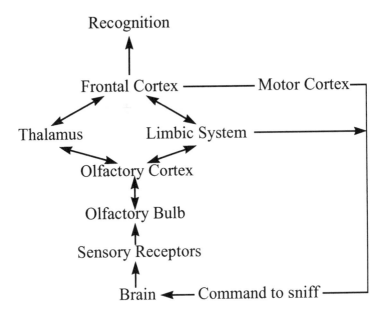

Figure 12.2 The major pathways of information processing in the brain for olfactory stimuli.

Panels of human noses continually assess the quality of raw materials, their processing and the quality of the final product in many industries. Because, human noses are fallible and can only cope with small numbers of samples in a given time, automated artificial odour sensing devices become attractive.

2.   Odour Evaluation Using Human Panels

The character of an odour is reported using "odour descriptors". There are no consistent ways of standardising the descriptors of odours and many individuals and companies have compiled lists of descriptors that may be used to train a human odour panel. Since the odour panel may be specialised to assess only certain products, or odour quality problems, the lists of relevant descriptors are often expanded or contracted as required. Table 12.1 illustrates a portion of a list compiled by Amoore [7]. It would be seen that in order to assess odour quality, a specialised vocabulary must be learnt, with each descriptor associated with specific notes, or nuance in a particular odour. In fact, an odour is normally comprised of many chemical species mixed together in different proportions that the receptors in the nose perceive all at once, without separation into individual chemical species.

Alcohol-Like	Eucalyptus	Oak Wood
Almond-Like	Fecal (Like Manure)	Cognac-Like
Animal	Fermented (Rotten)	Oily, Fatty
Anise (Licorice)	Fruit	Orange (Fruit)
Apple (Fruit)	Fishy	Paint-Like
Aromatic	Floral	Peach (Fruit)
Bakery (Fresh Bread)	Fragrant	Peanut Butter
Banana-Like	Fresh Green Vegetables	Pear (Fruit)
Bark-Like, Birch Bark	Fresh Tobacco Smoke	Perfumery
	Fried Chicken	Pineapple (Fruit)
Bean-Like	Fruity (Citrus)	Popcorn
Beery (Beer-Like)	Fruity (Other)	Putrid, Foul, Decayed
Bitter	Garlic, Onion	Raisins
Black Pepper-Like	Geranium Leaves	Rancid
Burnt Candle	Grainy (As Grain)	Raw Cucumber-Like
Burnt Rubber-Like	Grape-Juice-Like	Raw Potato-Like
Burnt, Smoky	Grapefruit	Rope-Like
Burnt milk	Green Pepper	Rose-Like
Buttery (Fresh)	Hay	Rubbery (New Rubber)
Cadaverous	Heavy	Sauerkraut-Like
Camphor Like	Herbal, Green, Cut	Seasoning (For Meat)
Cantaloupe Melon	Grass	Seminal, Sperm-Like
Caramel	Honey-Like	Sewer Odor
Caraway	Houseold Gas	Sharp, Pungent, Acid
Cardboard-Like	Incense	Sickening
Cat-Urine-Like	Kerosene	Soapy
Cedarwood-Like	Kippery (Smoked Fish)	Sooty
Celery	Laurel Leaves	Soupy
Chalky	Lavender	Sour Milk
Cheesy	Lemon (Fruit)	Sour, Acid, Vinegar
Chemical	Like Gasoline, Solvent	Spicy
Cherry (Berry)	Like Mothballs	Stale
Chocolate	Light	Stale Tobacco Smoke
Cinnamon	Like Ammonia	Strawberry-Like
Clove-Like	Like Blood, Raw Meat	Sulphidic
Coconut-Like	Like Cleaning Fluid	Sweaty
Coffee-Like	Like Burnt Paper	Sweet
Cologne	Malty	Tar-Like
Cooked Vegetables	Maple (As In Syrup)	Tea-Leaves-Like
Cool, Cooling	Meaty(Cooked, Good)	Turpentine (Pine Oil)
Cork-Like	Medicinal	Urine-Like
Creosote	Metallic	Vanilla-Like
Crushed Grass	Minty, Peppermint	Varnish
Crushed Weeds	Molasses	Violets
Dill-Like	Mouse-Like	Warm
Dirty Linen-Like	Mushroom-Like	Wet Paper-Like
Disinfectant, Carbolic	Musk-Like	Wet Wool, Wet Dog
Dry, Powdery	Musty, Earthy, Moldy	Woody, Resinous
Eggy (Fresh Eggs)	Nail Polish Remover	Yeasty
Etherish, Anaesthetic	Nutty (Walnut, Etc.)	

Table 12.1 Odour descriptors.

Determination of how much of an odour is present is often carried out by olfactometry. Odour samples are evaluated using an instrument called an "olfactometer" with a human odour panel, typically consisting of four to eight persons. Both European and American Standards for olfactometry exist, and odours are evaluated in accordance with defined standard practices. Odour panels consist of individuals (panelists) that are selected and trained. An odour panel is conducted using a panel of individuals and an olfactometer. Many olfactometer designs exist, but typically sample odour is delivered at a fixed rate from a Tedlar^{TM} gas-sampling bag to the olfactometer. A filtered air system delivers non-odourous dilution air to the olfactometer. The trained odour panelist sniffs the diluted odour sample as it is discharged from one of two or three sniffing ports. The panelist sniffs all sniffing ports and must select one that is different from the others. This is a forced choice approach. The panelist decides if the selection was a "guess", "detection" or "recognition". The panelist then sniffs the next set of ports, one of which also contains the diluted odour sample. However, this next set presents the odour at a higher concentration (i.e., two to three times). The panelist continues to additional sets of sniffing ports. This statistical approach is called "ascending concentration series". This method is a rapid means of determining sensory thresholds of any substance usually in an air medium. The threshold may be characterised as being either (a) only detection (awareness) that a very small amount of added substance is present but not necessarily recognisable, or (b) recognition of the nature of the added substance [8,9]. It is recognised that the degree of training received by a panel with a particular substance may have a profound influence on the threshold obtained with that substance, and that thresholds determined by using one physical method of presentation are not necessarily equivalent to values obtained by another method.

The number of binary dilutions of the odour sample to human threshold, gives an objective, numerical measure of 'odour units' present in the sample. While these are dimensionless units, dimensions of odour unit per unit volume are commonly applied. Odour concentrations have been previously reported in a variety of ways that are difficult to compare. These include detection threshold, recognition threshold, dilution to threshold, effective dose at 50 percentile, dilution ratio, odour units, odour dilution units, and best estimated threshold.

The use of a human panel for assessment of odour is highly subjective, costly and time consuming. However in industries that are associated with foods, beverages, cosmetics, soaps, and deodorants the human panel plays an invaluable role. However many industrial processes and agricultural operations also produce malodours as well as chemically toxic substances that need to be monitored and controlled. The human panel can only cope with limited numbers of samples, and has to be protected from exposure to potential hazardous substances. Hence, automated methods for odour measurement are desirable.

3. Analytical Procedures for Odour Measurement

Chemical mixtures are commonly analysed using gas chromatography (GC). Gas chromatography-mass spectrometry (GC-MS) combines the sample mixture separation obtainable with GC with the ease of sample identification associated with

MS. This technique can be highly selective and sensitive, but is time consuming and expensive. Other common methods used for odour identification include:

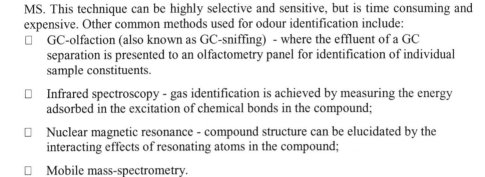

☐   GC-olfaction (also known as GC-sniffing) - where the effluent of a GC separation is presented to an olfactometry panel for identification of individual sample constituents.

☐   Infrared spectroscopy - gas identification is achieved by measuring the energy adsorbed in the excitation of chemical bonds in the compound;

☐   Nuclear magnetic resonance - compound structure can be elucidated by the interacting effects of resonating atoms in the compound;

☐   Mobile mass-spectrometry.

All of the methods mentioned above, despite their particular strengths, are inadequate to replace the human nose, either in terms of the cost and equipment needed or their limited discriminatory power. There is a great need for a low-cost, portable sensor system that is able to sense and identify a wide range of chemicals, not necessarily odourous, at low concentrations and with the ability to cope with discrimination of complex mixtures.

In the design of artificial odour sensing devices, much attention needs to be paid to the analysis of the requirements of the human user, and in what areas the instrument are likely to be used. An odour-sensing device will perform well in a particular defined context where a specific question is being asked of the apparatus. For such an instrument to be useful to a wide variety of users, it needs to be flexible enough to adapt to changing requirements of odour discrimination and intensity estimation. It is noted that the chemosensory system of humans and vertebrates consist of arrays of chemical transducers, where each type of sensing element has chemical specificity's that are broad, but distinct from each other. The signals from this system are processed in the brain to produce descriptors of the odour of individual chemicals, or mixtures of chemicals, as well as the intensity of the odour. It thus makes sense to design artificial chemical sensing arrays on this basis. Pattern recognition systems based on neural network odour classifier software and statistically based classifiers may then be applied to produce an analogy of the higher information processing mechanisms in the brain.

Biometric Aspects of Human Odour Discrimination

Mastiff Electronics (UK) demonstrated a system (Scentinel) on British Television in 1996 that verifies a person by their body odour, the makeup of which is genetically linked. The volatile chemicals that make up a person's smell are emitted from skin pores all over the body and it is claimed that the instrument can pick it up from a person's hand. The basis for the work stems from some fundamental studies on the composition of body odour carried out by Sommerville and co-workers [10] who demonstrated that the pattern of volatiles in human axillary sweat reflects a genetic influence and can be used to distinguish identical twins from unrelated people. These were later correlated to some common HLA Class 1 antigens. Practical difficulties exist, however. The odour profile of the human body may be affected by habits such as the use of deodorants or perfumes. Diet and medication also influence the body odour. There is the possibility of odour contamination or transfer from one person to the other. These aspects make body odour identification difficult. However it has been demonstrated that despite these variables, trained dogs are able to distinguish between persons on the basis of odour [11]. Hence there is research interest in the area of security applications where smell may be one parameter that may be monitored.

Currently, 'electronic nose' instruments are finding their way into applications such as the diagnosis of bacterial infections, and many medical diagnosis applications, where the normal odour is distinguished from an abnormal odour. In some cases the species of bacteria can be identified, as shown in a study of patients with bacterial vaginosis [12].

4.   Odour Sensing Instruments

An electronic system reproducing the characteristics of the human nose could utilise the approach taken by human evolution, i.e. use a large array of sensors, where each sensor element has broad but overlapping sensing characteristics with the other sensors. In order for the information from the artificial array-based odour sensing system to be applied practically, information is required from the biological system, in terms of odour/structure relationships, to provide the basis for translation criteria between the patterns produced by the sensor arrays and odour descriptions. More thought has to be given to the system as a whole and not just to the chemical sensing elements.

Although steps to design odour sensing systems had been taken as early as 1950 [13], one of the first practical instruments specifically designed to detect odours was developed in 1961 [14]. An electronic nose was reported in 1964 by Wilkens and Hartman [15] based on redox reactions of odorants at an electrode. In the following year, Buck *et al.* [16] and Dravnieks and Trotter [17] showed the potential of electronic noses based on the modulation of conductivity and the modulation of contact potential by odorants, respectively. Little progress was reported until 1982 when Persaud and Dodd [18] reported a successful discrimination of a wide variety of odours using plural semi-conductor transducers. The concept of an electronic nose, as an intelligent chemical array sensor system, for odour classification was introduced.

Array Based Odour Sensors

The first steps to viable odour sensing technology were taken when it was proved that discrimination between odours was possible using a small array of broad specificity semiconductor sensors [18,19]. The discriminatory power of a small array lies in the utilisation of cross-sensitivities between sensor elements. The responses of the individual sensors, each possessing a slightly different response towards the sample odours, when combined by suitable mathematical methods, can provide enough information to discriminate between sample odours. Since then many research groups, exploiting a variety of sensor technologies have joined in development of electronic multi-sensor systems that could eventually start to mimic some aspects of the biological olfactory system. These system have been given the terminology 'electronic nose' and consist of an array of chemical sensors possessing broad specificity, coupled to electronics and software that allow feature extraction - extraction of salient data for further analysis, together with pattern recognition - identification of sample odour. Software techniques and material science are important aspects of the development of the system. Advancement in software signal processing techniques, coupled with pattern recognition, enable optimum usage of sensor responses. The specificity and sensitivity of existing chemical sensors are constantly being developed, as well as new materials. Table 12.2 summarises some of the most popular technologies used currently. This list is expanding rapidly as more research groups become active in this exciting field.

Sensor Technology	Application
Sintered metal oxides (thick and thin film)	CO, H_2, anaesthetic gases, many 'nose' applications
Catalytic gas sensors	Combustible gases
MOS Field effect transistors	Combustible gases, organic vapours
Electrochemical cells	NH_3, CO, SO_2, others
Piezoelectric crystals	Broad range of vapours, nonpolar to polar
Surface acoustic wave devices	Broad range of vapours, nonpolar to polar
Conducting polymers	Mainly polar gases, many 'nose' applications
Fibre optic devices	NH_3, SO_2, wide variety of organic gases
Spectrophotometric devices, absorption, fluorescence, luminescence	Wide applications
Langmuir-Blodgett films	Organic vapours
Metal phtalocyanines	NO_x
Mass spectrometry based devices	Wide applications, including mixtures of vapours
Pt, Pd doped organic semiconductors	Combustible gases
Langmuir Blodget films	Various

Table 12.2 Sensor technologies currently in use for 'electronic nose' applications.

Many pioneering investigations into the use of sensor arrays, before intelligent gas sensing systems were developed, used various multisensor systems for identification of different types of gases, often in parallel with analytical methods. Zaromb and Stetter [19] provided a theoretical basis for the selection and effective use of an array of chemical sensors for a particular application. Bott and Jones [20] attempted to build a multisensor system to monitor hazardous gases in a mine using six sensors of three different types in combination with oxidising layers and absorbent traps. The system was able to distinguish between gases evolved from a fire and those evolved from diesel engines or explosives. Odour identifying systems containing two or more different types of sensors are often an attempt to enhance the different dimensional characteristics of the responses. Stetter [21-24], who was also involved in developing a portable device to detect hazardous gases and vapours to warn U.S. Coast Guard emergency response personnel using four different electrochemical sensors, demonstrated a combined system with a hydrocarbon sensor and an electrochemical sensor. He managed to get responses from both the hydrocarbon sensor and the electrochemical sensors after passing the vapours through a combustible gas sensor. The resulting data was successfully analysed using pattern recognition methods based on neural networks.

There are numerous types of gas sensor arrays used in electronic noses as shown in Table 12.2. The most popular types include metal oxide semiconductor (MOS) sensors, catalytic gas sensors, solid electrolyte gas sensors, conducting polymers, mass-sensitive devices, and fibre-optic devices based on Langmuir-Blodgett films. The oxide materials, which have been popular for use in electronic noses, operate on the basis of modulation of conductivity when the odorant molecules react with chemisorbed oxygen species. There are commercially available metal oxide sensors (*e.g.*, Figaro Inc., Japan) which operate at elevated temperatures, between 100-600°C, to help adsorption/desorption kinetics [25]. They are sensitive to combustible materials (0.1-100 ppm), such as alcohols, but are generally poor at detecting sulphur or nitrogen based odours and have a major problem of irreversible contamination with these compounds. Integrated thin-film metal oxide sensors have been designed using planar integrated microelectronic technology which has advantages of lower power consumption, reduction in size and improved reproducibility; however, they tend to suffer from poor stability. Although there are some oxide materials that show good specificity to certain odours [26], there are a number of advantages in employing organic materials in electronic noses. A wide variety of materials are available for such devices and they operate close to or at room temperature (20-60°C) with a typical sensitivity of around 0.1-100 ppm. Furthermore, functional groups that interact with different classes of odorant molecules can be built into the active material, and the processing of organic materials is easier than oxides.

Despite of a number of disadvantages including high power consumption, elevated operational temperature, poisoning effects from sulphur-containing compounds, and poor long-term stability, MOS gas sensors are the most widely used in gas and odour detection. The main reason is that MOS commercial products have been available for a number of years. Abe *et al.* examined an automated odour sensing system based on plural semiconductor sensors to measure 30 substances [27] and 47 compounds [28]. They analysed the sensor outputs using pattern recognition techniques: Karhunen-Loeve (K-L) projection for visual display output; and *k*-nearest neighbourhood (*k*-NN) method and potential function method for classification. Shurmer *et al.* worked on discrimination of alcohols and tobaccos using tin-oxide sensors based on the correlation coefficient method in their research [29]. Weimar *et al.* demonstrated the possibility of determining single gas components, such as H_2, CH_4 and CO, in air from specific patterns of chemically modified tin-oxide based sensors by using two different multicomponent analysis approaches [30]. Most methods applied to identification, classification and prediction of gas sensor outputs were based on conventional pattern recognition techniques until the late 1980's, when artificial neural networks were applied [31]. Gardner and co-workers implemented a three layer back-propagation network with 12 inputs and 5 outputs architecture for the discrimination of several alcohols, where they reported that it was better than the previous work [32] carried out using analysis of variance (ANOVA). Cluster analysis and principal component analysis (PCA) were used to test 5 alcohols and 6 beverages from 12 tin-oxide sensors. The results were presented by raw and normalised responses, and showed that the theoretically derived data normalisation substantially improved the classification of chemical vapours and beverages. Further investigations were carried out to discriminate the blend and roasting levels of coffees, the differences between tobacco blends in cigarettes, and three different types of beer. The result confirmed the potential application in an electronic instrument for on-line quantitative process control in the food industry. Hines and Gardner developed a stand-alone microprocessor-based instrument which can classify the signals from an array of odour sensitive sensors [33]. Data from the odour sensor array were used to train a neural network and then the neuronal weights were sent to an artificial neural emulator (ANE), which consisted of microprocessor, ADC chips, ROM (Read Only Memory) and RAM (Random Access Memory).

Another approach to odour sensing was studied using a quartz-resonator sensor array where the mechanism of odour detection based on the changes in oscillation frequencies when gas molecules are adsorbed onto sensing membranes. Nakamoto *et al.* employed neural network, including three layer back-propagation and principal component analysis, for the discrimination of several different types of alcoholic drinks using a selection of sensing membranes [34-36].

Persaud and Pelosi [37-40] proposed an odour sensing instrument using conducting polymers after investigating properties of a number of conducting polymers. They have found several organic conducting polymers that respond to gases with a reversible reaction of conductivity, fast recovery and high selectivity towards different compounds. In an experiment with an array of five different conducting polymer sensors and 28 odorants, they observed 20 different sets of responses and showed possible discrimination with 14 of the odorants by measuring changes in the electrical resistance. These results led Persaud *et al.* to produce arrays of 20 gas sensitive polymers that had reversible changes in conductivity and rapid adsorption/desorption kinetics at ambient temperatures when they were exposed to volatile chemicals. The concentration-response profiles of such sensors are almost linear over a wide concentration range to single chemicals. This is advantageous as simple computational methods may be used for information processing. The odour sensing system, developed at UMIST, and commercialised by Aromascan plc is the successful outcome of this research. The array of sensors have been expanded to thirty-two sensor elements, and may be expanded further.

Conducting Polymers

Conducting polymers based on aromatic or heteroaromatic compounds such as polypyrrole and polythiophene [41,42] are sensitive to many odorants and a reversible change in conductance is observed. Conducting polymer sensor arrays are being increasingly reported for use in odour detection and identification [43,44]. Although the understanding of chemical interaction with conducting polymers is still poor, it is believed that reversible characteristics of conformational and/or charge transfer take place between volatile odour chemical and polymer. Several advantages exist over other technologies such as little poisoning effects, rapid reversibility, use at room temperature (little power consumption and no breakdown of volatiles at the sensor surface from increased heating), rapid response absorption/desorption within seconds to most volatile chemicals, and a long sensor lifetime of several years. Polypyrrole is one of the most stable of the organic conducting polymers (OCPs) under ambient conditions and as such has been the focus for most of the research on OCPs. To develop conducting polymers that are useful for sensing volatile chemicals, a good understanding of the physical, chemical and electrical mechanisms is required. The problems of repeatable and reproducible manufacture are immense, and require very strict control of the synthetic conditions. The base unit of polypyrrole is pyrrole, a five-member heteroaromatic ring containing nitrogen Polypyrrole is synthesised from the pyrrole monomer by mild oxidation, using either chemical or electrochemical methods. The polymer formed is usually a black solid, the exact form depending on the nature of preparation, the counterions are usually incorporated in the conducting polymer during oxidation of the monomers. The stoichiometric ratio of monomer units to counterions is generally 3:1 to 4:1, decreasing slightly with molecular weight and negative charge of the anion [45].

The microstructure of the polypyrrole layer can be significantly altered by the introduction of counterions. Ions with more than one charge will tend to associate with the same number of charges on the polymer chain, causing distortion of an anisotropic (ordered) chain as the distance between the anion and the associated chain charges is minimised. As the polymer structure becomes more isotropic, this effect will diminish, as the disordered structure will generally possess more positive charges in the vicinity of the counterion.

In a semiconductor the band gap is narrow, allowing electrons to be energetically excited from one band to another, thereby enabling current to flow. The thermal excitation of an electron results in a 'hole' in the top of the valence band. The resulting positive charge is delocalised over the entire material, no local lattice distortion occurring within the crystalline material. Remaining electrons in the valence band are able to jump to this 'hole', leading to the appearance of metallic character. The addition of impurity (dopant) increases the number of charge carriers, which coupled with the high mobilities of the charge carriers in the crystalline lattice, leads to high conductivity's for doped semiconductors. However, this theory does not sufficiently explain conduction in OCPs [46]. The charge carriers in OCPs appear to be spinless, unlike the charge carriers considered above (i.e., electrons). This has led to the introduction of other theories, which are outlined below.

Conducting Polymer Mechanisms

When an electron is removed from the polymer chain by doping, a radical cation, possessing a spin of ½, is produced which partially delocalises over a small number of monomer units. This phenomenon is termed a polaron, derived from the fact that it polarises its surroundings in order to stabilise itself. This polarisation results in an electronically excited area, raising the local energy level above the valence band and into the band gap [47]. This polarisation does not occur in normal, 'intrinsic' semiconductors. OCPs do not possess the rigid lattice structure of metal and normal semiconductors, having a more amorphous, 'springy' structure, permitting local distortions in the chain.

If another electron is removed from the polymer by doping [48] it can either come from a separate site, in which case a second polaron is created, or the lone electron in the polaron can be removed creating a bipolaron. The formation of this spinless charge carrier would explain the general lack of spin observed in this semiconductor. The polarons are unstable compared to the bipolarons, and so tend to disappear at higher doping levels. This has been confirmed by ESR studies [49]. However, some spin character still remains in the polymer, and is thought to be due to polarons trapped in lattice defects or from thermal dissociation of bipolarons [50].

The actual mechanism of charge transport in polypyrrole can consist of two components; intrachain transport occurring along the polymer chains by rearrangement of the conjugated bond system, and interchain charge transport involving hopping of the charge carriers to neighbouring chains. Considering the chain length of polypyrrole is estimated to be up to 1000 monomer units, and the large degree of conjugation-limiting defects in the polymer chain, it would be expected that interchain conductivity would be the limiting factor in the conductivity.

The gas sensing properties of polypyrrole were first investigated by Nylander [51]. Exposure of a polypyrrole impregnated filter paper to ammonia vapour reversibly altered the resistance of the polymer. The performance of the sensor, operating at room temperature, was linear with higher concentrations (0.5% - 5%), responding within a matter of minutes.

The gas sensing abilities of polypyrrole may be altered in a number of ways. These include polymerisation of derivatised monomers [52] and the dopant ion used. Many types of gas sensitive conducting polymers have now been prepared, and these are typically based on functionalised pyrroles, thiophenes, anilines, indoles and others.

Data Acquisition

The software provided for the volatile chemical sensing system consists of three modules: data acquisition and instrument control; data manipulation for extraction of patterns; and pattern recognition.

The acquisition software samples the sensor array resistance at regular intervals, storing the resultant data in the computer. As the resistance of conducting polymers are inversely proportional to temperature, the temperature of the array is controlled and monitored (typically to 35°C ± 0.1°C). The sample temperature and sample humidity are also monitored, since these parameters are important for reproducible sampling. Individual sensors on the array may be deactivated or activated as required. The responses of the sensors are shown in real time in a strip chart display on screen, as seen in Figure 12.3. The signal is expressed as the percentage resistance change of each sensor compared to the initial sensor resistance.

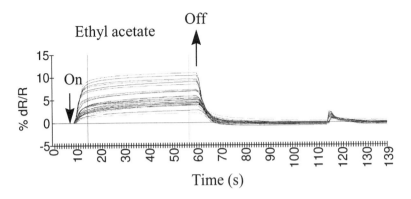

Figure 12.3 Response of a 32 sensor conducting polymer array to ethyl acetate vapour. The % change in resistance (dR/R) in response to the vapour is shown.

The response of a sensor is then normalised by expressing the fractional change of the individual sensor as a percentage of the fractional changes summed over the whole array, as denoted in Equation 12.1 for an array of n sensors:

$$N_x = 100 \cdot \frac{\Delta r_x / r_x}{\sum_{i=1}^{n} \left| \Delta r_i / r_i \right|}$$

(12.1)

where $1 \leq x \leq n$, N_x is the normalised response of sensor x, Δr_x is the resistance change of sensor x, r_x is the base (initial) resistance of sensor x, and $\Delta r_i / r_i$ is the fractional change in resistance of the i'th component of the array. The normalised data thus forms a pattern across the sensor array, as shown in Figure 12.4.

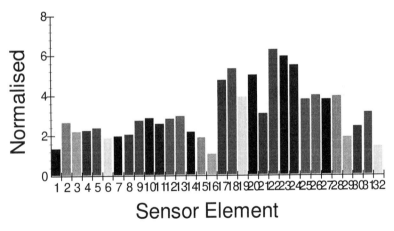

Figure 12.4. Normalised response pattern averaged between the cursors shown in Figure 12.3.

Data Processing

Any biometric application will rely heavily on the data processing and pattern recognition software associated with it. The response to a volatile odour of each sensor is proportional to the concentration, and is unique for each type of single or complex odour. With each sensor in an array having a certain response character, an array of sensors with broad but different chemical specificities provides a measurement pattern of broad overlapping selectivity. These responses, or signals, are processed to produce a set of descriptors for the input, which can be identified as a "fingerprint" for an odour, and then saved into a database for further manipulation within statistical pattern recognition methods, cluster analysis, and artificial neural networks. The input data into these methods is a normalised pattern of responses of each sensor relative to the whole array. For pure solvents the normalised patterns produced are almost concentration-independent for some, but not all of the technologies listed in Table 12.2; however, in general, when monitoring a complex odour or chemical mixture the patterns are non-linearly concentration dependent. This creates difficulties for information processing.

From the previous discussions, it would be seen that it is desirable to discriminate between odours, compare one odour sample with another, and get an estimate of intensity or concentration of the odour. Ideally, it is desirable to map these properties to the human perception using descriptors such as those described in Table 12.1. The use of neural networks within artificial sensory analysis has been growing in momentum in recent years. The ability to recognise pattern characteristics from relatively small pieces of information has led to growing interest in the possible applications and development within sensory recognition. A variety of pattern recognition techniques including neural networks may be applied to the classification of different odours, quantitative prediction and recognition of unknown gases and odours. Backpropagation, used for multilayer perceptron networks, is probably the most widely used neural network paradigm [52]. This algorithm has been used with good results with a wide variety of odour recognition problems. In this case an exemplar set of data is measured experimentally, and patterns for each odour class of interest are collected. The neural network is then used to associate these patterns with outputs that are descriptors. Once trained, such a system is able to generalise, and is able to recognise even noisy patterns. One disadvantage is a difficulty in classifying a previously unknown pattern that is not classified to any prototypes in the training set. We have investigated the characteristics of radial basis function (RBF) networks applied to odour classification problems. RBF networks train rapidly, usually orders of magnitude faster than backpropagation, while exhibiting none of backpropagation's training pathologies such as paralysis or local minima problems. A RBF network [53-57] is a two-layer network where the output units form a linear combination of the basis functions computed by hidden units. The basis functions in the hidden layer produce a localised response to the input and typically uses hidden layer neurones with Gaussian response functions.

The performance of radial basis function classifiers is highly dependent on the choice of basis centres and width. This has been a focus of our attention in order to optimise RBF networks for odour classification. For a minimum number of nodes, the selected centres should well represent the training data for acceptable classification. We have found that one optimum solution is to combine a learning vector quantizer (LVQ) [56,57] algorithm with RBF, the LVQ being used to find suitable cluster centres. The LVQ network is based on competitive learning and is used to both quantize input pattern vectors into reference values and to use these reference values for classification. The distance measurement between one vector and neighbouring vectors is a useful means of deciding whether correct classification is possible.

Performance of the RBF Network

In order to test the classifier based on RBF, the odour sensing system was used to collect representative patterns of pure chemicals as well as a mixture of methanol and ethanol in fixed ratios. Once validated, the network was then applied to a real world, complex odour measurement problem involving characterisation of the ageing process of a foodstuff.

The final output from the RBF network is presented in such a way that the recognition output values are directly displayed as a map instead of class labels and error values. The results of applying the radial basis function are seen in all figures with Gaussian non-linearities using a constant, fixed parameter and centres chosen by tentative classification using the network explained previously from the training patterns. The x-axis in Figures 12.5 and 12.6 depicts the target values (or the desired output); the y-axis represents the sequencing of individual patterns. Any symbol on (or closer to) a target line means the pattern is recognised as the class which holds the target value. Four mixtures of methanol and ethanol in ratio of 1:1, 1:2, 1:4, 1:8 were trained and tested, and the results are shown in Figure 12.5. Figure 12.5(b) also shows the output when untrained input data of mixture 1:4 is tested with a network previously trained with methanol, ethanol and the three previous mixtures. While the prediction does have some error associated with it, the performance of RBF network to previously unseen patterns is very robust.

Further studies were then carried out on real complex odour changes from one type of dry food material (biscuits). Data were collected over a thirteen week period on a particular sample batch of biscuits in order to assess if ageing effects on foods could be measured by the conducting polymer array and identified using radial basis function neural networks. Controls of water, methanol, ethyl acetate and butyl acetate, were used to assess the reliability and stability over this period. Little sensor drift and a high reliability in sensor response were observed. At the start of the experimentation period, a part of the biscuit sample was frozen to minimise any ageing effects and stored at -20^0C, and were used as control samples. The other part of the biscuit sample was kept at room temperature within the original food container. Each week an amount of control sample and aged sample was analysed by the thirty-two sensor conducting polymer array. In Figure 12.6(a) the radial basis function was trained with control and aged samples from weeks 1, 3, 5, 7, 9, 11, and 13. The radial basis function was then tested in Figure 12.6(b) with unknown control and aged data from weeks 2, 4, 8, 10, and 12.

The radial basis function worked very well in the prediction of unknown unaged and aged samples compared to known unaged and aged exemplar samples. In this case the change in odour being detected was due to oxidation of fats in the samples, leading to the onset of rancidity. A trend over the thirteen-week period of ageing is clearly seen. The capability to quantify and discriminate between complex odours has been demonstrated, by the use of the radial basis function network.

The results would demonstrate that by combining array based sensor systems with powerful pattern recognition methods, it is possible to produce odour-sensing instruments that have practical applications in industry. While the particular example illustrates the use of conducting polymer arrays, other technologies may have similar or better performance for a given application. Hence the user would need to evaluate which sensor technology has the best performance for the perceived needs.

5. Conclusions

It will be seen that it is very difficult to create a biomimetic device with the capabilities of the human nose. Improvements in sensor technologies combined with the revolution in microprocessor-based electronics, as well as the parallel evolution of pattern recognition algorithms, mean that practical devices with limited capabilities can now be realised. Difficulties remain however, due to the nature of the sensors that are available. Sensors may drift, or suffer from poisoning. In the biological systems, the sensors have a limited lifetime and are continually replaced. Somehow, the biological system is able to maintain invariant pattern recognition, even though the characteristics of the sensors involved may have changed. The new odour sensing instruments are not yet able to cover the full repertoire of the human nose, in terms of the diversity of odours that can be discriminated. As sensor arrays become larger, and the sensor types become more tuned to specific odour classes, some of these limitations will disappear. Sensitivity is also another problem, the sensors available are not as sensitive as the human nose. This may be overcome in the future by the combination of large numbers of sensors, carrying out signal averaging to increase the signal to noise ratio. For biometrics applications, discrimination between humans on the basis of body odour may be feasible. However this is complicated by the fact that humans often use deodorants, perfumes, and have diets that may affect their body odour. Instruments capable of distinguishing invariant components of human odour are still in the process of development.

Acknowledgements

This work was supported by grants from the EPSRC, BBSRC, and Aromascan UK.

References

1. G. Ohloff, "Chemistry of odor stimuli," *Experientia*, Vol. 42 pp. 271-279, 1986.
2. W. G. Freeman, "The Physiology of Perception," *Scientific American*, Vol. 264, No. 2, pp.78-85, 1991.
3. L. B. Buck, "Information coding in the vertebrate olfactory system," *Annual Review Of Neuroscience*, Vol. 19, pp. 517-544, 1996.

4. K. Bauer, D. Garbe, and H. Surburg, *Common fragrance and flavor materials*, VCH Verlagsgesselschaft, FRG, 1990.

5. M. K. McClintock, "Menstrual synchrony and suppression," *Nature*, Vol. 299, pp. 244-245, 1971.

6. F. Wortmann, "Receptor multispecificity and similarities between the immune system, the sense of smell, and the nervous system from the point of view of clinical allergology," *Ann. Allergy*, Vol. 59, pp. 65-73, 1987.

7. J. Amoore, *Molecular Basis of Odor*, Thomas, Springfield, Illinois, 1970.

8. F. H. H. Valentin and A. A. North, *Odour control - A concise guide*, Warren Springs Laboratory, Stevenage, 1980.

9. M. Hangartner, J. Hartung, M. Paduch, B. F. Pain, and J. H. Voorburg, "Improved recommendations on olfactometric measurements," *Environmental Technology Letters*, Vol. 10, pp. 231, 1989.

10. B. A. Sommerville, J. P. McCormick, and D. M. Broom, "Analysis of human sweat volatiles: an example of pattern recognition in the analysis and interpretation of gas chromatograms," *Pestic. Sci.*, Vol. 41, pp. 365-368, 1994.

11. B. A. Sommerville, M. A. Green, and D. J. Gee, "Using chromatography and a dog to identify some of the compounds in human sweat which are under denetic influence," *Chemical signals in vertebrates* Vol. 2 (Eds. D. W. MacDonald, D. Muller-Schwartze, S. E. Natynczug), Oxford University Press, Oxford, pp. 634-639, 1990.

12. S. Chandiok, B. A. Crawley, B. A. Oppenheim, P. R. Chadwick, S. Higgins, and K. C. Persaud, 1997, "Screening for bacterial vaginosis: a novel application of artificial nose technology," *J. Clin. Path.*, Vol. 50, No. 9, pp. 790-791, 1997.

13. N. N. Tanyolaç and J. R. Eaton, "Study of odors," *J. Am. Pharm. Assoc.*, Vol. 39, No. 10, pp. 565-574, 1950.

14. R. W. Moncrieff, "An instrument for measuring and classifying odors," *J. Appl. Physiology*, Vol. 16, pp. 742-749, 1961.

15. W. F. Wilkens and J. D. Hartman, 1964, "An electronic analogue for the olfactory processes," *Ann. N.Y. Acad. Sci.*, Vol. 116, pp. 608-612, 1964.

16. T. Buck, F. Allen, and M. Dalton, "Detection of chemical species by surface effects on metals and semiconductors", In: Bregman and Dravnieks (eds.), *Surface effects in detection*, Spartan Books Inc., USA, 1965.

17. A. Dravnieks and P. Trotter, "Polar vapour detection based on thermal modulation of contact potentials," *J.Sci., Instrum.*, Vol. 42, pp. 642, 1965.

18. K. C. Persaud and G. Dodd, "Analysis of discrimination mechanisms in the mammalian olfactory system using a model nose," *Nature*, Vol. 299, pp. 352-355, 1982.

19. S. Zaromb and J. Stetter, "Theoretical basis for identification and measurement of air contaminants using an array of sensors having partly overlapping selectivities," *Sensors and Actuators*, Vol. 6, pp. 225-243, 1984.

20. B. Bott and T. Jones, "The use of multisensor systems in monitoring hazardous atmospheres," *Sensors and Actuators*, Vol. 9, pp. 19-25, 1986.

21. J. Stetter, P. Jurs, and S. Rose, "Detection of hazardous gases and vapors : pattern recognition analysis of data from an electrochemical sensor array," *Anal. Chem.*, Vol. 58, pp. 860-866, 1986.

22. T. Otagawa and J. Stetter, "A chemical concentration modulation sensor for selective detection of airborne chemicals," *Sensors and Actuators*, Vol. 11, pp. 251-264, 1987.

23. J. Stetter, "Sensor array and catalytic filament for chemical analysis of vapors and mixtures," *Sensors and Actuators B*, Vol. 1 pp. 43-47, 1990.

24. J. Stetter, "Chemical sensor arrays: Practical insights and examples," In *Sensor and sensory systems for electronic nose*, (Eds. Gardner J. and Bartlett P.), NATO ASI Series E: Applied Sciences, Vol. 212, Springer-Verlag, Berlin, pp. 273-301, 1992.

25. P. Moseley, J. Norris, D. Williams (eds.), *Techniques and mechanisms in gas sensing*, Adam Hilger, Bristol., 1991.

26. M. Egashira, Y. Shimizu, and Y. Takao, "Trimethylamine sensor based on semiconductive metal oxides for detection of fish freshness," *Sensors and Actuators B*, Vol. 1, pp. 108-112, 1990.

27. H. Abe, T. Yoshimura, S. Kanaya, Y. Takahashi, Y. Miyashita, and S. Sasaki, "Automated odor-sensing system based on plural semiconductor gas sensors and computerized pattern recognition techniques," *Analytica Chimica Acta*, Vol. 194, pp. 1-9, 1987.

28. H. Abe, S. Kanaya, Y. Takahashi, and S. Sasaki, "Extended studies of the automated odor-sensing system based on plural semiconductor gas sensors with computerized pattern recognition techniques," *Analytica Chimica Acta*, Vol. 215, pp. 151-168, 1988.

29. H. Shurmer, J. Gardner, and H. Chan, "The application of discrimination techniques to alcohols and tobaccos using tin-oxide sensors," *Sensors and Actuators*, Vol. 18, pp. 361-371, 1989.

30. U. Weimar, K. Schierbaum, and W. Göpel, "Pattern recognition methods for gas mixture analysis: Application to sensor arrays based upon SnO_2," *Sensors and Actuators B*, Vol. 1, pp. 93-96, 1990.

31. R. Sleight, *Evolutionary strategies and learning for neural networks*, MSc. Dissertation, UMIST, 1990.

32. J. Gardner, E. Hines, and M. Wilkinson, "Application of artificial neural networks to an electronic olfactory system," *Meas. Sci. Technol.*, Vol. 1, pp. 446-451, 1990.

33. E. Hines and J. Gardner, "An artificial emulator for an odour sensor array," *Sensors and Actuators B*, Vols. 18-19, pp. 661-664, 1994.

34. K. Ema, M. Yokoyama, T. Nakamoto, and T. Moriizumi, "Odour-sensing system using a quartz-resonator sensor array and neural-network pattern recognition," *Sensors and Actuators*, Vol. 18, pp. 291-296, 1989.

35. T. Nakamoto, K. Fukunishi, and T. Moriizumi, "Identification capability of odor sensor using quartz-resonator arrays and neural-network pattern recognition," *Sensors and Actuators B*, Vol. 1, pp. 473-476, 1990.

36. T. Nakamoto, A. Fukuda, and T. Moriizumi, "Improvement of identification capability in an odor-sensing system," *Sensors and Actuators B*, Vol. 3, pp. 221-226, 1991.

37. P. Pelosi and K. C. Persaud, "Gas sensors : Towards an artificial nose," In *Sensors and sensory systems for advanced robots,* (Eds. Dario P. et al.), NATO ASI Series F: Computer and System Science, Springer-Verlag, Berlin, pp. 361-382, 1988.

38. K. C. Persaud, J. Bartlett, and P. Pelosi, "Design strategies for gas odour sensors which mimic the olfactory system," In *Robots and biological systems*, (Eds. Dario P. et al.), NATO ASI Series, Springer-Verlag, Berlin, 1990.

39. K. C. Persaud and P. Travers, "Multielement arrays for sensing volatile chemicals," *Intelligent Instruments and Computers*, pp. 147-153, July-August, 1991.

40. K. C. Persaud and P. Pelosi, "Sensor arrays using conducting polymers for an artificial nose," In *Sensors and sensory systems for an electronic nose*, (Eds. Gardner J. and Bartlett P), NATO ASI Series E: Applied Sciences, Vol. 212, Springer-Verlag, Berlin, pp. 237-256, 1992.

41. A. F. Diaz, K. K. Kanazawa, and G. P. Gardini, "Electrochemical polymerization of pyrrole," *J. Chem. Soc. Chem. Comms.*, pp. 635-636, 1979.

42. J. Roncali, "Conjugated poly(thiophenes): synthesis, functionalization, and applications," Chem. Rev., Vol. 92, pp. 711-738, 1992.

43. J. J. Miasik, A. Hooper, and B. C. Tofield, "Conducting polymer gas sensors," *J. Chem. Soc. Faraday Trans.* 1, Vol. 82, pp. 1117-1126, 1986.

44. R. Cabala, V. Meister, and K. Potje-Kamloth, 1997, "Effect of competitive doping on sensing properties of polypyrrole," *J. Chem. Soc. Faraday Trans.*, Vol. 93, 131-137, 1997.

45. G. Zott, G. Schiavon, and N. Comisso, "On effects on conductivity of isomorphous polypyrrole: charge pinning by nucleophilic anions," *Syn. Metals*, Vol. 40, pp. 309-316, 1991.

46. D. Bloor and B. Movaghar, 1983, "Conducting polymers," *IEE Proc. I*, Vol. 130, pp. 225-232, 1983.

47. G. Zotti, G. Schiavon, and N. Comisso,1990, "The charge-potential relationship in polyconjugated conducting polymers: Determination of E0 values and n-values for polypyrrole and polythiophene," *Electrochim. Acta*, Vol. 35, pp. 1815-1819, 1990.

48. H. S. Nalwa, "Phase-transitions in polypyrrole and polythiophene conducting plymers demonstrated by magnetic susceptibility measurements," *Phys. Rev. B - Condensed Matter*, Vol. 39, pp. 5964-5974, 1989.

49. J. L. Brédas, B. Thémans, J. G. Fripiat, J. M. André, and R. R. Chance, "Highly conducting polyparaphenylene, polypyrrole, and polythiophene chains: An ab initio study of the geometry and electronic-structure modifications upon doping," *Phys. Rev. B - Condensed Matter*, Vol. 29, pp. 6761-6773, 1984.

50. G. B. Street, S. E. Lindsey, A. I. Nazzal, and K. J. Wynne, "The structure and mechanical properties of polypyrrole," *Mol. Cryst. Liq. Cryst.*, 118: 137-148, 1985.

51. C. Nylander, M. Armgarth, and I. Lundström, In *Proceedings of the international meeting on chemical sensors, Fukuoka, 1983*. (Eds. Seiyama T, Fueki K, Shiokawa J, Suzuki S) Elsevier, Amsterdam, 1983.

52. J. Gardner and P. Bartlett, 1991, "Pattern recognition in gas sensing," In *Techniques and Mechanisms in Gas Sensing*, (Eds. Moseley P. et al.), Adam Hilger, Bristol, pp. 347-384, 1991.

53. J. Moody and C. Darken, "Fast learning in networks of locally-tuned processing units," *Neural Computation*, Vol. 1, pp. 281-294, 1989.

54. M. Musavi, W. Ahmed, K. Chan, K. Faris, D. Hummels, On the training of radial basis function classifiers, Neural Networks, Vol. 5, pp. 595-603, 1992.

55. T. Kohonen, *Self-Organization and Associative Memory*, Springer-Verlag, 3rd ed., pp. 199-202, 1989.

56. K. C. Persaud and H-G. Byun, 1998, "Classification of complex odours using conducting polymer arrays and neural networks," In *Industrial applications of neural networks*, Eds. Fogelman Soulié, World Scientific, Singapore, New Jersey, pp. 85-90, 1998.

57. D-H. Lee, J. S. Payne, and H-G. Byun, and K. C. Persaud, 1996, "Application of radial basis neural networks to odour sensing using a broad specificity array of conducting polymers," In *Lecture Notes in computer science* (Eds. C. Von der Malsburg, W. von Seelen, J. C. Vorbroggen, B. Sendhoff), Vol. 1112, pp. 299-304, 1996.

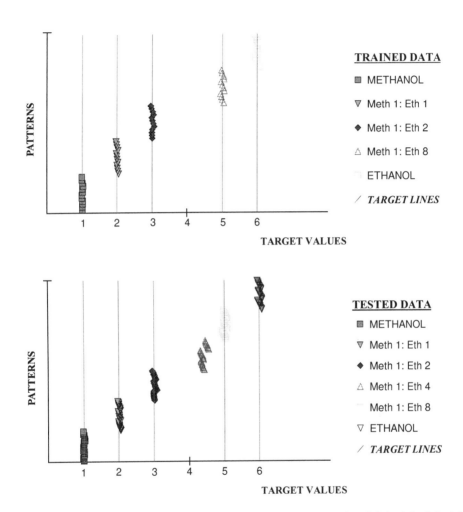

Figure 12.5 (a) Four mixtures of methanol and ethanol in ratio of 1:1, 1:2, 1:4, 1:8 were trained and tested using a radial basis function neural network.(b) shows the output when previously unseen input data of mixture 1:4 is tested with a network previously trained with methanol, ethanol and the three previous mixtures. The x-axis in figures depicts the target values (or the desired output); the y-axis represents the sequencing of individual patterns, offset for clarity.

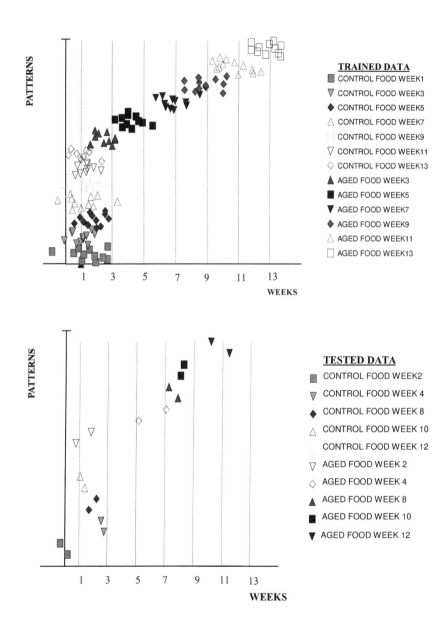

Figure 12.6 (a) The radial basis function was trained with control and aged samples of biscuits from weeks 1, 3, 5, 7, 9, 11, and 13. The radial basis function was then tested (b) with unknown control and aged data from weeks 2, 4, 8, 10, and 12.

13 EAR BIOMETRICS

Mark Burge and Wilhelm Burger
Johannes Kepler University
Linz, Austria
{burge,burger}@cast.uni-linz.ac.at

Abstract *A new class of biometrics based upon ear features is introduced for use in the development of passive identification systems. The availability of the proposed biometric is shown both theoretically in terms of the uniqueness and measurability over time of the ear, and in practice through the implementation of a computer vision based system. Each subject's ear is modeled as an adjacency graph built from the Voronoi diagram of its curve segments. We introduce a novel graph matching based algorithm for authentication which takes into account the erroneous curve segments which can occur due to changes (e.g., lighting, shadowing, and occlusion) in the ear image. This new class of biometrics is ideal for passive identification because the features are robust and can be reliably extracted from a distance.*

Keywords: *Ear biometrics, passive identification, structural image analysis, generalized Voronoi diagrams, error correcting graph matching.*

1. Introduction

Try a simple experiment, try to visualize what your ears look like. You were not able to? Well, then try to describe the ears of someone you see everyday. You will find that even if you are looking directly at someone's ears, they are still difficult to describe. We simply do not have the vocabulary for it; our everyday language provides only a few adjectives which can be applied to ears, all of which are generic adjectives like *large* or *floppy* and not ones which are solely[1] used to describe ears. On the other hand, we are all capable of describing the faces of even briefly glimpsed strangers with significant detail to allow police artists to reconstruct remarkable resemblances of them. Even though we apparently lack the means to recognize one another from our ears, we will see that the rich structure of the ear is unique and that it can be used as an effective biometric for passive identification.

---

[1] In fact, in English at least two ear specific adjectives do exist, namely "Vulcan" and "Charlesesque".

2. Automated Biometrics

Automating identification through biometrics [15], especially *face recognition* [8], has been extensively studied in machine vision. Despite extensive research, many problems in face recognition remain largely unsolved due to the inherent difficulty of extracting face biometrics. A wide variety of imaging problems (e.g., lighting, shadows, scale, and translation) plague the attempt for unconstrained face identification [13]. In addition to the many imaging problems, it is inherently difficult to collect consistent features from the face as it is arguably the most changing part of the body due to facial expressions, cosmetics, facial hair and hair styling. The combination of the typical imaging problems of feature extraction in an unconstrained environment, and the changeability of the face, explains the difficulty of automating face biometrics. Despite the attractiveness of face biometrics (e.g., they are easily verifiable by non-experts), other biometrics (e.g., fingerprint based) provide the basis for most commercial implementations.

Unlike facial biometrics, *fingerprint-based biometrics* have been shown to be highly amenable to automation by machine vision techniques [6]. The automation of fingerprint biometrics began in 1971 [14] and has culminated in a number of commercial machine vision based systems. Fingerprint imaging is done within a controlled environment, usually with a specially designed scanner, which eliminates the problem of localization and artifacts from shadowing and lighting variations. Physical changes, a bane of facial biometrics, is a miniscule problem as the finger, barring surgery, remains comparatively constant over time.   Machine vision techniques [11] have been successfully applied to create highly accurate and robust commercial systems which are in use worldwide.

3. Passive Biometrics

Fingerprints are not the only successful example of the application of machine vision techniques to automated biometrics; both the three dimensional *shape of the hand* and *retinal patterns* have also been used. All of the biometrics which have been successfully automated using machine vision techniques are inherently *invasive*. They require the subject to participate actively in both enrolling into the system and during subsequent identification. The willing participation of the subject in the controlled environment of these systems is intrinsic to the success of the identification.

One class of *passive* physiological biometrics are those based upon *iris scans*. Unlike retinal scans, which require close contact with the scanner, iris-based recognition has been reported to be successful at distances of up to 46 cm [17]. The unique collection of striations, pits, and other observable features of the iris along with the ease of segmenting the iris from the white tissue of the eye which serves as its background, make iris based biometrics attractive. The decided disadvantage is the small size of the iris which makes image acquisition from a distance, and therefore passive usage, problematic.

To summarize the two classes of passive physiological biometrics which have been researched in machine vision up to now: face and iris based techniques both have a number of drawbacks which make their usage in commercial applications limited. Facial biometrics fail due to the changes in features caused by expressions, cosmetics, hair styles, and the growth of facial hair as well as the difficulty of reliably extracting them in an unconstrained environment exhibiting imaging problems such as lighting and shadowing. Unlike facial biometrics, iris biometrics remain relatively consistent over time and are easy to extract, but acquisition of the image at the necessary resolution from a distance is difficult. Therefore, we propose a new class of biometrics for passive identification based upon ears which have both reliable and robust features which are extractable from a distance.

4. Ear Biometrics

In proposing the ear as the basis for a new class of biometrics, we need to show that it is viable (i.e., unique to each individual, and comparable over time). In the same way that no one can prove that fingerprints are unique, we can not show that each of us has a unique pair of ears. Instead, we will assert that this is probable and give supporting evidence by examining two studies from Iannarelli [10]. The first study compared over 10,000 ears drawn from a randomly selected sample in California, and the second study examined fraternal and identical twins, in which physiological features are known to be similar. The evidence from these studies supports the hypothesis that the ear contains unique physiological features, since in both studies all examined ears were found to be unique though identical twins were found to have similar, but not identical, ear structures especially in the Concha and lobe areas. Having shown uniqueness, it remains to ascertain if the ear provides biometrics which are comparable over time.

It is obvious that the *structure* of the ear does not change radically over time. The medical literature reports [10] that ear growth after the first four months of birth is proportional. It turns out that even though ear growth is proportional, gravity can cause the ear to undergo stretching in the vertical direction. The effect of this stretching is most pronounced in the lobe of the ear, and measurements show that the change is non-linear. The rate of stretching is approximately five times greater than normal during the period from four months to the age of eight, after which it is constant until around 70 when it again increases.

We have shown that biometrics based upon the ear are viable in that the ear anatomy is probably unique to each individual and that features based upon measurements of that anatomy are comparable over time. Given that they are viable, identification by ear biometrics is promising because it is passive like face recognition, but instead of the difficult to extract face biometrics, robust and simply extracted biometrics like those in fingerprints can be used.

5. Application Scenarios

Ear biometrics can be used as a supplementary source of evidence in identification and recognition systems, for example a system designed for face recognition already includes all the necessary hardware for capturing and computing ear biometrics. A typical example is supplementary identification at an automated teller machine (ATM). Most ATMs contain a camera mounted so that it records the face of the user during the transaction. These cameras could be supplemented with a simple optical mirror setup to allow simultaneous recording of the face and the ear of the user. While a user identifies himself to the ATM by inserting the bank card and keying in his personal identification number (PIN), the camera simultaneously records the face and ear of the user and uses ear biometrics to supplementary verify the identification of the user.

A second scenario, where passive ear biometrics are ideal, is where different levels of security are assigned to different groups of people. Typically, this situation is solved through the use of active badges. These badges, usually small identification cards containing passive transmitters, are automatically queried at each point of access (e.g., a door or elevator). Access to restricted areas (e.g., unlocking of the door) is automatically allowed or disallowed according to the identity supplied by the badge. A well known drawback of such a system is that someone can illegitimately obtain and use the active badge of another to gain access to restricted areas. For this reason, cameras are often installed to monitor the access points. The cameras record access and in the case that illegitimate access to an area is later discovered, the video tape can be examined to visually determine the perpetrator - but only *after* the fact. A prohibitively expensive solution, which defeats the purpose of automation, is to have a human analyze each request for access by comparing the identification reported by the active badge and the image provided by the camera to some stored image of the subject. Ear biometrics can be used in such a scenario to allow more secure automated access.

In the proposed solution, as the subject approaches the access point, the active badge is queried and the identification is ascertained. At the same time, an image of the subject's ear is acquired and biometrics are used to verify the identification provided by the active badge. In case the two identifications do not match, the video of the access point including the subject is displayed at the central security station along with an image retrieved from the database of the subject who should be in possession of that badge. The two images can then be visually compared by the security personnel and appropriate action taken.

6. Iannarelli's Ear Biometrics

An anthropometric technique of identification based upon ear biometrics was developed by Iannarelli [10]. The "Iannarelli System" is based upon the 12 measurements illustrated in Figure 13.1(b). The locations shown are measured from specially aligned and normalized photographs of the right ear. To normalize and align the images, they are projected onto a standard "Iannarelli Inscribed" enlarging easel which is moved horizontally and vertically until the ear image projects into a prescribed space on the easel. The system requires the exact alignment and normalization of the ear photos as is explained by Iannarelli:

> Once the ear is focused and the image is contained within the easel boundaries, adjust the easel carefully until the *oblique guide line* is parallel to the outer extreme tip of the tragus flesh line.... The oblique line should now be barely touching the tip of the tragus. (The left

white lines in Figure 13.1(a)) Move the easel slightly, keeping the oblique line touching the tip of the tragus, until the upper section of the oblique guide line intersects the point of the ear image where the start of the inner helix rim overlaps the upper concha flesh line area just below the slight depression or hollow called the triangular fossa....

When the ear image is accurately aligned using the oblique guide line, the ear image has been properly positioned. The technician must now focus the ear image to its proper size. The short vertical guide line (The right white line in Figure 13.1(a)) on the easel is used to enlarge or reduce the ear image to its proper size for comparison and classification purposes. [10, pp. 83-84]

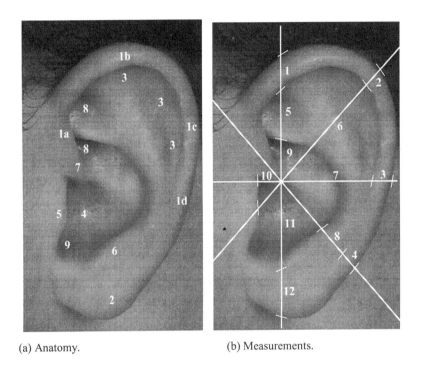

(a) Anatomy.   (b) Measurements.

Figure 13.1 Ear Biometrics: (a) 1 Helix Rim, 2 Lobule, 3 Antihelix, 4 Concha, 5 Tragus, 6 Antitragus, 7 Crus of Helix, 8 Triangular Fossa, 9 Incisure Intertragica. (b) The locations of the anthropometric measurements used in the "Iannarelli System".

Since each ear is aligned and scaled during development, the resulting photographs are normalized, enabling the extraction of comparable measurements directly from the photographs. The distance between each of the numbered areas in Figure 13.1(a) is measured in units of 3 mm and assigned an integer distance value. These twelve measurements, along with information on sex and race, are then used for identification. The system as stated provides for too small of a classification space as within each sex and race category a subject is classified into a single point in a 12 dimensional integer space, where each unit on an axis represents a 3 mm measurement difference. Assuming an average standard deviation in the population of four units (i.e., 12 mm), the 12 measurements provide for a space with *less* than 17 million distinct points.

Though simple remedies (e.g., the addition of more measurements or using a smaller metric) for increasing the size of the space are obvious, the method is additionally not suited for machine vision because of the difficulty of localizing the anatomical point which serves as the origin of the measurement system. All measurements are relative to this origin which, if not exactly localized, results in all subsequent measurements being incorrect. In fact, Iannarelli himself was aware of this weakness as he states on page 83, "This is the first step in aligning the ear image.... and it must be accurate or the entire classification of the ear will be inaccurate".

In the next section we present a proof of concept implementation which avoids the problem of localizing anatomical points and the frailty of basing all subsequent feature measurements on a single such point.

7. Automating Ear Biometrics

The goal in *identification* is to verify that the biometric extracted from the subject sufficiently matches the previously acquired biometric for that subject. Let s' be the subject at the time of identification and s the subject at the time of enrollment, further let $G_s = f(s)$ represent a function which extracts some biometric from a subject s as a graph G_s, and let $d(G_s, G_{s'})$ compute some previously defined distance metric between these two graphs. Identification is then the task of determining if $d(G_s, G_{s'}) <$ t, where t is a given acceptance threshold.

Since the subject and the environment change over time, a certain tolerance in the matching criterion must be permitted. This tolerance can be defined in terms of the *false reject rate* (FRR) and the *false acceptance rate* (FAR) exhibited by the system. A system is usually designed to be tunable to minimize either the FAR or the FRR (i.e., in the given formulation by lowering or raising t, respectively) depending upon the type of security which is required.

The problem of *recognition* is harder than that of identification since the system must determine if the subject's identity can be verified against *any* previously enrolled subject. If the system's enrolled identities are the set $I = \{G_0, G_1, ..., G_n\}$ then recognizing some subject s' is equivalent to finding the member of the set $\{G_i | G_i \in I \wedge d(G_{s'}, G_i) < t\}$ with the smallest distance. We have developed a machine vision system as a proof of concept of the viability of ear biometrics for passive identification. The system implements $f(s')$ using the following steps:

1. *Acquisition*: A 300 by 500 grayscale image is taken of the subject's head in profile using a CCD camera. Next the location of the ear in the image must be found. Fortunately, a number of techniques from face localization are applicable. Two particularly promising methods for still images are the application of Iconic Filter Banks [16] and Fischerface [1]. When sequences of color images are available then color and motion based segmentation [18] can be used to locate the subject before applying ear localization. Since our goal was to construct a proof

of concept system, we used a relatively simple method based on deformable contours.

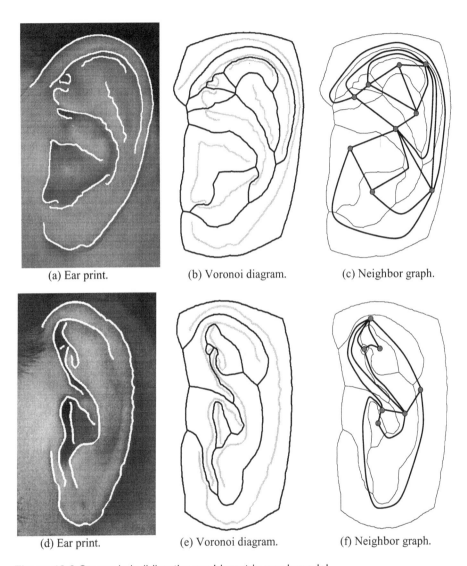

(a) Ear print.   (b) Voronoi diagram.   (c) Neighbor graph.

(d) Ear print.   (e) Voronoi diagram.   (f) Neighbor graph.

Figure 13.2 Stages in building the ear biometric graph model.

2. *Localization*: The ear is located by using deformable contours [12] on a Gaussian pyramid representation of the image gradient.

3. *Edge extraction*: Edges are computed using the Canny [7] operator (i.e., $\sigma = 3.0$) and thresholding with hysteresis using upper and lower thresholds of 46 and 20 (Figure 13.2(b)).

4. *Curve extraction*: Edge relaxation [9] is used to form larger curve segments, after
    which the remaining small curve segments (i.e., length less than 10) are removed
    as is shown in Figure 13.3(b). We could attempt to perform identification at this
    stage by trying to match features computed from the extracted curves to those
    computed from the model. Differences in lighting and positioning would render
    such a method very unreliable. What is needed is a description of the relations
    between the curves in a way which is first invariant to affine transformations and
    secondly invariant to small changes in the shape of the curves resulting from
    differences in illumination. To achieve invariance under affine transformations,
    we turn to the neighboring relation, and construct a Voronoi neighborhood graph
    of the curves and use it as our model.

5. *Graph model*: A generalized Voronoi diagram [4] of the curves is built and a
    neighborhood graph is extracted (Figure 13.2(c)).

Using the above steps results in a high FRR due to variations in the graph models
due to underlying differences in the spatial relations of the extracted curves [3]. To
improve the FRR rate, we first eliminate some of the erroneous curves and then
develop a new matching process which takes into account broken curves.

Error Correcting Graph Matching

Let $G(V,E)$ denote the graph model with each vertex $v \in V$ containing unary features
of a curve and edges $e \in E$ containing binary features between two neighboring
curves. Matching is done by searching for subgraph isomorphisms between the
subject's stored graph G_s and the extracted graph $G_{s'}$. If the distance $d(G_{s'}, G_s)$ between
them is less than the established acceptance threshold t then identification is verified.

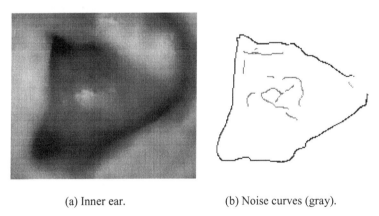

(a) Inner ear.   (b) Noise curves (gray).

Figure 13.3 Removal of noise curves in the inner ear.

In the case where $G_{s'}$ and G_s belong to the same subject, erroneous curves can rise
from differences in lighting and orientation. From our analysis [4], most of these
false curves occur within the inner cavity (Figure 13.3(a)) of the ear. The main
reasons are that areas of high specularity arising from oil and wax buildup and
shadowing caused by the Tragus and Antitragus (Figure 13.1 (a) parts 5 and 6) create

edges in Step 3 which are built into false curves in Step 4. These false curves (Figure 13.3(b)) are removed by first segmenting the inner cavity and then removing small, high curvature, and closed curves occuring within it.

This removes many of the false curves while preserving those arising from ear structures. Unfortunately, due to imaging problems, many of the remaining curves may be broken even after Step 4. To compensate for this, we have developed Algorithm 1 for computing subgraph isomorphisms between G_s and $G_{s'}$ which considers the possibility of broken curves in $G_{s'}$. The idea is to merge neighboring curves in $G_{s'}$ if their Voronoi regions indicate that they are possibly part of the same underlying feature.

Algorithm 1 Calculate $d(G_s, G_{s'})$

1	**While** $d(G_s, G_{s'}) \cdot c < t$ and $|V| \le |V'|$ **do**
2	**for all** $v \in V'$ **do**
3	**for all** a adjacent to v **do**
4	**if** $d_v(v,a) < \gamma$ **then** {see Equation 13.1}
5	contract (v,a)
6	**end if**
7	**end for**
8	**end for**
9	increase t and decrease c
10	**end while**

Let the boundary of the Voronoi region of a curve c_j be represented by ∂Vc_j such that $\partial Vc_j = \{p \,|\, d(p,c_j) = d(p,c_k), j \ne k\}$, where $d(p,c_j)$ is the distance $\min_{q \in c_j} d(p,q)$ between a point p and any point on the curve c_j. Then the adjacent vertices v and a are contracted (i.e., all incident edges of a are added to v and self-loops removed) when

$$d_v(v,a) = \frac{|\partial V(c_a) \cap \partial V(c_v)|}{|\partial V(c_a)| + |\partial V(c_v)|} \tag{13.1}$$

is less than some threshold, γ, the contraction threshold ($\cap$ denotes set intersection operation and $|\,.\,|$ is set cardinality operator). We continue in this way to change the topology of $G_{s'}$ until either we have a match or the number of vertices in $G_{s'}$ is less then that in G_s and since curves may be erroneously merged, we decrease our confidence in the match each time by a factor c.

Decreasing the FAR

As only the topological relations between the extracted curve segments are used during the matching process, there is the possibility of a false acceptance since there exists a set of ears having the same topology. By measuring physical features of the ear curves, we can significantly decrease the FAR. We have found that measurements based on the length of the ear curves are not reliable since small changes occur due to lighting. More reliable is the width of an ear curve, in particular we have found that the width of the curve corresponding to the upper Helix rim (Figure 13.3) can be reliably extracted and normalized against the height of the ear (i.e., the distance from the top of the upper Helix rim to the lowest point on the Lobule as shown in Figure 13.1(a)) as found during the Localization step.

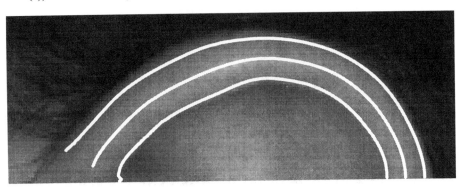

Figure 13.4 Improving the FRR with ear curve widths.

Thermograms and Occlusion by Hair

The main drawback of ear biometrics is that they are not usable when the ear of the subject is covered. In the case of active identification systems, this is not a drawback as the subject can pull his hair back and proceed with the authentication process. The problem arises during passive identification as in this case no assistance on the part of the subject can be assumed.

In the case of the ear being only partially occluded by hair, it is possible to recognize the hair and segment it out of the image. This can be done using texture and color segmentation, or as we have implemented it, using thermogram images. A thermogram image is one in which the surface heat (i.e., infrared light) of the subject is used to form an image. Figure 13.5 is a thermogram of the external ear. The subject's hair in this case has an ambient temperature between 27.2 and 29.7 degrees Celsius, while the pinna (i.e., the external anatomy of the ear) ranges from 30.0 to 37.2 degrees Celsius. Removing partially occluding hair is done by segmenting out the low temperature areas which lie within the pinna.

The Meatus (i.e., the passage leading into the inner ear) of the ear is easily localizable using the thermogram imagery. In a profile image of a subject, if the ear is visible, then the Meatus will be the hottest part of the image, with an expected 8 degree Celsius temperature differential between it and the surrounding hair. In Figure

13.5, the Meatus is the clearly visible section in the temperature range of 34.8 to 37.2 degrees Celsius. By searching for this high temperature area, it is possible to detect and localize ears using thermograms.

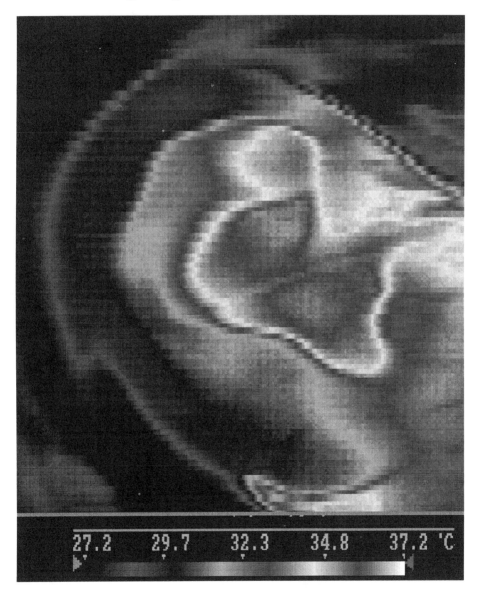

Figure 13.5 Thermogram of an ear. Image provided by Brent Griffith, Infrared Thermography Laboratory, Lawrence Berkeley National Laboratory.

8. Conclusions

The proof of concept system discussed lends support to the theoretical evidence that ear biometrics are a viable and promising new passive approach to automated human identification. They are especially useful when used to supplement [4] existing automated methods. Though ear biometrics appear promising, additional research needs to be conducted to answer important questions like:

- *Feature or appearance based*: Can primitives (e.g., curves) be extracted under varying imaging conditions with sufficient reliability for a feature based approach or will appearance based approaches (e.g., Fisherimages) be necessary?

- *Occlusion by hair*: In the case of the ear being completely occluded by hair, there is no possibility of identification using ear biometrics. It remains to be seen with what degree of partial occlusion identification is possible and if thermogram imagery can resolve this problem.

In conclusion, we have shown that ear biometrics can be used for passive identification. For the further development, testing, and comparison of ear biometric algorithms, the creation of a database of ear images and a set of standardized tests must be the next step.

References

[1] P. Belhumeur, J. Hespanha, and D. Kriegman, "Eigenfaces vs. fisherfaces: Recognition using class-specific linear projection," *IEEE Trans. Pattern Analysis and Machine Intelligence*, Vol. 19, No. 7, pp. 711-720, 1997.

[2] M. Burge, "The Representation and Analysis of Document Images," *volume 100 of Computer Vision and Graphics Dissertations*. Austrian Computer Society Vienna, 1998.

[3] M. Burge and W. Burger, "Ear biometrics for machine vision," In *21st Workshop of the Austrian Association for Pattern Recognition*, Hallstatt. ÖAGM, R. Oldenbourg Verlag, 1997.

[4] M. Burge, and W. Burger, "Identification using ear biometrics," In *22nd Workshop of the Austrian Association for Pattern Recognition*, pp. 195-204. ÖAGM, R. Oldenbourg Verlag, 1998.

[5] M. Burge, and W. Burger, "Using ear biometrics for passive identification," In *14th International Information Security Conference*. Kluwer Academic. To appear.

[6] G. T. Candela, and R. Chellappa, "Comparative performance of classification methods for fingerprints," In *NISTIR*, 1993.

[7] J. Canny, "A computational approach to edge detection," *IEEE Trans. Pattern Analysis and Machine Intelligence*, Vol. 8, No. 6, pp. 679-698, 1986.

[8] R. Chellappa, C. L. Wilson, and S. Sirohey, "Human and machine recognition of faces: A survey," *Proceedings of the IEEE*, Vol. 83, No. 5, pp. 705-740, 1995.

[9] E. R. Hancock, and J. V. Kittler, "Edge-labeling using dictionary-based relaxation," *IEEE Trans. Pattern Analysis and Machine Intelligence*, Vol. 12, No. 2, pp. 165-181, 1990.

[10] A. Iannarelli, *Ear Identification*. Forensic Identification Series. Paramont Publishing Company, Fremont, California, 1989.

[11] K. Karu and A. K. "Fingerprint classification," *Pattern Recognition*, Vol. 29, No. 3, pp. 389-404, 1996.

[12] K. F. Lai, and R. T. Chin, "Deformable contours: Modeling and extraction," *IEEE Trans. Pattern Analysis and Machine Intelligence*, Vol. 17, No. 11, pp. 1084-1090, 1995.

[13] S. Lawrence, C. L. Giles, A. C. Tsoi, and A. D. Back, "Face recognition: A convolutional neural network approach," *IEEE Transactions on Neural Networks*, Vol. 8, No. 1, pp. 98-113, 1997

[14] B. Miller, "Vital signs of identity," *IEEE Spectrum*, Vol. 83, pp. 22-30, 1994.

[15] W. C. Shen, M. Surette, and R. Khanna, "Evaluation of automated biometrics-based identification and verification systems," *Proceedings of the IEEE*, Vol. 85, No. 9, pp. 1464-1478, 1997.

[16] B. Takacs, and H. Wechsler, "Detection of faces and facial landmarks using iconic filter banks," *Pattern Recognition*, Vol. 30, No. 10, pp. 1623-1636, 1997.

[17] R. P. Wildes, "Iris recognition: An emerging biometric technology," *Proceedings of the IEEE*, Vol. 85, 9, pp. 1348-1363, 1997.

[18] C. Wren, A. Azarbayejani, T. Darrell, and A. Pentland, "Pfinder: Real-time tracking of the human body," *IEEE Trans. Pattern Analysis and Machine Intelligence*, Vol. 19, No. 7, pp. 780-785, 1997.

14 DNA BASED IDENTIFICATION

Norah Rudin
Forensic DNA Consulting
Richmond, CA
nrbiocom@uclink4.berkeley.edu

Keith Inman
Forensic Scientist
Berkeley, CA
kinman@ix.netcom.com

Gustavo Stolovitzky, Isidore Rigoutsos
IBM T. J Watson Research Center
Yorktown Heights, NY
{gustavo,rigoutso}@us.ibm.com

Abstract *No single biometric technique provides an optimal identification method in all cases. However, depending on the biological system, a particular biometric may stand out as the preferred method of identification. DNA profiling clearly emerges as the most powerful and reliable system for measuring genetic traits. DNA typing provides valuable information in such diverse applications as medical science, environmental science, historical research, and, of course, forensic science. In this chapter, we review the most common genetic typing systems and the laboratory techniques employed to analyze the markers. We discuss the precautions that must be taken in collecting samples and the consequences of analyzing non-optimal material. Of particular interest from a societal standpoint is the creation of DNA databanks and the privacy issues associated with them. Finally, when two samples appear to be indistinguishable by the tests conducted, the significance of the association must then be* *determined. The methods by which the rarity of a genetic profile is determined provide the DNA typing community with it's most fertile ground for debate. We review the major elements of the discussion here. With the advent of solid state and automated methods for DNA typing, the technique will soon become not only indispensable, but practical.*

Keywords: *DNA, DNA typing, DNA testing, DNA profiling, genetic typing, genetic testing, genetic profiling.*

1. Introduction

No single biometric technique provides an optimal identification method in all cases. However, depending on the biological system, a particular biometric may stand out as the preferred method of identification. **DNA profiling** clearly emerges as the most powerful and reliable system for measuring genetic traits [1].

That genetic traits, displayed at their most fundamental level in **DNA (deoxyribonucleic acid),** provide an excellent tool for both species and individual identification is not a coincidence. All living organisms result from complex developmental interactions between their original genetic blueprints and both the cellular and external environment. Although different cells may express only various portions of genetic information, the DNA in each cell (excluding gametes) contains all the genetic information inherent in the organism. Additionally, when a cell's DNA complement is copied during **mitosis** (cell division), the cellular machinery performs the process with incredible fidelity. These attributes result in one of the greatest advantages to DNA analysis, the fact that the DNA profile throughout an organism is homogeneous.

But to understand how DNA analysis is helpful in differentiating between individual organisms, it is the variability that we must explore. Because the normal physical and biological specifications of humans leave little room for modification, all humans share about 99.5% to 99.9% of their DNA; only about 0.1% to 0.5% may vary. However, even that small percentage of the **human genome** contains millions of **base pairs**, the building blocks of DNA. Because those regions are rarely responsible for specifying physical functions, evolutionary mutations may accumulate without detriment to the organism or the species. This provides the genetic basis for variability in DNA that is so useful in individual identification.

As with any technique, **DNA typing** has its advantages and disadvantages. Compared to some of the other biometrics discussed in this book, it is slow, labor intensive and difficult to automate. However, when the fingerprints, eyeballs, and even teeth are lost or destroyed, and all that is left is a few cells of biological material, DNA techniques still hold the power to identify that person.

2. A Brief History of DNA Identification

In 1944 Oswald Avery defined the role of the cellular component known as DNA as the vehicle of generational transference of heritable traits. In 1953, James Watson and Francis Crick elucidated the structure of the DNA molecule as a **double helix**. In science, as in art, form follows function; the very nature of the molecule provided an explanation for its unique properties, including the ability to propagate itself faithfully from generation to generation.

> "...Our model for deoxyribonucleic acid is, in effect, a pair of templates, each of which is complementary to the other. We imagine that prior to duplication the hydrogen bonds are broken, and the two chains unwind and separate. Each chain then acts as the template for the formation onto itself of a new companion chain, so that eventually we shall have two pairs of chains, where we only had one before. Moreover, the sequences of the pairs of bases will have been duplicated exactly." (Watson and Crick)[2]

In 1980, David Botstein and coworkers were the first to exploit the small variations found between people at the genetic level as genetic landmarks to construct a human gene map. The particular type of variation they used is called **Restriction Fragment Length Polymorphism** or **RFLP**. In 1984, while searching for disease markers in DNA, Jeffreys et al. [3] discovered a unique application of RFLP technology to the science of personal identification. His method, which he termed a "DNA fingerprint", has been modified since its original inception. Scientists generally agree that a more descriptive and inclusive term for the process as currently applied **is DNA typing or DNA profiling.** In 1986, the **polymerase chain reaction** (**PCR**) was invented by Mullis [4], who received a portion of the 1993 Nobel prize in chemistry for his discovery. PCR, more than any other scientific advance since the elucidation of the structure of DNA, has changed the face of molecular biology. RFLP and PCR technology together form the cornerstone of forensic DNA typing.

Applications of DNA Identification Technology

DNA analysis enjoys a broad usage across many disciplines, including human health, forensic sciences, parentage, missing persons, anthropology and animal sciences.

Medical science
An initial and ongoing application is the search for genes implicated in disease. Due to the efforts, in large part, of the **human genome project**, all of the information contained in the human genetic code is rapidly being deciphered and catalogued. In addition to improved diagnosis and drug therapies, actual gene replacement is already in clinical trials for some of these diseases. Because the genetic information collected in this application contains information about disease predisposition, it poses a serious ethical dilemma that has yet to be resolved at a societal level.

Forensic science
Wherever an individual leaves a **nucleated** cell (most cells other than red blood cells or surface skin cells), she has also left her complete DNA signature. Forensic laboratories may use that information to link suspects of a crime with the biological evidence left at the crime scene. DNA typing is particularly well suited to evidence left as a result of violent crimes (e.g., blood, semen, hair, skin), and like the time-honored biometric of **dermatoglyphic fingerprinting**, provides a direct association to

an individual. Because the markers used in forensic DNA analysis are typically located in regions of DNA that produce no physical traits, ethical considerations are somewhat reduced.

The first forensic use of DNA took place in England and made use of Alec Jeffrey's original method of "DNA fingerprinting". In conjunction with a police investigation, the Home Office was able to identify the murderer responsible for killing two young girls in the English Midlands. Significantly, an innocent suspect was the first accused murderer to be freed based on DNA evidence. The exclusion of a suspect by DNA is absolute; there is no statistical probability associated with it. Numerous wrongly convicted men continue to be freed from prison after old evidence is re-tested using the newer DNA techniques.

Parentage and missing persons
Because the DNA molecule contains the hereditary information that is transmitted generationally from parent to child, it is an obvious choice for parentage determination. Immigration and paternity disputes were among the first legal arenas in which DNA typing was used. DNA testing can provide valuable clues about the identity of missing persons when remains are tested and compared to living relatives. Recently, DNA testing has been used to identify the biological families of children who were abducted as newborns during the military dictatorship in Argentina, enabling grandparents to be reunited with grandchildren they had never known. The U.S. Armed Forces DNA Identification Laboratory (AFDIL) now routinely collects and stores samples from all members of the armed forces for later comparison to remains recovered from combat, much in the same way fingerprints have been collected for decades. DNA typing systems are also invaluable in the identification of bodies and body parts from mass disasters.

Anthropology and animal sciences
Because of its relative stability as compared to other biological molecules, DNA, and in particular a certain type of DNA, mitochondrial DNA, has become an important tool in the study of anthropology and ancient history. Animal geneticists are typing DNA in endangered species, such as cheetahs and whales, to track migration and breeding patterns. Inroads have also been made into poaching and illegal trafficking in animal parts.

DNA testing for identification is now ubiquitous around the world. It is worth noting here that, with the exception of the use of DNA to link a suspect to the scene of a crime, none of the other uses of DNA typing detailed above has been seriously contested by the legal or scientific communities.

3.   Molecular Biology and Genetics

Biophysics of DNA

The DNA molecule consists of a linear arrangement of basic units called **nucleotides** or **bases**. These chemical moieties come in four types: **Adenine (A)**, **Cytosine (C)**, **Guanine (G)** and **Thymine (T)**. The specific sequence of the bases determines all the genetic attributes of a person. The genetic and chemical properties of the DNA molecule are directly related to its physical structure. DNA in nature takes the form of a double helix. Two ribbon-like entities (the sugar-phosphate backbone) are entwined around each other and held together by ladder rungs composed of hydrogen bonded **base pairs**. Only specific pairings between the four bases are chemically compatible. **A** always pairs with **T**, and **G** with **C**. This obligatory pairing, called **complementary** base pairing, is exploited in all DNA typing systems. When the double helix is intact, the DNA is called **double-stranded**; when the two halves of the helix come apart, either in nature or in the test tube (*in vitro*), the DNA is called **single-stranded**. Thus, in a double-stranded DNA molecule, the sequence AGTCGGTCA is paired with its complementary sequence to form the double-stranded structure:

AGTCGGTCA
TCAGCCAGT

In nature, complementary base-pairing is responsible for the ability to accurately replicate the DNA molecule, with its encoded genetic information, and pass it on to the next generation. The double helix is unzipped by special enzymes, and new building blocks (nucleotides) are brought in. Using each half of the original helix as a template, a second half is created, resulting in two molecules identical to the original. Because each original base captures a complementary replacement to complete the base pair, the order of the bases in the new strands is specified by the existing strands. This process can be recreated *in vitro* to a limited extent, and is the basis of the polymerase chain reaction (PCR) to be discussed in detail in a later section.

Under appropriate conditions, even short segments of complementary DNA will find each other and **reanneal** or **hybridize**. If the sequence at a particular location in the genome is of interest, single-stranded fragments can be artificially synthesized to target that location. These single-stranded fragments of known sequence are called DNA **probes** or DNA **primers**, depending on their intended function. Complementary base-pairing is essential to the detection of the genetic variations described in *Subsection Sources of Genome Variability* below.

Genome Organization

Nuclear genome
In eukaryotes (all organisms except bacteria), most of the cellular DNA resides in the cell **nucleus** and is organized into linear structures called **chromosomes**. This is called the **nuclear genome**. Human cells (except for gametes) contain 23 pairs of matched chromosomes. One pair comprises the **sex chromsomes** (either XX or XY) and the remaining 22 pairs are termed **autosomes**. Because the chromosomes are physical entities, genetic markers residing on the same chromosomes are inherited together; they exhibit **genetic linkage**. In contrast, markers on different chromosomes are generally inherited independently of one another. This principle is called **random assortment**.

The genetic information contained in the chromosomes can be compared both between the paired chromosomes of an individual and between the corresponding chromosomes of different people. If the DNA sequence (or those portions detected in a specific test) of two paired chromosomes within an individual is the same at a particular location (**locus**; pl. **loci**), the situation is termed **homozygous**; if the sequence is different, the chromosomes are **heterozygous** at that locus. Different forms of the same **gene** or **genetic marker** at a locus are called **alleles**.

Mitochondrial genome

A small amount of cellular DNA is found outside of the nucleus in several cell **organelles**. Of particular interest in DNA typing is **mitochondrial DNA (mtDNA)**. Mitochondrial DNA is physically organized into a small circular molecule (16.5 Kb), and exists in 100s to 1,000s of copies within the mitochondria of each cell. Mitochondrial DNA has a unique inheritance pattern. In contrast to nuclear DNA, which is inherited in equal parts from both parents, genetic material from mitochondria is inherited only from the egg cell of the mother. Thus mtDNA is said to exhibit maternal inheritance. Therefore, the mtDNA type of an individual cannot be heterozygous, or exhibit two different types. Because the mitochondrial chromosome lacks a matched partner, it is termed **hemizygous**.

Sources of Genome Variability

Through scientific investigation, mostly as a by-product of disease research, standard loci in the human genome have been established where the sequence varies more than usual between people. The existence of multiple alleles of a genetic marker at a single locus is called **polymorphism**. When such loci exhibit extreme numbers of variants (as many as hundreds), they are called **hypervariable**. Variations, or polymorphisms, can occur either in the sequence of bases at a particular locus or in the length of a DNA fragment between two defined endpoints. **Sequence polymorphisms** are like different spellings for the same word in British English and American English. When you see *analyze* spelled as *analyse*, the word and meaning are still recognizable. In DNA, a sequence mutation would be manifest by small changes in the sequence. For example, a point mutation might look like this:

<u>AGTCGGTCA</u> to <u>AGGCGGTCA</u>
TCAGCCAGT to TCCGCCAGT

   Length polymorphisms are most easily analogized to a train that can accommodate different numbers of boxcars. The engine and caboose define the end of the train; the total length may vary according to the number of cars attached between them at any one time. Each boxcar contains the same small DNA sequence. In genetic terminology, the boxcars are termed **tandem repeats**. A locus that shows variation in the number of tandem repeats is called a **variable number tandem repeat (VNTR)** locus. A particular number of tandem repeats, for instance, 35, defines a VNTR allele at that locus. The following is an example of a double-stranded tandem repeat region:

······[AGTCGGTCA][AGTCGGTCA][AGTCGGTCA]······
······[TCAGCCAGT][TCAGCCAGT][TCAGCCAGT]······

4. Forensic DNA Typing Systems

The space limitations of this chapter do not permit us to provide detailed laboratory protocols for DNA analysis. Such protocols can be found in [5] and [6]. We will concentrate here on giving a general description of the techniques such that the reader will be in a position to appreciate the advantages and limitations of different typing systems.

RFLP Systems

RFLP analysis determines variation in the length of a defined DNA fragment. When complete, an RFLP pattern looks like a very simple supermarket bar code. In looking at two samples, the pattern of bars on the autoradiograph is compared to determine if they could have originated from the same source (Figure 14.1).

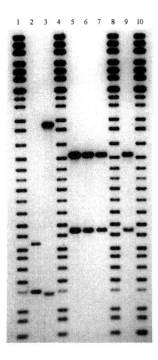

Figure 14.1 RFLP autorad. The end result of RFLP analysis at one locus. Each vertical lane contains a different DNA sample. Lanes 1, 4, 8, and 10 contain synthetic molecular weight standards from which the unknown band sizes can be measured. The samples in lanes 5-7, 9 all exhibit the genetic profiles at the locus tested. The donors of these samples cannot be distinguished by this genetic test. The samples in lanes 2 and 3 show different patterns. The donors of these samples could not have contributed the samples in lanes 5-7, 9 and also could not be the same individual. All of the human samples are heterozygous (show two bands) for this locus.

RFLP, at this writing, still provides the highest degree of discrimination *per locus*. If two samples originate from different sources, RFLP is the technique most likely to differentiate them. Two circumstances contribute to the huge **power of discrimination (P_d)** of RFLP. One is that many loci have been established for RFLP analysis; the more places you look, the greater the chance of finding a difference between two people. Forensic DNA laboratories now have access to probes for over 15 different loci. Second, forensic workers have chosen RFLP loci that have as many as hundreds of variations at each locus, increasing the chance that samples from different individuals will be differentiated.

Although the number of loci for newer systems, namely **STR** (**short tandem repeat**) analysis, are rapidly increasing, simple genetics will always maintain RFLP as the most discriminating technique *per locus*. Because of this, RFLP remains the method of choice for distinguishing among the contributors of mixed samples. Both contributors will likely be detected clearly on the final readout. Unfortunately, the RFLP technique requires more and better quality DNA than some of the newer PCR techniques. Because forensic evidence is often old, degraded and of limited quantity, RFLP analysis is sometimes not possible.

PCR Amplification

PCR-based techniques have gained preference because of all the other advantages they provide. At least some information may be gleaned from samples that might otherwise be refractory to analysis because of limited or degraded starting material. While the sample preparation time is the same for both RFLP and PCR methods, a PCR-based analysis is much more rapid than an RFLP analysis and also more amenable to automation.

The DNA samples prepared using PCR are analyzed in a variety of different ways. Although not yet ready for forensic laboratory applications, researchers are even working on applying optimized PCR techniques to the larger size DNA fragments generated by RFLP that have traditionally been recalcitrant to reliable and consistent amplification.

Of the genetic systems now in use for forensic analysis of PCR-amplified DNA, each locus tends to show less variation than for RFLP loci. Therefore, results may be obtained for a sample of limited quantity and quality, but the P_d will be lower. Because the combined P_d, at least for single-source samples, increases with each additional locus, the recent development of numerous **short tandem repeat** (**STR**) loci has greatly increased the usefulness of PCR-based typing. The simultaneous amplification of multiple loci (as many as 8 or 9) in a single "cocktail" (**multiplexing**) most efficiently utilizes a limited or degraded sample. However, the relatively smaller number of alleles at each locus will always limit the application of current PCR-based systems in mixed samples.

HLA DQA1 and AmpliType® PM ("Polymarker")

The first system adapted for forensic analysis of PCR-amplified DNA is from the HLA DQA1 locus. The type of variation present at this locus resides in the DNA sequence and is detected using specially designed molecular probes, synthetic fragments of DNA designed to be complementary to, and thus target, particular subregions within this locus. Since only one locus with limited variability is analyzed in this system, the P_d is not nearly as high as for RFLP. The chief advantages of HLA DQA1 are the ability to investigate very small samples and the rapidity of analysis. The final results are seen as a series of blue dots on a paper-like strip. A comparison of the pattern of the dots between typing strips indicates whether two samples may have originated from the same source (Figure 14.2).

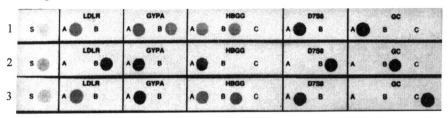

Figure 14.2 AmpliType™ PM (polymarker) dot blot. The end result of PCR-PM analysis at 5 different loci. Each horizontal strip represents one DNA sample. The vertical demarcations on each strip separate the individual loci. LDLR, GYPA, HBGG, D7S8 and GC are the names of the loci analyzed by this test. The S dot at the far left of each strip is a threshold control. All three strips show different genetic profiles, indicating that they were contributed by different individuals. The LDLR locus in sample 1 is homozygous; the GYPA locus in sample 1 is heterozygous.

The AmpliType® PM system, commonly known as **polymarker,** is just an expansion of the technique used in HLA DQA1 analysis. This was the first forensic system to exploit multiplexed amplification of DNA. Although each of the five additional markers do not contain as much individual variation as HLA DQA1, the combined result, along with that from HLA DQA1, increases the power of the test considerably. A disadvantage of this test is that it is often more difficult to interpret the results from samples containing DNA from more than one contributor.

D1S80

The first length-based PCR system that was implemented in forensic laboratories is called D1S80. Like RFLP, D1S80 is a VNTR; however, because of its relatively smaller size, this fragment is amenable to amplification by PCR. The DNA fragments produced by the D1S80 system can be counted in hundreds of base pairs, about an order of magnitude smaller than the fragments normally analyzed in RFLP typing. Thus D1S80 analysis combines the advantages inherent in any PCR system (specifically the ability to analyze samples of limited quantity and quality) with the greater variation generally seen in length-based systems. Again, because only one locus is analyzed in this system, the power of discrimination is not as high as RFLP. Also, the D1S80 locus in particular contains two alleles that are common among many people in some racial groups. If one of these two alleles is present in the sample being analyzed, the significance of the test result may be reduced.

In RFLP analysis, all of the DNA is processed, and the regions of interest are detected with molecular probes. In D1S80 analysis, the regions defined by the PCR amplification are effectively purified before the DNA is analyzed. Thus no special probes are needed to visualize the final result. This is like buying a package of 1-inch nails which are pre-selected for you instead of buying a package of assorted nails and having to fish out the ones you need for your fence. While subtle variation exists within all tandem repeat polymorphisms, the D1S80 system has been designed so that variants are detected as discrete alleles (digital system) and thus can be compared directly to a standard ruler made up of all possible alleles (allelic ladder) run on the same gel. The DNA is commonly detected using a silver stain, and the final result looks much like the simplified supermarket bar- code often used to describe RFLP. The D1S80 locus is also amenable to the automated detection and analysis techniques described later in this chapter, although the laboratories that can afford such automated systems tend to use them to run more powerful systems, such as STRs.

STRs

Short Tandem Repeats (STRs) are similar to the D1S80 system described above, except that the repeat units are shorter. The loci chosen for forensic use generally have a tandem repeat unit of 3 or 4 bp and may be repeated from a few to dozens of times. The number of alleles present in the population varies from about 5 to 20, depending on the locus. Like D1S80, STR loci are detected as discrete alleles and thus can be compared directly to an allelic ladder run on the same gel, simplifying comparison and analysis. Including flanking sequences amplified by the primers, the size of the DNA fragments produced by amplification of STR loci tends to be in the range of hundreds of base pairs, rather than the thousands of base pairs found in RFLP fragments. This makes STRs an ideal choice for degraded DNA.

Although each locus is only moderately polymorphic (i.e. fewer alleles are found), many such loci exist and can be analyzed simultaneously. In this respect, the system is similar to RFLP. In fact, PCR amplification of several different loci is often performed simultaneously in the same tube (multiplexing), conserving time, materials, and most important, sample. Because the human genome contains an almost unlimited choice of STR loci, it is possible to choose those in which the alleles in any given population tend to be reasonably well distributed, another advantage over the D1S80 system. Laboratories are increasingly employing semi-automatable systems, in particular capillary electrophoresis, for the detection and analysis of STR loci.

Gender ID

It is often useful to know if male or female components are present in a forensic sample. The **amelogenin** locus, which is coincidentally the gene for tooth pulp, shows a length variation between the sex chromosomes. Analysis of this locus is often appended to another PCR system, such as DQA1 or a multiplex STR system. Then no additional sample need be expended to make this determination, which in and of itself might eliminate only 50% of the population.

Mitochondrial DNA

Mitochondrial DNA analysis tends to be used more often to answer identification questions that arise outside the criminal justice system. Because mtDNA is inherited maternally, it is particularly useful in tracking families and populations. Also, because of its relatively small size in comparison to nuclear chromosomes and the presence of numerous multiple copies in a single cell, it is often the last typable DNA present in a small, old or badly degraded sample, so is particularly useful for anthropological research. Researchers have even been successful in typing mtDNA embedded in the anucleated cells of hair shafts, as well as in bones and teeth. However, because mtDNA constitutes only a single locus and presents some technical challenges, this analysis is best reserved for cases where nuclear DNA analysis has failed due to minimal quality or quantity. At present, only a few forensic DNA laboratories in the world possesss mtDNA analysis capabilities.

5. Acquisition and Storage of DNA Evidence

Collection and Preservation

From the moment biological material is out of the body, it is in a foreign environment and changes begin to take place. DNA is subject to degradation and that degradation can have an effect on the ability to obtain a useful result from DNA typing, particularly RFLP. Factors leading to the degradation of DNA include time, temperature, humidity, light (both sunlight and UV light), and chemical or biological contamination. Numerous studies have been conducted to determine the effects of these conditions, which, with a few exceptions, tend to degrade the samples into smaller fragments.

An important outcome of these studies is the finding that these environmental factors will not change DNA from one type into another; in other words, there is no danger that environmental degradation will produce a complete DNA pattern that would include someone who is not the donor of the sample (when interpreted by an experienced analyst). It is true that only a partial type may remain, but wholesale change of types is not seen. Degradation limits the usefulness of DNA typing, but does not invalidate it. DNA, under normal environmental conditions, can remain stable and typable for years. This is especially true of the PCR systems, which can tolerate a large amount of degradation and still yield readable types.

An important goal in collecting and preserving biological evidence is to halt any degradative process already in progress and limit any future deterioration. In general, biological processes are slowed by removing moisture and lowering the temperature. Thus the goal of the crime scene investigator is to dry a sample, and freeze it as soon as it is practical.

Contamination

Just as important as preserving the biological integrity of the sample is the consideration of any contamination that might interfere in the analysis. In fact, there are different types of contamination, and the final effect on evidence, if any, varies. *Non-biological* contamination (e.g., dyes, soaps and other chemicals) may affect the sample by interfering in the analytical procedures. This type of interference typically produces an inconclusive result or no type at all. *Non-human* biological contamination includes the physiological material or DNA from other organisms. Although cross-typing is occasionally seen in some systems, it generally does not interfere with interpretation of the final result. A particular concern is *microorganismal* contamination. Crime scene samples such as blood and semen provide a fertile environment for the growth of bacteria and fungi. As they grow, these microorganisms secrete biochemicals that degrade the human DNA in the sample.

Even so, the DNA type will simply go away, as opposed to being magically converted into someone else's type.

The most significant type of contamination is that from a *human* source. In this sense, contamination is defined as the inadvertent addition of an individual's physiological material/DNA during or after collection of the sample as evidence. It is important to differentiate between a "mixed sample" and a "contaminated sample". A mixed sample is one that contains DNA from more than one individual, and where the mixture occurred before or during the commission of the crime. A contaminated sample is one in which the foreign material was deposited during collection, preservation, handling, or analysis.

PCR-type testing, inherently a more sensitive technique than RFLP, is more likely to detect traces of a second type, whatever the source. Safeguards are set up not only to guard against contamination with non-evidence DNA, but to detect it should it occur.

Evaluation of Evidence

Before an evidence item is analyzed for DNA type, presumptive tests are sometimes performed to establish the type of biological material that is present. It would be wasteful to run a full spectrum of DNA tests with no result only to find that ketchup or shoe polish was being analyzed. Presumptive color tests for various fluids such as blood, semen, or saliva may be performed at the scene before a sample is collected, or in the laboratory.

Once the identification of a sample as a particular biological substance is established, preliminary tests are conducted to establish the "state of the DNA" contained in the sample. It is possible to run tests that will reveal the quality of the DNA (how much degradation is present) in an item of evidence, how much total DNA is present, and how much of the total DNA is human. An evaluation of the "state of the DNA" is crucial in making decisions about what might be accomplished with any particular sample, for example, whether RFLP is possible or whether a PCR method might be more suitable.

DNA and the Databank

Direct comparison of a sample to a suspect utilizes only a small fraction of the potential of DNA typing. The storage of DNA profiles from convicted criminals in a databank engenders the possibility to search for possible perpetrators of suspectless and serial crimes. Most people are familiar with the databanks now in use to track latent fingerprints. The Automated Fingerprint Identification System (AFIS) contains millions of people's fingerprints in computer files.

DNA profiles are particularly suited for computer storage and automated searches because information can be stored as a set of numbers, requiring very little in the way of sophisticated technology. It is essential to realize that an initial "cold hit" in a databank is only used as probable cause to obtain a sample from a suspect for further testing. The markers used to identify the suspect are then retested, and the samples are compared further, using yet additional markers. This system provides a safeguard against any clerical errors or sample switches that may have occurred in generating the databank.

In order for databanks to be most effective, especially on a national level, the DNA system used to create them must be standardized. The FBI is leading the effort to create a national databank, to which all states will contribute information. PCR-type systems that are just now coming on-line are beginning to be included and will greatly streamline the task of processing the tens of thousands of backlogged samples sitting in some state freezers.

Useful databanks may be created from groups of individuals other than crime suspects. For example, a databank composed of DNA profiles from evidence samples in suspectless cases could be searched against itself, linking cases to a common (unknown) perpetrator. A databank containing DNA profiles from unidentified bodies would obviously be useful in identifying them as new information is uncovered. The U.S. Armed Forces DNA Identification Laboratory (AFDIL) has already instituted a sample collection program from military personnel. This will greatly assist in the identification of victims of war, particularly when extremities containing fingerprints are missing.

Privacy considerations in general are a larger issue for DNA testing than for other personal identification methods. Additional information pertaining to diseases, relatedness and physical traits are contained in the DNA sample. These are not traits, however, that are tested in forensic labs, and any data stored in a computer would consist only of information about specific forensic markers. Nevertheless, great care must be taken to protect the privacy of the individual, convict or not, as well as the security of the sample. Recommendations to this effect have been made by national committees, such as the National Research Council (NRC), and are being followed [7,8].

6. Procedures for Forensic DNA Analysis *Isolation and Evaluation of DNA*

Before any type of testing can be performed, DNA must be isolated from the rest of the cellular components, as well as from any non-biological material that might be present. The isolation or extraction procedure varies somewhat according to the type of biological evidence present (e.g., blood, semen, saliva, hair etc.), the amount of evidence (which influences the type of test that is subsequently performed), and the kinds of cells that are present. These determinations are made by the visual inspection, microscopic examination, presumptive, and confirmatory testing.

A special situation involves samples in which sperm are present along with other types of cells, often vaginal in origin. Many cells found in forensic evidence fall into a category called **epithelial cells** or **e.cells**. This includes saliva, skin, buccal, and vaginal cells, as well as those found in urine and feces. The different properties of epithelial cells from sperm cells are exploited in order to separate them from each other before DNA is isolated. This simplifies the final interpretation, as the victim's and suspect's types may be analyzed and compared separately.

Before any analysis proceeds, it is imperative to determine not only how much DNA is present, but how much of it is human and how **degraded** (broken up) it is.

DNA remaining in relatively large pieces is said to be of **high molecular weight (HMW)**. Specific tests are performed to define these parameters before proceeding with any analysis.

RFLP Procedures

RFLP analysis measures the size of the DNA fragments produced by **restriction enzymes**. Restriction enzymes recognize specific short DNA sequences, and cleave the strand when this sequence is encountered. When a VNTR occurs between two restriction enzyme sites, the resulting fragment sizes will vary depending on the number of repeat units between the restriction sites. After restriction enzyme digestion, the cut DNA is separated according to fragment length in an **agarose** gel to which an electric field is applied. Because DNA carries an overall negative (−) charge, all of the DNA fragments, which have been loaded at the negative pole, will start to migrate towards the positive (+) pole. At the end of the run, the fragments are arrayed from largest to smallest in parallel lanes. The DNA fragments in the gel are then **denatured** (separated into single strands) and transferred by capillary action to a piece of nylon membrane. This procedure is called **Southern blotting** (Figure 14.3).

In order to detect DNA originating from designated locations in the genome, short single-stranded fragments of DNA, now used as probes, are labeled with a radioactive or chemiluminescent tag. These probes are designed to match specific places in the genome that are well-characterized as highly polymorphic. Under the right conditions, DNA strands that match will reunite into a double-stranded form. This process is called **hybridization**. The labeled fragments signal where they have hybridized and this signal is recorded on a sheet of X-ray film. Each piece of exposed film is called an **autorad**, short for autoradiogram or autoradiograph (Figure 14.1).

In the RFLP loci chosen for forensic use, the two chromosomes inherited from each parent will frequently contain different numbers of repeat units (heterozygous), producing DNA fragments of different sizes. Thus, two bands are generally detected in each sample lane. If a person is homozygous for a particular locus (that is, they have inherited the same length allele from both parents), only one band will be detected on the autorad.

Once the information from probe number 1 is recorded, it is **stripped** (removed) and the nylon membrane is exposed to the next probe in the series. The process is repeated for every additional probe. In forensic DNA analysis, as many as 5 or 6 different loci are commonly analyzed so that 10 to 12 bands are ultimately detected in each lane. The exposed films, or autorads, from each particular probe provide a permanent record of the results of the analysis. The band locations are compared from lane to lane in order to identify any similar patterns. Samples which look visually similar are then subjected to computer imaging and analysis. In order to aid in this analysis, an artificially constructed molecular ruler is run on each gel; the computer then has an objective internal standard from which to calculate band (fragment) sizes. If two samples are suspected to have originated from the same source, a calculation is also performed in order to estimate how often the evidence profile occurs in any population.

PCR Amplification

PCR amplification faithfully replicates a defined segment of DNA millions of times, a process dependent on the enzyme *Taq* polymerase. An essential feature of this particular enzyme is that it can survive high temperatures and still keep working. As will be seen, this is key to the "chain reaction" used to replicate the DNA.

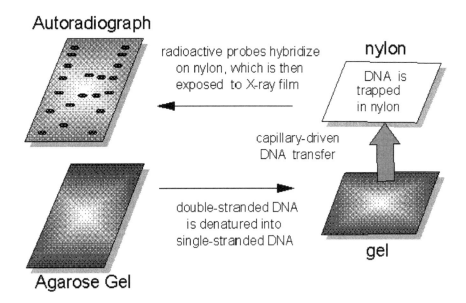

Figure 14.3 Southern blot. After DNA fragments are separated by size in an agarose gel, they are denatured into single strands and transferred to a nylon membrane by capillary action. The membrane is then exposed to tagged DNA probes for the region of interest. The fragments corresponding to the genomic location specified by the probe are visualized on X-ray film. An autorad, the end result of this process, is depicted in Figure 14.1.

The three main steps of 1) **denaturing** the double helix into single strands, 2) **annealing** DNA **primers** to define the amplification region, and 3) **extension** to create new DNA strands, are repeated dozens of times. In each cycle, the number of DNA copies is doubled, resulting in millions of copies identical to the original (Figure 14.4).

Analysis of PCR Product

Depending on the type of polymorphism being investigated, the product of PCR reactions, henceforth called "PCR product" is analyzed in one of two ways. Sequence polymorphisms are detected using a hybridization procedure, or less commonly, by direct sequence analysis. Length polymorphisms are most commonly detected using various procedures similar to the gel used in RFLP analysis.

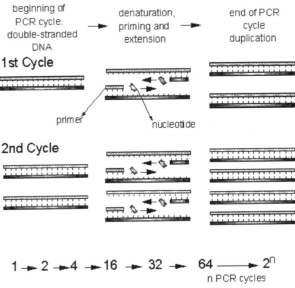

Figure 14.4 Polymerase chain reaction (PCR). PCR allows a selected portion of DNA to be replicated many times over the rest of the genome. The fragment is doubled in each cycle of the reaction, resulting in a geometric increase. After several dozen PCR cycles, the region of interest is increased by a factor of millions.

Sequence Polymorphisms

DQA1, Polymarker. Complementary base pairing forms the scientific basis for the detection of sequence polymorphisms. A nylon strip, to which DNA probes have been attached, is challenged with the PCR product. The strips contain specific DNA sequences originating from the same locus in the genome as the DNA that has been amplified by PCR. Each probe contains a specific sequence of DNA that defines an allele. These types of probes are known as **sequence specific oligonucleotides (SSO).** The SSO probes on the strip (in the shape of dots) define a finite number of variations (types) seen in that particular region. The type of the sample is revealed by the hybridization of the amplified DNA to a specific immobilized probe on the strip. Hybridization is detected via a chain reaction that ultimately results in a blue color appearing on those dots where PCR product has hybridized, hence the term "**dot-blot**". The pattern of dots corresponds to the alleles present in the sample (Figure 14.2).

Mitochondrial DNA. A number of analysis and detection systems for the polymorphisms found in mitochondrial DNA are currently in use or development. The variations exhibited in the two hypervariable regions are generally point mutations, although small deletions and insertions may also occur. The most common method currently in use involves direct DNA sequencing. Other methods, based on defined hypervariable regions within the mtDNA control region are under development.

Length polymorphisms (D1S80, STRs, Gender ID)

Similar procedures are employed in the analysis of all three PCR-amplified, length-based marker systems used in forensic testing. The PCR product is loaded into either an **acrylamide** gel or capillary column to be separated on the basis of length. Because no extraneous DNA is present, the bands may be visualized directly, negating the need for the secondary detection method of probing and hybridization used in RFLP. In manual detection, a silver stain is used to visualize the separated DNA bands, and the gel is then dried to be kept as a permanent record. In automated detection, a fluorescent tag is incorporated into the fragments via the amplification primers; the bands may then be detected, statically, after the run is complete, or dynamically, during the run.

In fluorescent detection, only the primer complementary to one strand is tagged, thus eliminating any confusion resulting from reading doublets (resulting from the resolution of single DNA strands) at each allele. Additionally, the use of multiple colored fluorescent tags allows the combination of STR loci (multiplexing) in which the lengths of some alleles overlap. These systems can be run in the same gel lane, and still be clearly distinguished by color. The possibility of in-lane sizing standards mean that the computer will calculate the size of a particular band against a ruler that has been subjected to exactly the same electrophoretic micro-environment as the sample, rendering extremely reliable results.

Because of its high sensitivity and relative ease of use, partially automated capillary electrophoresis has become the preferred method of detecting and analyzing STRs for forensic use. The use of mass spectrometry to detect and analyze the STR-PCR products is currently under development and it is expected that other new systems, such as mass-array chip technology will eventually also be adapted for forensic use.

7. Significance of Results

The entire purpose of DNA typing is to test the hypothesis that a particular person is the source of an item of biological evidence. An attempt is made to ascertain whether an association exists between an evidence sample and a reference sample taken from an individual. The evidence sample (a biological fluid or tissue) and reference (typically a blood sample) are subjected to a battery of DNA tests. Upon completion, the analyst is able to render a determination as to the genetic similarity of the samples. Three conclusions are possible.

1. **exclusion**. The types are different and therefore must have originated from different sources. This conclusion is absolute and requires no further analysis or discussion.

2. **inconclusive**. It is not possible to be sure, based on the results of the test, whether the samples have similar DNA types. This might occur for

a variety of reasons including degradation, contamination, or failure of some aspect of the protocol (e.g., inhibition of restriction enzyme). Various parts of the analysis might then be repeated with the same or a different sample in an attempt to obtain a more conclusive result. One way of thinking about an inconclusive result is that there is no more information after the analysis than before; it is as if the analysis had never been performed.

3. **inclusion**. The types are similar, and could have originated from the same source. If the samples are determined to be similar, the question becomes: What is the significance of this similarity?

Determination of Similarity

Frequently the word "match" is used to describe the genetic similarity between the evidence and reference samples. Scientists are careful to limit the term match to mean that no significant differences were observed between the two samples in the particular test(s) conducted. It is certainly possible that two samples may be different, but that the test used has failed to reveal those differences. Because DNA tests currently sample a relatively small percentage of the entire human genome, further analysis might reveal differences that would lead to a exclusion. In contrast, the perception of the general public is that the word match connotes an absolute "individualization". A conclusion of **genetic similarity** or **genetic concordance** merely describes the fact that no differences were seen between the two samples in the particular tests conducted. Having said that, however, the strength of DNA typing lies in its immense powers of discrimination; samples that show genetic concordance over several highly discriminating DNA loci approach, and in some cases reach, individuality.

In determining whether two samples have similar types, it is important to know the kind of marker system used. Typing systems may be divided into those detecting **continuous alleles** and those detecting **discrete alleles**.

Evaluation of Results—The Strength of the Association

Samples may show genetic similarity under three circumstances:

a) The samples come from a common source. This means that the evidence sample (blood stain, semen sample, saliva stain, etc.) comes from the same person who provided the reference sample.

b) The similarity is a coincidence. This means that the evidence sample comes from someone other than the person who provided the reference sample. The genetic similarity results from two individuals, the reference donor and the true donor, who share the same genetic profile for the particular markers examined.

c) The similarity is an accident (erroneous). This means that the evidence sample comes from someone other than the reference donor but that some collection/analytical/clerical error has occurred to make the evidence and reference samples appear to have the same DNA profile.

We want to evaluate which of these three alternatives is the correct one for the case under consideration. The likelihood of each alternative provides insight into the strength of the association between the evidence and reference sample.

The samples come from a common source

Keep in mind that if the evidence sample is, in fact, from the reference donor (common source), then the probability of finding the evidence profile is one (1).

The similarity is a coincidence

If many individuals have this type, then the significance is minimal, because it means that there is some reasonable chance that anyone taken at random from the population (e.g., the wrong suspect chosen by the detective) will have the same type. If, on the other hand, only a low probability exists that the types found are from someone other than the reference donor, then the association is strong.

The question then becomes: What is the probability of a "match" if someone other than the reference donor is the true donor? (or what is the probability of a random match?) The answer has typically been provided in the form of a profile frequency, that is, the number of times that this profile is seen in some reference population. An alternate form of that question is: What is the probability of finding this profile if the reference donor were the true donor, compared to the probability of finding this profile if someone other than the reference donor were the true donor? While this seems like a more complicated question, in fact it is a more complete statement of the first question.

Frequency estimate calculations. In simple terms, we want to express how many people might possess the profile seen in the biological evidence. The only way to determine this is by testing a representative number of people from a reference population and counting the number of times each genotype occurs. For genetic marker systems with just a few alleles, it is likely that all of the genotypes will be seen several times in a relatively small population such as 10 or 20 people. The problem becomes more complex with additional alleles. We may have to test 50 to 100 people to have confidence that the genotype frequencies are a true representation of the population makeup.

Some of the hypervariable RFLP loci have 50 or more alleles, such that each locus can produce approximately 1275 genotypes (49x50/2=1225 heterozygotes and 50 homozygotes). Typing enough people to find how often each of these types occurs would be a daunting task. Considering four hypervariable loci, each with 50 alleles; the number of possible allele combinations at four loci is (1275) x (1275) x (1275) x (1275), or about 2.6 trillion possible genotypes. Testing everyone in order to obtain a fair representation of all these types is clearly impossible. Additionally, since there are only about 6 billion people alive on earth at this time, most of these combinations do not even exist.

The solution to this dilemma (how to estimate frequencies when there are a large number of alleles, each at low frequency) is to invoke population genetics theory, particularly two principles called **Hardy-Weinberg equilibrium (H-W)** and **linkage equilibrium (LE)**. These principles allow for the estimation of genotypes based on individual **allele frequencies**, rather than observed **genotype frequencies**. Since the total number of alleles is much smaller than the possible combinations of those alleles into genotypes, this is clearly a much more practical proposition. The principles may be summarized as follows:

> The **Hardy-Weinberg** model states that there is a predictable relationship between allele frequencies and genotype frequencies at a single locus. This is a mathematical relationship that allows for the estimation of genotype frequencies in a population even if the genotype has not been seen in an actual population survey.

> **Linkage equilibrium** is defined as the steady-state condition of a population where the frequency of any multi-locus genotypic frequency is the product of each separate locus. This allows for the estimation of a DNA profile over several loci, even if the profile has not been seen in an actual population survey.

Population substructure. Theoretical application of the Hardy-Weinberg principle rests on several assumptions. Mating must be random, the mating population large, and migration negligible. However, it can be reasonably argued that mating is not random in most human populations, that some mating populations are not large, and that migration is variable among mating populations throughout the world. In fact, it is well accepted that the United States population is a mixture of people of various origins. For instance, in New York City, it is well known that neighborhoods exist of, for example, Italians, Germans, and Russians. It is also commonly accepted that people tend to mate among those with similar ancestry. This results in matings among people who are more closely related to each other than to people outside of their common ancestry. If a suspect comes from such a group, a greater number of people with similar DNA types may exist in this particular community than we would estimate from a survey of the general population. The phrases used to express this existence of smaller populations within a larger group include population subgroups, subpopulations, population substructure, and structured populations.

Given that the U.S. population is structured to some extent, and the assumptions for Hardy-Weinberg cannot be met, how is it possible to use these principles and arrive at useful frequency estimates? In actual fact, imperfect adherence to Hardy-Weinberg and linkage equilibrium does not invalidate the use of these principles in estimating frequencies for DNA profiles. This is substantiated by both scientific theory and empirical testing.

Research has shown that the effects of substructuring are predictable. Relative to theoretical Hardy-Weinberg proportions, the effect is to increase the occurrence of homozygotes and to reduce the number of heterozygotes at a single locus. The effect on linkage equilibrium is to increase the correlation between some loci, while decreasing the correlation between others. These deviations can be taken into account and accommodated statistically. Once sufficient data has been gathered for a specific population, departures from both Hardy-Weinberg and linkage equilibrium can be estimated. This allows for an evaluation of the extent and direction of the error that might occur if frequency estimates are calculated using H-W and LE calculations.

The concern over the lack of knowledge regarding the effects of population substructure has instigated two major studies by the National Research Council (NRC) of the National Academy of Sciences. The first study [7], concluded that insufficient knowledge existed to substantiate use of the H-W and LE calculations. Several recommendations were made that resulted in an extremely conservative method of estimating the frequency of a DNA profile (termed the "**ceiling principle**"). The second NRC report (known as NRC II) [8] concluded that enough information had been collected since the original study to eliminate the most conservative recommendations (including the ceiling principle) as unnecessary. This second report is an excellent source for a deeper understanding of the issues and the solutions presented here.

Finally, it is imperative to emphasize that frequencies are estimated for the evidence profile, not the suspect profile. The race/ethnicity of the suspect is irrelevant when interpreting test results. It is erroneous to assume that the suspect was at the crime scene to determine if the suspect was at the crime scene! Most of the time, general population frequencies (often limited to a particular geographical region, such as a state) can be employed, and racial/ethnic frequencies are used as comparisons or limits. In the case of a mixture where alleles cannot be reliably paired into genotypes, the correct frequency is calculated by the sum of all possible genotypes that could be present in the sample. The significance of the evidence can then be assessed under several different scenarios or assumptions, and the scenarios then compared via likelihood ratios. Most analysts agree that likelihood ratios are the only way to handle the complexities of mixed samples.

Estimating Frequencies. The goal in deriving frequencies for any DNA profile is to provide an estimate that is scientifically conservative; that is, it should not overstate the strength of the association. Forensic scientists, geneticists, and biostatisticians have devised several methods to accomplish this goal. First, population studies must be performed for the specific loci that will be used and the data evaluated for H-W and LE. Once data for a specific case has been generated, genotype frequencies must be calculated for each locus, then the frequencies for each locus multiplied to obtain a profile frequency.

Different workers have formulated a variety of correction factors that ensure conservative frequency estimates. NRC II provides guidance on reasonable corrections that prevent over-estimating the significance of the concordance without seriously compromising the power of the tests. Because of the ultimate destination of forensic DNA results (often a court of law), the way in which the results are presented are often limited by court rulings and case law. Thus, sometimes, a more accurate scientific calculation must be sacrificed to a legal ruling. It must be remembered that forensic scientists work solely at the behest of the legal system. The most recent example of this compromise of scientific accuracy for legal conservatism is California v. Venegas, 1998, which adopted the (admittedly) absurdly conservative recommendations of the first National Research Council committee report (NRC I).

The calculations outlined above are for random unrelated individuals. A special case exists for related individuals. Siblings potentially share more genetic material with each other than anyone else. This is because they inherit their genes from the same two people, Mom and Dad. This idea can be extended to more distant relationships such as children, grandchildren, and cousins. In these relationships some genetic material is shared, but the more distant the relationship, the fewer the genes that are held in common. For the highly variable DNA regions that are used in forensic testing, this means that even siblings are unlikely to share the same profiles when several highly variable DNA regions are analyzed. Special calculations, usually based on Bayesian statistics, are applied in those circumstances where relatives might be involved.

The power of DNA testing is such that the information provided by several highly variable loci is often sufficient to convince us of the source of a sample. With adequate data, it might be concluded that two samples originate from a single common source to the exclusion of all other individuals. We suggest that when close relatives can be eliminated (including identical twins), the appropriate quality control measures have been followed, and the conservative frequency estimates reach one thousand times the population of the earth (presently 6 billion people), individuality can be concluded.

The similarity is an accident
When the genetic similarity is an accident (through sample switch, contamination, or clerical error), the reference donor only appears to be the true donor, when, in fact, the evidence DNA is from someone else. A few authors have suggested that the power of DNA testing is limited to the error rate for the industry, the lab, or the individual conducting the test, whichever is available. NRC II and others have suggested that the risk of error can only be evaluated on a case-by-case basis.

Quality assurance and quality control are ways in which laboratories have attempted to prevent and detect errors. While not fool-proof, accreditation of laboratories, certification and proficiency testing of individual scientists, and the proper use of standards and controls, greatly reduce the risk of error in lab analysis and reporting.

8. Strengths and Limitations of DNA Typing

The immense diversity (the great capacity to discriminate between individuals) of the hypervariable RFLP loci that makes them so attractive for forensic work also presents analytical challenges and renders them susceptible to environmentally induced alterations. A great amount of work has been performed to determine whether these artifacts constitute fatal flaws, or merely complications that can be factored into the conclusions drawn from analytical results on a case-by-case basis. The exquisite sensitivity of PCR-based systems, both sequence-based and length-based, and the relatively low variation per locus, confers upon them another set of interpretational issues. These issues have also been addressed, both empirically and theoretically. In particular, the recent success in multiplexing STR systems promises discrimination of single-source samples approaching that of RFLP systems. The enormous amount of validation and proficiency testing performed using forensic DNA analysis systems has led to the consensus among crime labs and academic experts (with a few exceptions) that the difficulties can be adequately addressed and accounted for in most cases, and that the DNA results can be properly and conservatively applied to the case or situation as a whole.

The "wet chemistry" necessitated by current techniques limits both the portability and rapidity of DNA typing. This stands to change as the chemistry involved in the reactions is adapted to solid-state technologies already in development for basic research. It is not unreasonable to expect that, in the foreseeable future, the DNA typing laboratory might be reduced to a small mobile package that could produce a DNA result in minutes or hours instead of the days or weeks now necessary.

References

[1]   K. Inman, and N. Rudin, *An Introduction to Forensic DNA Analysis*. CRC Press, Boca Raton, Florida, 1997.

[2]   J. D. Watson, and F. H. C. Crick, "Genetic Implications of the structure of deoxyribonucleic acid," *Nature*, Vol. 171, No. 964, 1953.

[3]   J. Jeffreys, V. Wilson, and S. L. Thein, "Individual Specific 'Fingerprints' of Human DNA," *Nature*, Vol. 316, No. 76 , 1985.

[4]   K. B. Mullis and F. Faloona, "Specific synthesis of DNA *in vitro* via a polymerase-catalyzed chain reaction", *Methods Enzymol.*, Vol. 155, No. 335, 1987.

[5]   L. T. Kirby, *DNA Fingerprinting, An Introduction*, Oxford University Press, New York, 1992.

[6]   M. A. Innis, D. H. Gelfand, and J. J. Sninsky, (editors), *PCR strategies*, Academic Press, San Diego, CA, 1995.

[7]   National Research Council, *DNA Technology in Forensic Science*, National Academy Press, Washington D.C., 1992.

[8]   [8]   National Research Council, *The Evaluation of Forensic DNA Evidence*, National Academy Press, Washington D.C., 1996.

15 LARGE SCALE SYSTEMS

Robert S. Germain
IBM T. J. Watson Research Center
Yorktown Heights, NY
rgermain@us.ibm.com

Abstract *This chapter describes some of the fundamental issues related to biometric identification (one-to-many matches) on a large scale as well as some details of a particular implementation of a scalable fingerprint matcher. The decomposition of system error rates into expressions in terms of component error rates and the extrapolation of identification accuracy from verification accuracy are explained. Details of the Flash fingerprint matcher are presented including assumptions about allowable transformations connecting instances of model fingerprints and the structure of the invariant index used to form tentative model correspondences. Finally, the pose clustering stage used to filter tentative correspondences and accumulate evidence is described. Results using this preliminary implementation on a database containing 100,000 models are presented and interpreted in the context of the above-mentioned extrapolation framework described above.*

Keywords: Flash, fingerprint, performance evaluation, geometric hashing, indexing.

1. Introduction

There are two general classes of problems which a fingerprint matcher is expected to address. The first class of problems involves situations for which it is necessary to verify or authenticate an individual's identity. This is a one-to-one matching problem which is of interest here primarily as a conceptual basis for one-to-many matching.

The second, more challenging, problem occurs when it is important to ensure that a particular database contains only a single entry for any given individual. This occurs in the case of social services wherein one wishes to prevent individuals from collecting welfare under multiple aliases or in the case of identity card issuance. This identification problem requires that one search a large database of individuals and determine whether a person is already in the database.

Previous work in automatic fingerprint identification systems has concentrated on criminal justice applications. In the criminal justice application arena the cost of missing a potential match is quite high, e.g., a wanted criminal is released, and it is acceptable to require the employment of trained fingerprint officers to inspect large number of candidate matches. Also, criminal fingerprint identification queries almost always involve a large amount of filtering, effectively reducing the size of the database that is actually searched. This filtering may involve classification based on general ridge pattern, but also includes demographic filtering based on age, race, geographic location, etc. Finally, criminal justice fingerprinting systems retain images of all ten fingers.

This work was undertaken to address the requirements of a non-criminal identification application. The challenges for this application are to support the large throughput required for enrollment of a large population over a limited period of time and to minimize the time a clerk in a social service agency must spend in investigating ambiguous cases. Additionally, to simplify the enrollment process and minimize storage requirements, the system must achieve acceptable performance with as few fingerprint impressions from each individual as possible. A common requirement is to take imprints of the two index fingers only.

Large scale social service or national identity registry applications will require searches of databases containing imprints from a large fraction of the total population of a state or country; the ability to search databases of tens of millions of people is required. An understanding of the systematic changes in the error rates of a fingerprint identification system with database size is also needed as part of a framework for extrapolating measurements from small benchmarking or sample databases. Criminal justice fingerprint systems are not characterized in a manner that allows extrapolation of measured performance data to large database sizes. In particular, there is a focus on characterization in terms of the frequency with which the correct result appears in the top-ranked position or in the top ten positions, metrics which are useful for comparing results achieved on a particular database, but which are ill-suited for making estimates of the identification error rates on larger databases. The existing ANSI/IAI standard [1] for benchmarking of fingerprint identification systems is focused on criminal justice applications and on *relative* performance of competing systems. Some of the issues in measuring accuracy in law enforcement applications are discussed in [11].

System Performance vs. Matcher Performance

When architecting a large-scale identification system based on any biometric, there are several universally applicable strategies for improving throughput that do not depend on any details of the matcher. That is, the matcher can be treated as a "black box", characterized only by its error rates for verification and the time required to perform a match as a function of database size. These strategies include front-end filtering (a.k.a. classification) and fusion of multiple biometric data, such as using fingerprint impressions from multiple fingers. Before estimating the system performance of a particular configuration of filtering and data-fusion, characterization of the component performance is needed In particular, the performance of the classifier and the matcher operating on a single instance of a biometric must be measured individually. From these building blocks, the system performance for a particular choice of filtering and data fusion can be derived. The very simple data fusion methods described here are only a small subset of the available possibilities.

Flash and Geometric Hashing

The straightforward approach to searching a large database is to scan the entire database and to compare the query fingerprint against each reference model. The increased efficiencies obtained from generating index tables to speed access are well known in the database community [13]. An index can be formed from a subset of the feature points in a model instance and if multiple indices are generated for a single model instance from subsets that redundantly include feature points; the indexing scheme allows retrieval of models that differ from the query by one or more feature points. One of the attractions of redundant indexing schemes in computer vision applications, the earliest example of which is geometric hashing [8], is their robustness in the presence of partial occlusion.

The Flash algorithm uses a higher dimensional indexing scheme than geometric hashing by adding additional invariant properties of the feature subset to the index. Scalar properties such as color might be appropriate in some vision applications, while in fingerprint recognition, the relationship of the chosen subset of features to the local ridge pattern provides additional distinguishing power. The second stage of the Flash algorithm uses transformation parameter clustering to accumulate evidence [2].

Object instances are represented by a collection of feature points, which might be points of maximum curvature in a vision application, minutiae in a fingerprint application, or an ASCII character in a string matching application. When a model is added to the database, invariant information computed from each subset of feature points is used to form a key or index labeling an entry which is added to a multi-map or bag [13], a variant of associative memory which permits more than one entry to be stored with the same key value. This entry minimally contains the identifier of the model that generated the key and may also contain information concerning the feature subset as shown in Figure 15.1.

When servicing a query, each key generated by the query object instance is used to retrieve any items in the multi-map which are stored under the same key. Each item retrieved represents a hypothesized match between subsets of features in the query object instance and the reference model instance which created the item stored in the multi-map. This hypothesized match is labeled by the reference model identifier and possibly, parameters characterizing the geometric transformation which bring the two subsets of features into closest correspondence.

Votes for these hypothesized matches are accumulated in another associative memory structure, keyed by the model fingerprint identifier and the transformation parameters as shown in Figure 15.2. This structure is a map or keyed set, a container which permits only a single item to be stored under a given key. Each time that a hypothesis is constructed, the program checks to see if a hypothesis with the same label already exists in the hypothesis table (map container). If the hypothesis already exists, the score of the existing hypothesis entry is updated appropriately. If there is no hypothesis with the same label, a new hypothesis entry is added to the hypothesis table with its score set to an initial value. Finally, a sorted list of hypotheses whose scores exceed some threshold can be used to determine whether a match to the query object instance exists in the database or as input to another stage of matching machinery.

If a globally parameterizable transformation can be computed from each set of local feature correspondences, then when a good match exists, many of the local feature subset correspondences located during the index lookup phase will generate the same parameters for the geometric transformation and a large number of votes for that hypothesized match will be accumulated. Examples of globally parameterizable transformations include affine, similarity, or rigid transformations in one, two, or three dimensions. The computation of transformation parameters and accumulation of evidence based on binned values for these parameters is an instance of the "pose-clustering" [12] technique. This is related to the alignment technique used by other workers [6], but the Flash algorithm [10] permits parallel accumulation of evidence at a finer granularity. It is only necessary to compute transformations or relative poses for pairs of corresponding feature subsets that are mapped to the same invariant index. Verification of consistency in the correspondence of different local feature sets is implicit in the evidence accumulation process since large numbers of consistent relative poses will only be generated when the relative positions of many local feature sets are consistent in both the query and model object instances.

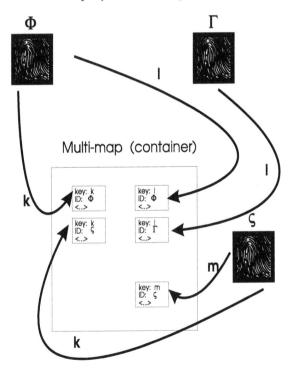

Figure 15.1 The extracted features from each fingerprint are used to generate keys or indices. For each key generated, an entry is added to the multi-map data structure. For example, fingerprint Φ generates keys **k** and **l**, fingerprint Γ generates key **l**, and fingerprint ς generates keys **k** and **m**.

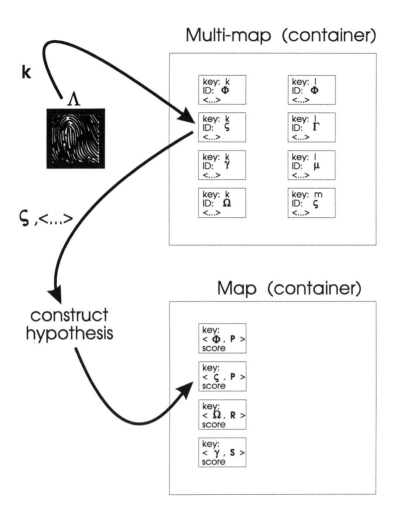

Figure 15.2 Retrieval.

2. Application to Fingerprint Matching

In the fingerprint application, the class of transformations that connects different object instances is assumed to be that of two-dimensional distance preserving (rigid) transformations. A least squares estimation methodology is used to solve the over-constrained pose estimation problem for each hypothesized local correspondence generated by the index look-up process.

Data Abstraction and Index Generation

Both automatic and manual fingerprint recognition schemes use the feature points determined by singularities in the finger ridge pattern. Unlike the general shape recognition problem [2], in fingerprint matching, the singularities in the ridge pattern known as *minutiae* provide a natural choice for feature points. These features, which consist of points where a ridge either ends or splits into two ridges, form the basis of most fingerprint matching applications. Each feature point is represented by a triplet of numbers (X, Y, θ) as shown in Figure 15.3. A typical "dab" impression has approximately 40 minutiae which are recognized by the feature extraction software, but the number of minutiae can vary from zero to over one hundred depending on the finger morphology and imaging conditions. Not all of these minutiae will be reproducible from imprint to imprint and therefore the redundancy in the combinatorial index formation process described below is essential.

One additional piece of information is utilized by the Flash matcher. Part of the output of the feature extraction process is a skeletonized version of the ridge pattern on the finger. If a line is drawn between each pair of minutiae, the number of ridges crossed by this line may be counted as shown in Figure 15.4. This ridge-counting procedure is carried out for each pair of minutiae in the fingerprint and the results are used as part of the Flash index.

The Flash algorithm uses redundant combinations of three feature points when forming indices to give some immunity against noise consisting of insertions and deletions of feature points as well as to provide more uniquely descriptive information than is available from a single feature point. An exhaustive listing of the possible combinations of three feature points requires $\binom{n}{3}$ entries where n is the number of minutiae. To keep the number of indices generated within bounds, restrictions are placed on the "acceptable" combinations of feature points to be used when forming an index. Only triplets for which the distances separating each pair of points fall into the specified range are used.

Even the restriction on pairwise separations does not prevent large variations in the number of indices generated by different fingerprints. In order to guarantee a relatively constant number of generated indices, a deterministic selection process is used to select a sampling of those indices whose generating triangles satisfy the imposed side length constraints.

The search engine requires the generation of indices used for table look-up that are simultaneously descriptive of the objects stored in the database and invariant under the transformations to which an object might be subjected. Each component of the index is invariant under rotation and translation.

While the model used here assumes that mated fingerprint impressions may be mapped onto each other by a rigid transformation, the realities of the imaging and feature extraction process are such that a certain amount of uncertainty is associated with the coordinates of a minutia. The implementation of the Flash algorithm requires that the index take on discrete values, hence some binning mechanism must be used; the bin size can be used to allow appropriate tolerance for irreproducibility in minutiae positions.

A reproducible choice for the ordering of the sides is made by traversing the triangle in a consistent sense (clockwise in this implementation) as shown in Figure 15.5. This procedure is invariant under rotation and translation, but *not* under reflection. The full index consists of nine components: the length of each side, the ridge-count between each pair, and the angles measured with respect to the fiducial side.

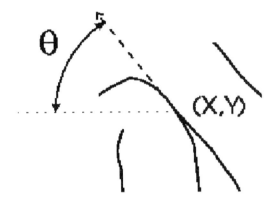

Figure 15.3 An example of a minutia point in a fingerprint. The (X,Y) coordinates are the location of the minutia in the reference frame of the print while θ is the angle that the ridge makes with respect to the x-axis of the reference frame of the print. With the appropriate convention for choosing this direction, θ has an unambiguous value in the range $[0,2\pi)$ radians.

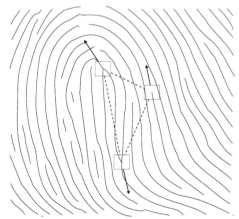

Figure 15.4 Triplet of minutiae on skeletonized image of fingerprint with the direction of ridges at minutiae (minutiae angles) shown.

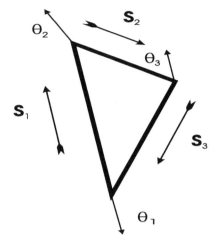

Figure 15.5 Geometry of ordered triangle without background of fingerprint. The ordering of the sides allows the expression of the minutiae angle direction with respect to an ordered side without any ambiguity. S_i are the lengths of the three sides. θ_i are the minutiae angles encoded in an transformation invariant fashion. RC_i are the number of ridges crossed by a line connecting a pair of minutiae. The sides are ordered so that the largest side appears first. Successive sides are enumerated proceeding in a defined orientation (e.g. in a clockwise fashion). The order used for the ridge counts RC_i is the same as that of the sides, while any quantity such as the minutia angle θ is ordered using the convention for the ordering the minutiae described here. The order of the minutiae will be the same as that of the sides with the first minutia being in a well-defined orientation with respect to the first side. For example, the first minutia is always taken to be the most counter-clockwise point on the first side.

Evidence Accumulation

In the Flash framework, the index generated by the procedure described above is used as the key to identify triangles that "resemble" one another. During the storage phase, each one of the indices generated by a fingerprint cause the storage of a data object labeled by the index and containing the identity of the fingerprint generating this index as well as information concerning the triplet of feature points that generated the index.

During the query phase, each index generated by the query fingerprint is used to retrieve all model objects stored in the table which are labeled with the same index. Each of these retrieved model objects represents a hypothesized correspondence between three points in the query print and three in the model print. Given this correspondence, the coordinate transformation that best maps the query triplet onto the model triplet is computed. The algorithm that computes the coordinate transformation does so in a way that minimizes the sum of the squared distances between the transformed query points and their corresponding model points. The essentials of the retrieval procedure are outlined in Figure 15.6.

The computed transformation parameters, X and Y translation and rotation θ, are binned and along with the reference fingerprint ID, used to form a key that indexes the map (keyed set) used for evidence accumulation pictured in Figure 15.2.

If a large number of feature points can be brought into correspondence by a rigid transformation of the coordinate system, all of the indices generated by the combinations of three feature points belonging to this set will generate the same coordinate transformation parameters. Hence a large number of votes for a correct match will be tabulated. There may be a number of random correspondences between triplets of points in the query print and some arbitrary reference print, but the likelihood of a number of *consistent* transformation parameters being generated by such random correspondences is quite small.

After all of the indices from the query fingerprint have been generated and all of the relevant hypotheses have been computed, the entries in the hypothesis table exceeding some threshold are sorted by score with the highest scores appearing first. This ranked list of scores can be provided in response to the original query or used as input to other decision-making machinery that might combine the results from queries using imprints taken from additional fingers.

3. Characterizing Accuracy for Verification and Identification

First consider the problem of determining whether or not two fingerprints were made by the same finger (verification). This problem amounts to assigning the pair to either of the mated or non-mated pair populations. The objective is to find a test that assigns a pair of prints to one of these two populations while making the smallest number of mistakes in large a number of trials. This problem in statistical decision making has a very long history [9]. The decision framework described here is similar in spirit to that used to describe recent work on iris identification [3]. In the case where one of two mutually exclusive hypotheses, H_0 and H_1, must be selected, two classes of errors can be made. Suppose that H_0 is the hypothesis that the pair of prints belongs to the non-mated population and that H_1 is the hypothesis that the pair of prints belongs to the mated population. Four scenarios are possible:

test says H_0 is true and H_0 is true	Test says H_0 is true and H_0 is false
test says H_1 is true and H_1 is true	test says that H_1 is true and H_1 is false

The test breaks down in two of the four scenarios; two distinct types of errors can be made:

- False Negative or Miss: incorrectly assigning a mated pair to the non-mated population

- False Positive or False Alarm: incorrectly assigning a non-mated pair to the mated population

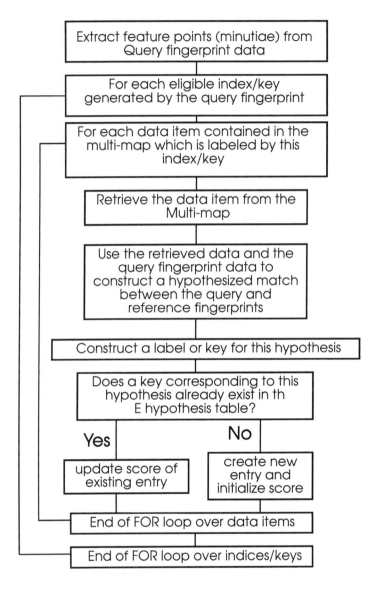

Figure 15.6 The main portion of the query phase is given in broad outline. Hypothesis key construction involves the estimation of the rigid transformation parameters, rotation and translation, using the hypothesized correspondence between triplets of minutiae in the query and model fingerprint. A description of the algorithm used for this sort of estimation can be found in the literature [5].

The number of matching triangles, henceforth referred to as the score, that generate a consistent rigid transformation between two prints can be used as the basis for a test that assigns a pair of fingerprints to the mated pair population or to the non-mated pair population. Histograms of the scores achieved by the matcher on the two test populations can be used as estimates for the conditional probability densities of the score, $f_{mated}(x)$ and $f_{non-mated}(x)$. It is natural to use a threshold x_{th} to assign a pair of imprints to one of the two possible populations. Any pair whose score exceeds x_{th} will be assigned to the mated pair population and other pairs will be assigned to the non-mated pair population. Note that there is a tradeoff between the false positive error rate (FPR) and the false negative error rate (FNR). The FNR can be reduced to an arbitrarily small value by decreasing x_{th} sufficiently, but a large number of false alarms will result with a corresponding increase in the FPR.

If this decision criterion is used, it is straightforward to compute the two error rates from the conditional probability densities computed from the test populations. The error rate for incorrectly assigning a mated pair to the non-mated population (false negative rate) is given by the distribution function defined below:

$$FNR = F_{mated}(x_{th}) = \int_0^{x_{th}} f_{mated}(t)dt. \tag{15.1}$$

Similarly, the error rate for incorrectly assigning a non-mated pair to the mated pair population (false positive rate) is given by the following function of the conditional probability distribution function:

$$FPR = 1 - F_{non-mated}(x_{th}) = 1 - \int_0^{x_{th}} f_{non-mated}(t)dt. \tag{15.2}$$

Insofar as the mated and non-mated pair test populations form representative samples of the real populations, the estimates may be used to extrapolate to behavior on the real populations. The measured accuracy of a matcher is a strong function of the database from which estimates of the error rates are derived, most importantly, the variation of the false negative error rate with threshold depends on the care with which fingerprint images were acquired. Note that these estimates of the error rates are independent of the size of the test database used although the uncertainties in the estimates depend on the sizes of the sample pair populations.

Now consider a one-to-many identification query, which may be viewed as a series of one-to-one verifications executed against every print in the database. With the assumption that at most one mate to the query is present, and the assumption that the candidate list of hypothesized matches is formed by taking all prints from the reference database whose verification matching scores with the query print exceed some fixed threshold, the false positive rate (FPR) and the false negative rate (FNR) for an identification search against a database of N individuals is as follows:

$$FNR = FNR(1)$$
$$FPR(N) = 1 - (1 - FPR(1))^N \tag{15.3}$$
$$\approx N \times FPR(1) \quad \text{for} \quad FPR(1) << \frac{1}{N}$$

The false positive rate increases drastically with database size because each additional entry in the database provides another opportunity to randomly achieve a high score. A matcher operating at a point where its false positive verification rate is 1% may be satisfactory in a verification application, but in even a small scale identification application, the error rate will become unacceptable. For example, when used on a ten person database, this matcher will generate false matches at a rate of 1 - 0.99^{10} or 9.5%. On a hundred person database this matcher's false positive error rate will be 1 - 0.99^{100} or 63%. Figure 15.7 shows the extrapolated false positive rate versus population size for a variety of one-to-one error rates. In order to keep the false positive rate within reasonable bounds when operating on large population sizes, a matcher must be operating in a mode for which its false positive rate for verification is in the range of 10^{-9}-10^{-6}. To make model-independent estimates of false positive rates in this range requires a correspondingly large sample population of mismatched pairs of prints. Because of the tradeoff between FPR and FNR, the need to operate at very small values of the false positive rate in identification applications may lead to unacceptable miss rates (FNR) when using only a single finger. The system miss rate can be reduced dramatically by executing searches using two different query fingers and considering a match on either finger to be a hit while causing a modest increase in the false positive rate. More sophisticated schemes involving larger numbers of fingers and requiring matches on multiple fingers can be used to design identification systems with less stringent requirements on the false positive rate of the matcher component.

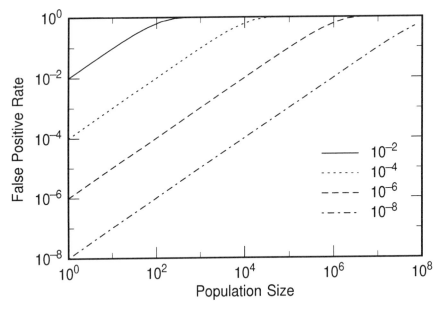

Figure 15.7 Extrapolated false positive identification error rate (FPR) plotted versus population size for a series of verification false positive rates.

4. Results

There are two aspects of the system to be characterized, the accuracy and the matching speed. Results for these two matcher characteristics are presented in Figures 15.8 and 15.9.

In order to characterize the accuracy of the system, a reference database of model prints was constructed from approximately 100,000 inked dab images (actually 97,492) acquired in 1995 which were processed by feature extraction code developed by the Exploratory Computer Vision Group at the IBM Thomas J. Watson Research Center. A description of this class of feature extraction algorithms can be found in [7]. A set of 657 queries were executed against this database. The query set of prints were a subset of the 100,000 models. Conceptually, $657 \times 97,492$ comparisons of pairs took place. These pairs can be divided into three groups:

1. pairs consisting of identical fingerprints (657)

2. pairs consisting of different impressions of the same finger (768)

3. pairs consisting of impressions of different fingers (64,050,819)

The pairs in the first group are excluded from the analysis of results because they represent an experimental artifact. The reason that there are 768 pairs in the second group is because some query prints had a single mate while others had two or even three mates in the reference database.

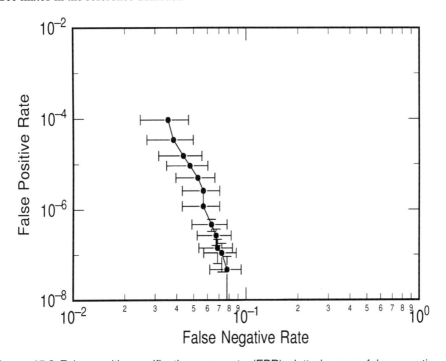

Figure 15.8 False positive verification error rate (FPR) plotted versus false negative verification error rate (FNR) for a variety of decision thresholds. The error bars represent the 90% confidence intervals for the estimates of the corresponding error rates obtained from this set of experiments. This presentation is similar to the Receiver Operating Curve (ROC) used to characterize the ability of a statistical test to distinguish between two alternative hypotheses [9].

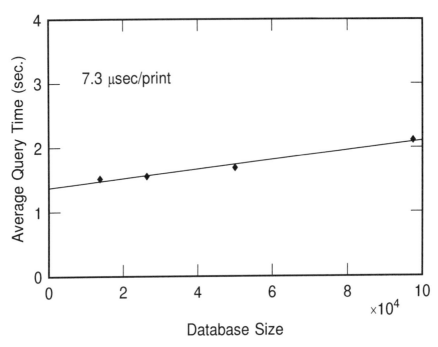

Figure 15.9 Measured average query times for a series of database sizes sizes: 13726, 26317, 50047, and 97492. Each point represents the average of 657 queries and the line is a least squares fit to the data. No front-end filtering was used in these tests--each data-point represents a similarity search of the entire database. The hardware and software configuration used for each test was the same. The non-zero intercept is a consequence of the requirement for doing index look-ups. In this series of runs, 32 disks were used to spread out the i/o burden.

The distributed Flash algorithm as implemented for fingerprint matching requires a few thousand i/o operations to look up the indices generated by each query. These i/o operations can be spread over a large number of disks to keep the elapsed time down. With 32 disks distributed over an 8 node IBM SP2 system, the incremental addition to the average query time caused by an additional print in the database is approximately 7 microseconds as shown in Figure 15.9. Thus, the system configuration used for these trials could search a database of 10 million prints in approximately 70 seconds. Additional experiments indicate that the load balancing of the i/o is such that disk parallelism can be used effectively to reduce the i/o contribution to the query time. Identification searches of very large databases of fingerprints without pre-filtering are thus possible through indexing subsets of features on the fingerprints.

The work presented here comprises the preliminary findings [4] of an IBM Research Division project initiated in early 1995 and done in collaboration with A. Califano, S. Colville, and the members of the Exploratory Computer Vision Group at the Watson Research Center. A large scale identification system based on Flash technology has been deployed in Peru as part of a voter registration and national identity program. No inferences about the current state of technology resulting from continued research and development for commercial application should be drawn from the work reported here.

References

[1] ANSI/IAI "Automated fingerprint identification systems – benchmark tests of relative performance," *Technical Report ANSI/IAI 1-1988*, American National Standards Institute, 1988.

[2] A. Califano and R. Mohan, "Multidimensional indexing for recognizing visual shapes," *IEEE Trans. Pattern Analysis and Machine Intelligence*, Vol. 16, No. 4, pp. 373-392, 1994

[3] J. Daugman, "High confidence visual recognition of persons by a test of statistical independence," *IEEE Trans. Pattern Analysis and Machine Intelligence*, Vol. 15, No. 11, pp. 1148-1161, 1993

[4] R. Germain, A. Califano, and S. Colville, "Fingerprint matching using transformation parameter clustering," *IEEE Computational Science and Engineering*, Vol. 4, No. 4, pp. 42-49, 1997.

[5] R. Haralick, H. Joo, C. Lee, X. Zhuang, V. Vaidya, and M. Kim, "Pose estimation from cooresponding point data," *IEEE Trans. Systems, Man and Cybernetics*, Vol. 19, No. 6, pp. 1426-1446, 1989.

[6] D. Huttenlocher and S. Ullman, "Object recognition using alignment," In *Proceeding of the First International Conference on Computer Vision*, pp. 102-111. IEEE Computer Society Press, 1987.

[7] A. Jain, L. Hong, S. Pankanti, and R. Bolle, "An identity authentication system using fingerprints," *Proceedings of the IEEE*, Vol. 85, No. 9, pp. 1365, 1997.

[8] Y. Lamdan and H. Wolfson, "Geometric hashing: A general and efficient model-based recognition scheme," In *Proceedings of the Second International Conference on Computer Vision*, pp. 238-249, 1988.

[9] E. Lehmann, *Testing Statistical Hypotheses, Second Edition*, Springer-Verlag, 1986.

[10] I. Rigoutsos and A. Califano, "Searching in parallel for similar strings (biological sequences)," *IEEE Computational Science and Engineering*, Vol. 1, No. 2, pp. 60-75, 1994.

[11] M. Sparrow, Measuring AFIS matcher accuracy, *The Police Chief*, pp. 147-151, 1994.

[12] G. Stockman, "Object recognition and localization via pose clustering," *Computer Vision, Graphics, and Image Processing*, Vol. 40, pp. 361-387, 1987.

[13] J. Ullman, and J. Widom, *A First Course in Database Systems*, Prentice Hall, 1997.

16 MULTIMODAL BIOMETRICS

Lin Hong and Anil K. Jain
Michigan State University
East Lansing, MI
{honglin,jain}@cse.msu.edu

Abstract *A biometric system based solely on one biometrics is often not able to meet the desired performance requirements. Identification based on multiple biometrics represents an emerging trend. We introduce a decision fusion framework which integrates two biometrics (faces and fingerprints) which complement each other in terms of identification accuracy and identification speed. This framework takes advantage of the capabilities of each individual biometrics. Therefore, it can be used to overcome, to a certain extent, both the speed and accuracy limitations of a single biometrics in making a personal identification. We have also implemented a multimodal biometric system which integrates information in both faces and fingerprints using our decision fusion framework. The system operates in the identification mode with an admissible response time. Experimental results demonstrate that the identity established by the integrated system is more reliable than the identity established by a face recognition system as well as by a fingerprint verification system.*

Keywords: *Multimodal biometrics, fingerprint matching, face recognition, integration, fusion.*

1. Introduction[1]

Accurate *automatic* personal identification is critical in a wide range of application domains such as national ID card, electronic commerce, and automated banking [14]. *Biometrics*, which refers to automatic identification of a person based on her physiological or behavioral characteristics [14], is inherently more reliable and more capable in differentiating between an authorized person and a fraudulent impostor than traditional methods such as passwords and PIN numbers. A *biometric system* is essentially a pattern recognition system which makes a personal identification by determining the authenticity of the specific physiological or behavioral characteristic possessed by the user. It can be based on either a (or one snapshot of a) single biometric characteristic or multiple biometric characteristics (or multiple snapshots of

---

[1] This chapter is based on our earlier paper based on this topic [8].

a single biometric characteristic) to make a personal identification. We define a biometric system which uses only a single biometric characteristic as a *unimodal biometric system* and a biometric system which uses multiple biometric characteristics as a *multimodal biometric system*.

A unimodal biometric system is usually more cost-efficient than a multimodal biometric system. However, it may not always be applicable in a given domain because of (*i*) unacceptable performance and (*ii*) inability to operate on a large user population. A multimodal biometric system can overcome, to a certain extent, these limitations. First of all, identification using multiple biometrics is essentially a sensor fusion problem, which utilizes information from multiple sensors to increase fault-tolerance capability, to reduce uncertainty, to reduce noise, and to overcome incompleteness of individual sensors [5,17]. A multimodal approach can increase the reliability of the decisions made by a biometric system [1,3,6,10]. Although a necessary requirement for a biometric characteristic is that each individual possess it, it is not necessary that a particular biometric characteristic of a specific individual is suitable for an automatic system. By using multiple biometric characteristics, the system will be applicable on a larger target population. Finally, a multimodal biometric system is generally more robust to fraudulent technologies, because it is more difficult to forge multiple biometric characteristics than to forge a single biometric characteristic.

In designing a multimodal biometric system, a number of issues need to be considered: (*i*) what is the main purpose of utilizing multiple biometrics? (*ii*) what is the operational mode? (*iii*) which biometrics should be integrated? and (*iv*) how many biometrics are sufficient? Since the applicable population and system robustness depend mainly on the characteristics of the selected biometrics, the main problem in designing a multimodal biometric system is the integration of individual biometrics to improve the performance in making a personal identification. Typically, performance refers to (*i*) *accuracy* and (*ii*) *speed*. System accuracy indicates how reliable and confident a biometric system is in differentiating between a genuine individual and an impostor. System speed refers to the time taken by a biometric system in making a personal identification. By properly incorporating those biometrics that are relatively fast, the overall speed of a biometric system can be improved.

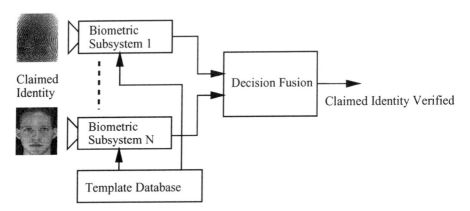

Figure 16.1 A generic multimodal verification system.

A biometric system can operate in either a *verification mode* or an *identification mode*. The integration schemes for these two modes are very different. Since only an one-to-one comparison is performed in a verification system, multimodal biometrics cannot really improve the verification speed. Therefore, integration of multiple biometrics in a verification system is mainly intended to improve the accuracy of the system. The block diagram of a generic multimodal verification system is shown in Figure 16.1. In a typical identification system, a large number of matchings need to be performed to identify an individual. A biometrics that has a large discriminating power can improve the identification accuracy, while a biometrics that is computationally efficient can improve the identification speed. The block diagram of a generic multimodal identification system is shown in Figure 16.2.

Which biometrics and how many of them should be integrated depend very much on the application domain. It is difficult to establish a systematic procedure to determine which biometrics should be used. Intuitively, the larger the number of integrated biometrics, the higher the system accuracy, but more expensive the system. In this paper, we mainly concentrate on improving the system performance by integration of two specific biometrics, namely face and fingerprint.

Multimodal Biometrics for Verification

Integration of multiple biometrics for a verification system may be performed in the following scenario: (*i*) integration of multiple snapshots of a single biometrics, for example, a number of fingerprint images of the same finger in fingerprint verification (Figure 16.3) and (*ii*) integration of a number of different biometrics (Figure 16.4). In this sense, multimodal biometrics is a conventional decision fusion problem - to combine evidence provided by each biometrics to improve the overall decision accuracy. Generally, multiple evidences may be integrated at one of the following three different levels [2]: (*i*) Abstract level; the output from each module is only a set of possible labels without any confidence value associated with the labels; in this case, the simple majority rule may be employed to reach a more reliable decision [20];

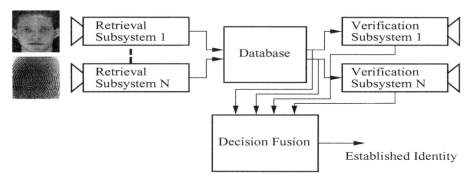

Figure 16.2 A generic multimodal identification system.

(*ii*) Rank level; the output from each module is a set of possible labels ranked by decreasing confidence values, but the confidence values themselves are not specified; (*iii*) Measurement level; the output from each module is a set of possible labels with associated confidence values; in this case, more accurate decisions can be made by integrating different confidence values.

Dieckmann *et al.* [6] have proposed an abstract level fusion scheme: "2-from-3 approach" which integrates face, lip motion, and voice based on the principle that a human uses multiple clues to identify a person. This approach uses a simple voting algorithm to find whether the decision made by each individual classifier is consistent with the other two classifiers. Brunelli and Falavigna [2] have proposed two schemes to combine evidence from speaker verification and face recognition. The first scheme is a measurement level scheme in which the outputs of two different speech classifiersand the outputs of three different face classifiers are normalized and combined using geometric average. The second scheme is a hybrid rank/measurement level scheme which uses HyperBF network to combine the outputs of these five classifiers. The authors have demonstrated that the system accuracy can be improved by using these fusion schemes. Kittler *et al.* [10] have demonstrated the efficiency of an integration strategy which fuses multiple snapshots of a single biometrics using a Bayesian framework. In this scheme, the *a posteriori* class probabilities for each individual are estimated and the decision is made based on the average or max or median of the *a posteriori* class probabilities for a given set of snapshots. Bigun *et al.* [1] have proposed a Bayesian integration scheme to combine different evidences based on the assumption that the evidences are independent of one another. Their scheme results in an improved recognition accuracy by combining voice and face as well as voice and lip motion. Maes and Beigi [12] have proposed to combine biometric data (*e.g.*, voice) with non-biometric data (*e.g.*, password).

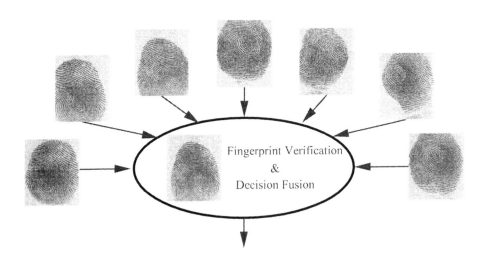

Figure 16.3 Integration of multiple snapshots of a single biometric characteristic.

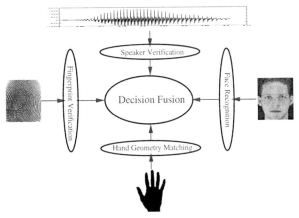

Figure 16.4 Integration of different biometric characteristics.

Multimodal Biometrics for Identification

All the decision fusion schemes mentioned above can be used to improve the identification accuracy in a multimodal identification system. However, since an identification system needs to perform one-to-many comparisons to find a match, the average computational complexity for each comparison should be as low as possible to enable a reasonable response time, especially for a large database. However, the integration schemes mentioned above increase the computational complexity for each comparison. Therefore, it is not practical to directly apply these schemes to an identification system; an integration scheme that is able to improve both the speed and the accuracy should be used. In this paper, we introduce a multimodal biometrics scheme which integrates two biometrics (in particular, face and fingerprint) which complement each other in terms of identification speed and identification accuracy: a biometric approach (*e.g.*, face recognition) that is suitable for database retrieval is used to index the template database and a biometric approach (*e.g.*, fingerprint verification) that is reliable in deterring impostors is used to ensure the overall system accuracy. In addition, since each biometric approach provides a certain confidence about the identity being established, a decision fusion scheme which exploits all the information at the output of each module can be used to make a more reliable decision.

In our multimodal scheme, a biometrics that is efficient in database retrieval is used to retrieve the top *n* matches from the database, where *n* is a design parameter. We define such a biometrics as a *retrieval biometrics*. Since retrieval biometrics is not necessarily very reliable, a large value of *n* can be used to guarantee that the true match is retrieved from the database. After the top *n* matches have been retrieved, a biometrics which has a very high identification accuracy is used to find which of the top *n* retrievals is the correct match. We define this biometrics as a *verification biometrics*. So far, the above integration has addressed only the identification speed. We can also improve the identification accuracy by combining the confidence values associated with the decisions made by each individual biometrics. In order to implement such a decision fusion scheme, we need to define (*i*) a confidence measure for each decision and (*ii*) a decision fusion criterion. Note that the confidence values associated with the top *n* matches obtained using the retrieval biometrics have a different distribution than the confidence values associated with the verification biometrics. In this paper, we will give a general framework to derive the distributions of these two different confidence values.

Remainder of the chapter is organized as follows. Section 16.2 formulates the decision fusion scheme. A prototype integration system which combines evidence provided by face and fingerprint using the proposed fusion scheme is described in Section 16.3. Experimental results on the MSU fingerprint databases and public domain face database are also reported in Section 16.3. Finally, the summary and conclusions are given in Section 16.4.

2. Decision Fusion

A biometric system makes a personal identification by comparing the similarity between input measurements and stored templates. Due to *intraclass variations* inherent in the biometric characteristics, the decision made by a biometrics has an associated confidence value. A decision fusion scheme should utilize all these confidence values associated with individual decisions to reach a more reliable decision. As we mentioned early, in order to derive a decision fusion scheme, we need to define (*i*) a confidence measure for each individual biometrics and (*ii*) a decision fusion criterion. The confidence of a given biometrics may be characterized by its *false acceptance rate* (FAR), which is defined as the probability of an impostor being accepted as a genuine individual. In order to estimate FAR, the *impostor distribution* which is defined as the distribution of similarity between biometric characteristic(s) of different individuals needs to be computed.

Impostor Distribution for Retrieval Biometrics

Due to the relatively low discrimination capability of the retrieval biometrics, the top n matches need to be retrieved from the database to guarantee that the genuine individual will be identified if he or she is in the database. In order to retrieve the top n matches, N comparisons need to be performed explicitly (in the linear search case) or implicitly (in organized search cases such as the tree search), where N is the total number of templates in the database. The comparisons of a query against templates from different individuals essentially provide an indication of interclass variations, which can be used to refine the confidence of a genuine match.

Let Φ_1, Φ_2, ..., Φ_N be the N templates stored in the database. For simplicity, let us assume that each individual has only one template and that a distance measure is used to indicate the similarity between a query and a template. So, the smaller the distance value, the more likely it is that the match is a correct match. Let Φ'_1, Φ'_2, ..., Φ'_n be the top n matches obtained by searching through the entire database using the retrieval biometrics. Let us further assume that the top n matches are arranged in the increasing order of distance values. Since the relative distances between consecutive matches are invariant to the mean shift of the distances, we will use relative distance values to characterize the impostor distribution. The probability that a retrieval in the top n matches is incorrect is different for different ranks. The probability that the first match is incorrect tends to be smaller than the probability that the second match is incorrect, the probability that the second match is incorrect tends to be smaller than the probability that the third match is incorrect, and so on. Thus, the impostor distribution is a function of both the relative distance values, Δ, and the rank order, i, which has the following form: $F_i(\Delta)P_{order}$ (i), where $F_i(\Delta)$ represents the probability that the relative distance between impostors and their claimed identity at rank i is larger than a value Δ and $P_{order}(i)$ represents the probability that the retrieved match at rank i is an impostor. In practice, $F_i(\Delta)$ and $P_{order}(i)$ need to be estimated from empirical data.

Let I_1, I_2, ...,I_N denote the identity indicators of the N individuals in the template database. Let X^α denote the distance between an individual and her own template which is a random variable with density function $f^\alpha(X^\alpha)$. Let X^β_1, X^β_2, ..., X^β_{N-1} denote the distances between an individual and the templates of the other individuals in the database, which are random variables with density functions, $f^\beta_1(X^\beta_1)$, $f^\beta_2(X^\beta_2)$, ..., $f^\beta_{N-1}(X^\beta_{N-1})$, respectively. Assume, for simplicity of analysis, that X^α and X^β_1, X^β_2, ..., X^β_{N-1} are statistically independent and $f^\beta_1(X^\beta_1) = f^\beta_2(X^\beta_2) = \ldots f^\beta_{N-1}(X^\beta_{N-1}) = f^\beta(X^\beta)$. For an individual, Π, which has a template stored in the database,$\{\Phi_1, \Phi_2, ..., \Phi_N\}$, the rank, R, of X^α among X^β_1, X^β_2, ..., X^β_{N-1} is a random variable with probability

$$P(R = i) = \frac{(N-1)!}{i!(N-1-i)!} p^i(1-p)^{(N-1-i)}, \tag{16.1}$$

where

$$p = \int_{-\infty}^{\infty} \int_{-\infty}^{X\alpha} f^\alpha(X^\alpha) f^\beta(X^\beta) dX^\beta dX^\alpha. \tag{16.2}$$

When $p \ll 1$ and N is sufficiently large, P(R) may be approximated by a Poisson distribution [15],

$$P(R = i) \doteq \frac{e^{(-a)}a^i}{i!}, \tag{16.3}$$

where $a \doteq np$. Obviously, $P(R=i)$ is exactly the probability that matches at rank R=i are genuine individuals. Therefore,

$$P_{order}(i) = 1 - P(R = i). \tag{16.4}$$

Although the assumption that X^β_1, X^β_2, ..., X^β_{N-1} are *i.i.d.* may not be true in practice, it is still reasonable to use the parametric form in Eq. (16.4) to estimate the probability that retrieved matches at rank i are impostors.

Without any loss of generality, we assume that, for a given individual, Π, X^β_1, X^β_2, ..., X^β_{N-1} are arranged in increasing order of values. Define the non-negative distance between the $(i+1)th$ and ith distance values as the *ith* relative distance,

$$\Delta_i = X^\beta_{i+1} - X^\beta_i, \quad 1 \leq i < N-1. \tag{16.5}$$

The distribution, $f_i(\Delta_i)$, of the ith relative distance, Δ_i, is obtained from the joint distribution $g_i(X^\beta,\Delta_i)$ of the ith value, X^β, and the ith relative distance, Δ_i,

$$f_i(\Delta_i) = \int_{-\infty}^{\infty} g_i(X^\beta,\Delta_i)dX^\beta,$$

(16.6)

(16.7)

$$\bar{F}^\beta(\Delta_i) = [1 - F^\beta(X^\beta + \Delta_i)],$$

$$g_i(X^\beta,\Delta_i) = CF^\beta(X^\beta)^{i-1}\bar{F}^\beta(\Delta_i)^{N-i}f^\beta(X^\beta + \Delta_i),$$

(16.8)

$$C = \frac{(N-1)!}{(i-1)!(N-2-i)!},$$

(16.9)

where [7]

$$F^\beta(X^\beta) = \int_{-\infty}^{X^\beta} f^\beta(X^\beta)dX^\beta.$$

With the distribution, $f_i(\Delta_i)$, of the ith relative distance defined, the probability that the relative distance of the impostor at rank i is larger than a threshold value, Δ, is

$$F_i(\Delta) = \int_{\Delta}^{\infty} f_i(\Delta_i)d\Delta_i.$$

(16.10)

Let us assume that $X^\beta{}_1, X^\beta{}_2, ..., X^\beta{}_{N-1}$ have a Gaussian distribution with unknown mean and variance. Note that $F_i(\Delta)$ depends on the number of individuals, N, enrolled in the database. However, it does not mean that F_i has to be recomputed whenever a new individual is enrolled in the database. If $N \gg 1$, the distributions of F_is for different values of N are quite similar to one another. On the other hand, the decision criterion still satisfies the FAR specification when N increases, though it may not be able to take a full advantage of the information contained in the N comparisons. In practice, an update schema which recomputes the decision criteria whenever the number of added individuals is larger than a pre-specified value can be used to exploit all the available information.

Impostor Distribution for Verification Biometrics

Depending on the availability of the statistical models or properties of the verification biometrics, *i.e.* a probability distribution function that precisely characterizes the decision making process, the problem of impostor distribution estimation may fall into one of the following three situations: (*i*) known model with known parameters, (*ii*) known model with unknown parameters, and (*iii*) unknown model. In situation (*i*), there exists a *statistical model* with known parameter values which exactly characterizes the verification biometrics. Unfortunately, there is no biometrics which can be precisely formulated by a known statistical model. In both situations (*ii*) and (*iii*), the impostor distribution of the verification biometrics can not be precisely determined. Instead, they can only be approximated by an empirical estimate.

However, if it is possible to obtain a general model of the overall impostor distribution by making some simplifying assumptions, then the impostor distribution can be reliably estimated from a set of test samples.

Fingerprint matching depends on the comparison of the two most prominent local ridge   characteristics (*minutiae*) and their relationships to determine whether two fingerprints are from the same finger or not [9]. A model that can precisely characterize the impostor distribution of a fingerprint matching algorithm is not easy, since (*i*) the minutiae in a fingerprint are distributed randomly in the region of interest, (*ii*) the region of interest for each input fingerprint may be different, (*iii*) each input fingerprint tends to have a different number of minutiae, (iv) there may be a significant number of spurious minutiae and missing minutiae, (*v*) sensing, sampling, and feature extraction may result in errors in minutiae positions, and (*vi*) sensed fingerprints may have different distortions. However, it is possible to obtain a general model of the overall impostor distribution by making some simplifying assumptions.

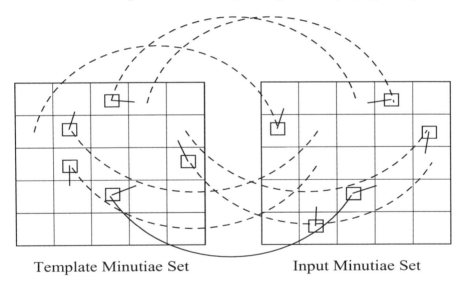

Template Minutiae Set   Input Minutiae Set

Figure 16.5 Minutiae matching model. A solid line indicates a match and a dashed line indicates a mismatch.

Let us assume that the input fingerprint and the template have already been registered and the region of interest of both the input fingerprint and the template is of the same size, a $W \times W$ (for example, 500×500) region. The $W \times W$ region is tessellated into small cells of size $w \times w$ which is assumed to be sufficiently large (for example, 40×40) such that possible deformation and transformation errors are within the bound specified by the cell size. Therefore, there are a total of $(W/w) \times (W/w)$ $(=N_c)$ different cells in the region of interest. Further, assume that each fingerprint has the same number of minutiae, N_m ($\leq N_c$), which are distributed randomly in different cells and each cell contains at most one minutiae. Each minutiae is directed towards one of the D (for example, 8) possible orientations with equal probability. Thus, for a given cell, the probability, P_{empty}, that the cell is empty with no minutiae present is (N_m/N_c) and the probability, P, that the cell has a minutiae that is directed towards a specific orientation is $(1-P_{empty})/D$. A pair of corresponding minutiae between a template and an input is considered to be identical if and only if they are in the cells at the same position and directed in the same direction (Figure 16.5). With the above simplifying assumptions, the number of corresponding minutiae pairs between any two randomly selected minutiae patterns is a random variable, Y, which has a Binomial distribution with parameters N_m and P [15]:

$$g(Y) = \frac{N_m!}{Y!(N_m - Y)!} P^Y (1-P)^{(N_m - Y)}.$$

(16.11)

The probability that the number of corresponding minutiae pairs between any two minutiae patterns is less than a given threshold value, y, is

$$G(y) = g(Y < y) = \sum_{k=0}^{y-1} g(k).$$

(16.12)

The decision made by a minutiae matching algorithm for an input fingerprint and a template is generally based on the comparison of the "normalized" number of corresponding minutiae pairs against a threshold. Therefore, under the assumption that minutiae in the region of interest of fingerprints of different individuals are randomly distributed, the probability that an impostor is accepted is $(1-G(y))$.

Decision Fusion

The impostor distribution for retrieval biometrics and the impostor distribution for verification biometrics provide confidence measures for each of the top n matches. In order to combine these confidence values to generate a more reliable decision about the genuine identity of a query, a joint impostor distribution of retrieval biometrics and verification biometrics is needed. It is reasonable to assume that the retrieval biometrics is statistically independent of the verification biometrics, because the similarity of one biometric characteristics between two individuals does not imply the similarity of a different biometric characteristics and vice versa. Let $F_i(\Delta)P_{order}(i)$ and $G(Y)$ denote the impostor distributions at rank i for the retrieval biometrics and the verification biometrics, respectively. The joint impostor distribution of retrieval biometrics and verification biometrics at rank i may be defined as

$$H_i(\Delta, Y) = F_i(\Delta)P_{order}(i)G(Y).$$

(16.13)

A decision criterion can then be derived from the joint impostor distribution. Since we are interested in deriving a decision criterion which satisfies the FAR specification, the decision criterion must ensure that the FAR of the multimodal system should be less than the given value. Without a loss of generality, we assume that at most one of the top n identities established by the retrieval biometrics for a given query is the genuine identity of the individual. The final decision by the multimodal system either rejects all the n possibilities or accepts only one of them as the genuine identity. Let $\{I_1, I_2, ..., I_n\}$ denote the n possible identities established by retrieval biometrics, $\{X_1, X_2, ..., X_n\}$ denote the corresponding n distances, $\{Y_1, Y_2, ..., Y_n\}$ denote the corresponding n distances for the verification biometrics, and FARo denote the specified value of FAR. The identity decision, $ID(\Pi)$, for a given individual Π can be determined by the following criterion:

$$ID(\Pi) = \begin{cases} I_k, & if \begin{cases} H_k(\Delta_k, Y_k) < FAR_o, and \\ H_k(\Delta_k, Y_k) = H_*(\Delta_*, Y_*) \end{cases} \\ impostor, & otherwise. \end{cases} \quad (16.14)$$

where

$$H_*(\Delta_*, Y_*) = \min\{H_1(\Delta_1, Y_1), ..., H_n(\Delta_n, Y_n)\}, \quad (16.15)$$

and $\Delta_i = X_{i+1} - X_i$. Since $H_i(\Delta, Y)$ defines the probability that an impostor is accepted at rank i with relative distance, Δ, and fingerprint matching score, Y, the above decision criterion satisfies the FAR specification.

3. Experimental Results

The proposed decision fusion scheme has been used in our prototype multimodal biometric system which integrates face recognition and fingerprint verification to improve the identification performance [8]. The block diagram of our prototype system is shown in Figure 16.6.

Face Recognition

In personal identification, *face recognition* refers to static, controlled full frontal portrait recognition [4]. There are two major tasks in face recognition: (*i*) face location and (*ii*) face recognition. Face location is to find whether there is a face in the input image and the location of the face in the image. Since the background is controlled or almost controlled in biometric applications, face location is not considered to be an extremely difficult problem. Face recognition finds the similarity between the located face and the stored templates to determine the identity. A number of face recognition approaches have been reported in the literature. The performance of some of the proposed face recognition approaches is very impressive. Two of the well-known commercial face recognition systems are Faceit [19] and Trueface [13]. Phillips *et al.* [16] concluded that "face recognition algorithms were developed and were sufficiently mature that they can be ported to real-time experimental/demonstration system." In our system, the eigenface approach [18] is used for the following reasons: (*i*) the background, facial transformations (*e.g.,* scale and rotation), and illumination can be controlled, (*ii*) it has a compact representation, (*iii*) it is feasible to index an eigenface-based template database using different indexing techniques such that the retrieval can be conducted efficiently [4], and (*iv*) the eigenface approach is a generalized template matching approach which was demonstrated to be more accurate than the attribute-based approach [2].

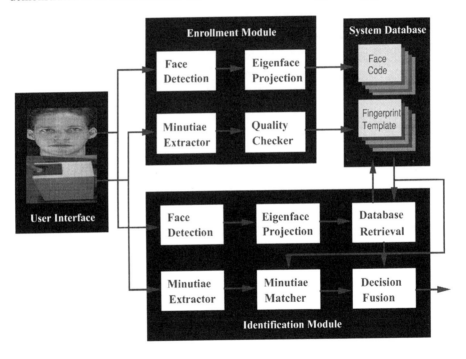

Figure 16.6 Block diagram of the prototype multimodal biometric system.

Fingerprint Verification

A *fingerprint* is the pattern of ridges and furrows on the surface of a fingertip (Figure 16.7). The uniqueness of a fingerprint is exclusively determined by the local ridge characteristics and their relationships. Fingerprint matching generally depends on the comparison of local ridge characteristics and their relationships [9,11]. The two most prominent ridge characteristics, called minutiae, are (*i*) *ridge ending* and (*ii*) *ridge bifurcation* (Figure 16.7). Automatic fingerprint verification mainly consists of two stages [9]: (*i*) minutiae extraction and (*ii*) minutiae matching. Minutiae extraction module extracts minutiae from input fingerprint images and the matching module determines whether two minutiae patterns are from the same finger or not. We have developed an automatic fingerprint verification system which has been evaluated on a large database [9].

Databases

The integrated biometric system was tested on the MSU fingerprint database and three public domain face databases. The MSU fingerprint database contains a total of 1,500 fingerprint images (640 × 480) from 150 individuals with 10 images per individual, which were captured with an optical scanner manufactured by Digital Biometrics. When these fingerprint images were captured, no restrictions on the finger position, orientation, and impression pressure were imposed. As a result, the fingerprint images vary in quality. Approximately 90% of the fingerprint images in the MSU database are of reasonable quality, similar to the images shown in Figures 16.9 (b) and (d). Images of poor quality with examples shown in Figures 16.9 (f) and (h) are mainly due to large creases, smudges, dryness of the finger, and high impression pressure. The face database contains a total of 1,132 gray level images of 86 individuals; 400 images of 40 individuals with 10 images per individual are from the Olivetti Research Laboratory, 300 images of 30 individuals with 10 images per individual are from the University of Bern, and 432 images of 16 individuals with 27 images per individual are from the MIT Media Lab. The images were re-sampled from the original sizes to a fixed size of 92 × 112 and normalized to zero mean.

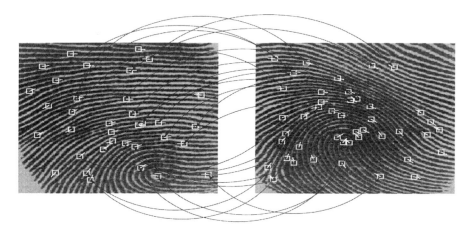

Figure 16.7 Fingerprint matching.

Test Results

We randomly selected 640 fingerprints of 64 individuals as the training set and the remaining as the test set. The mean and variance of the impostor distribution (Figure 16.10(a)) were estimated to be 0.70 and 0.64, respectively, from the 403,200 (640 × 630) impostor matching scores of "all against all" verification test by fitting a binomial model [8]. A total of 542 face images were used as training samples. Since variations in position, orientation, scale, and illumination exist in the face database, the 542 training samples were selected such that all the representative views are included. Eigenfaces were estimated from the 542 training samples and the first 64 eigenfaces associated with the 64 largest eigenvalues were saved. This resulted in about 95% of the total variance to be retained. The top $n = 5$ impostor distributions were approximated. Generally, the larger the value of n, the lower the false reject rate of face recognition. However, as n increases, more candidates need to be verified by fingerprint verification. This illustrates the trade-off between the accuracy and speed of a multimodal biometric system. Figure 16.10(b) shows the impostor distribution for face recognition at rank no. 1. We randomly assigned each of the remaining 86 individuals in the MSU fingerprint database to an individual in the face database (see Figure 16.9 for some examples). Since the similarity between two different faces is statistically independent of the fingerprint matching scores between the two individuals, such a random assignment of a face to a fingerprint is reasonable. One fingerprint for each individual is randomly selected as the template for the individual. To simulate the practical identification scenario, each of the remaining 590 faces was paired with a fingerprint to produce a test pair. In the testing phase, with a pre-specified confidence value (FAR), for each of the 590 fingerprint and face pairs, the top 5 matches were retrieved using face recognition. Then fingerprint verification was applied to each of the resulting top 5 matches and a final decision was made by the decision fusion scheme.

The typical FAR for a biometric system is usually very small (<0.0001). In order to demonstrate that a given biometric system does indeed meet such a specification, a very large set of representative samples (>100,000) is needed. Unfortunately, obtaining such a large number of test samples is both expensive and time consuming. To overcome this hurdle, we reuse the individual faces by different assignments - each time, a different fingerprint is assigned to a given face to form a face and fingerprint probe pair. Obviously, such a reuse schema might result in an unjustified performance improvement. In order to diminish the possible gain in performance due to such a reuse schema, we multiplied the estimated impostor distribution for face recognition by a constant of 1.25, which essentially reduces the contribution of face recognition to the final decision by a factor of 1.25. On the other hand, fingerprint verification operates in the one-to-one verification mode, so different assignments may be deemed as different impostor forgeries. Therefore, the test results using such a random assignment schema are able to reasonably estimate the underlying performance numbers. In our test, 1,000 different assignments were tried. A total of 590,000 (590 × 1,000) face and fingerprint test pairs were generated and tested. The FRR of our system with respect to different pre-specified FAR values, as well as the FRR obtained by "all-to-all" verifications using only fingerprints (2,235,000 = 1500 × 1490 tests) or faces (342,750 = 350 × (590-5) + 240 × (590-15) tests) are listed in Table 16.1. Note that the FRR values in fusion column include the error rate (1.8%) of genuine individuals not present in the top 5 matches established by face recognition. The receiver operating curves are plotted in Figure 16.11, in which the authentic acceptance rate (the percentage of genuine individuals being accepted, *i.e.* 1-*FRR*) is plotted against FAR. We can conclude from these test results that an integration of fingerprint and face does result in an improvement of identification accuracy.

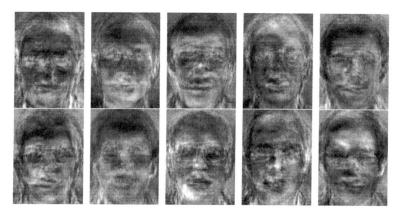

Figure 16.8 First ten eigenfaces obtained from 542 training images of size 92 × 112; they are listed, from left to right and top to bottom, in decreasing values of the corresponding eigenvalues.

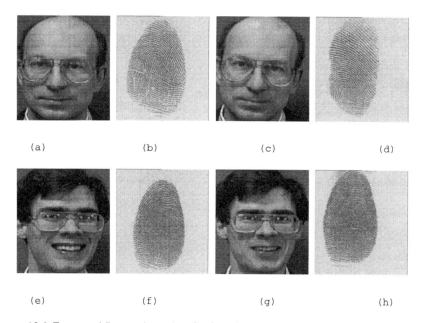

(a)   (b)   (c)   (d)

(e)   (f)   (g)   (h)

Figure 16.9 Face and fingerprint pairs; the face images (92 × 112) are from the Olivetti Research Lab. And the fingerprint images (640 × 480) are captured with a Digital Biometrics scanner; without any loss of generality, face and fingerprint pairs were randomly formed.

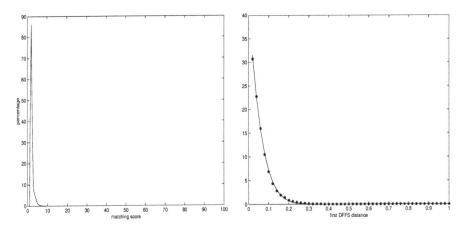

Figure 16.10 Impostor distributions; (a) impostor distribution for fingerprint verification; the mean and variance of the impostor distribution are estimated to be 0.70 and 0.64, respectively; (b) the impostor distribution for face recognition at rank no. 1, where the stars (*) represent empirical data and the solid curve represents the fitted distribution; the mean square error between the empirical distribution and the fitted distribution is 0.0014.

FAR (%)	Face FRR (%)	Fingerprint FRR(%)	Fusion FRR (%)
1.000	15.8	03.9	1.8
0.100	42.2	06.9	4.4
0.010	61.2	10.6	6.6
0.001	64.1	14.9	9.8

Table 16.1 False reject rates (FRR) on the test set with different values of FAR.

Face Recognition	Fingerprint Verification	Total
0.9 seconds	1.2 seconds	2.1 seconds

Table 16.2 Average CPU time per identification on a Sun UltraSPARC 1 workstation.

In order for the prototype identification system to be acceptable in practice, the response time of the system needs to be within a few seconds. Table 16.2 shows that our multimodal system does meet the response time requirement.

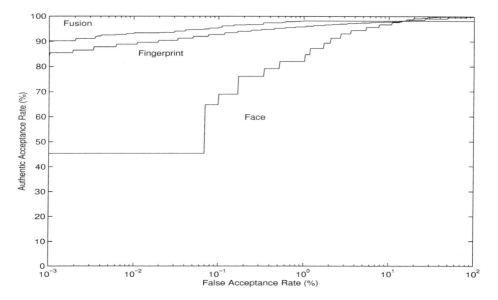

Figure 16.11 Receiver Operating Curves; the horizontal axis is FAR and the vertical ais is (1-FRR).

4. Conclusions

A biometric system which is based only on a (or one snapshot of a) single biometric characteristic may not always be able to achieve the desired performance. A *multimodal biometrics* technique, which combines multiple biometrics in making an identification, can be used to overcome the limitations. Integration of multimodal biometrics for an identification system has two goals: (*i*) improve the identification accuracy and (*ii*) improve the identification speed (throughput). We have developed a decision fusion scheme which integrates two different biometrics (face and fingerprint) that complement each other. In this scheme, a biometrics that is suitable for database retrieval is used to index the template database and a biometrics that is reliable in deterring impostors is used to ensure the overall system accuracy. In addition, a decision fusion scheme which exploits all the information in the decisions made by each individual biometrics is used to make a more reliable decision.

In order to demonstrate the efficiency of our decision fusion scheme, we have developed a prototype multimodal biometric system which integrates faces and fingerprints in making a personal identification. This system overcomes many of the limitations of both face recognition systems and fingerprint verification systems. Experimental results demonstrate that our system performs very well; it meets the response time as well as the accuracy requirements.

References

[1] E. S. Bigun, J. Bigun, B. Duc, and S. Fischer, "Expert conciliation for multi modal person authentication systems by Bayesian statistics," In *Proceedings 1st Int. Conf. On Audio Video-Based Personal Authentication*, pp. 327-334, Crans-Montana, Switzerland, 1997.

[2] R. Brunelli, and D. Falavigna, "Personal identification using multiple cues," *IEEE Trans. on Pattern Analysis and Machine Intelligence*, Vol. 17, No. 10, pp. 955-966, 1995.

[3] R. Brunelli and T. Poggio, "Face recognition: Features versus templates," *IEEE Trans. on Pattern Analysis and Machine Intelligence*, Vol. 15, No. 10, pp. 1042-1052, 1993.

[4] R. Chellappa, C. Wilson, and A. Sirohey, "Human and machine recognition of faces: A survey, "*Proceedings IEEE*, Vol. 83, No. 5, pp. 705-740, 1995.

[5] J. Clark and A. Yuille *Data Fusion for Sensory Information Processing Systems*. Kluwer Academic Publishers, Boston, 1990.

[6] U. Dieckmann, P. Plankensteiner, and T. Wagner, "SESAM: A biometric person identification system using sensor fusion," Pattern Recognition Letters, Vol. 18, No. 9, pp. 827-833, 1997.

[7] E. J. Gumbel, *Statistics of Extremes*. Columbia University Press, New York, 1958.

[8] L. Hong and A. Jain, "Integrating faces and fingerprints for personal identification," In *Proceedings 3rd Asian Conference on Computer Vision*, pp. 16-23 Hong Kong, China, 1998. To also appear in *IEEE Trans. Pattern Analysis and Machine Intelligence*.

[9] A. Jain, L. Hong, and R. Bolle, "On-line fingerprint verification," *IEEE Trans. Pattern Analysis and Machine Intelligence*, Vol. 19, No. 4, pp. 302-314, 1997.

[10] J. Kittler, Y. Li, J. Matas, and M. U. Sanchez, "Combining evidence in multimodal personal identity recognition systems," In *Proceedings 1st International Conference On Audio Video-Based Personal Authentication*, pp. 327-344, Crans-Montana, Switzerland, 1997.

[11] H. C. Lee and R. E. Gaensslen, *Advances in Fingerprint Technology*. Elsevier, New York, 1991.

[12] S. Maes and H. Beigi, "Open sesame! speech, password or key to secure your door?" In *Proceedings 3rd Asian Conference on Computer Vision*, pp. 531-541, Hong Kong, China, 1998.

[13] Miros, *Miros Homepage*. http://www.miros.com, 1998.

[14] E. Newham.,*The Biometric Report*. SJB Services, New York, 1995.

[15] A. Papoulis, *Probability, Random Variables, and Stochastic Processes*. McGraw-Hill, New York, 1965.

[16] P. J. Phillips, P. J. Rauss, and S. A. Der, *FERET (Face Recognition Technology) Recognition Algorithm Development and Test Results*, U. S. Government Publication, ALR-TR-995, Army Research Laboratory, Adelphi, MD, 1996.

[17] Solberg, T. Taxt, and A. Jain, "A Markov random field model for classification of multisource satellite imagery," IEEE Trans. Geoscience and Remote Sensing, Vol. 34, No. 1, pp. 100-113, 1996.

[18] M. Turk and A. Pentland, "Eigenfaces of recognition," Journal of Cognitive Neuroscience, Vol. 3, No. 1, pp.71-86, 1991.

[19] Visionics. Visionics homepage. http://www.visionics.com, 1988.

[20] Y. A. Zuev, and S. K. Ivanov, "The voting as a way to increase the decision reliability," In *Proceedings Foundations of Information/Decision Fusion with Applications to Engineering Problems*, pp. 203-210, Washington, D.C., 1996.

17 TECHNICAL TESTING AND EVALUATION OF BIOMETRIC IDENTIFICATION DEVICES

James L. Wayman
National Biometric Test Center
San Jose State University
San Jose, CA
biomet@email.sjsu.edu

Abstract *Although the technical evaluation of biometric identification devices has a history spanning over two decades, it is only now that a general consensus on test and reporting measures and methodologies is developing in the scientific community. By "technical evaluation", we mean the measurement of the five parameters generally of interest to engineers and physical scientists: false match and false non-match rates, binning error rate, penetration coefficient and transaction times. Additional measures, such as "failure to enroll" or "failure to acquire", indicative of the percentage of the general population unable to use any particular biometric method, are also important. We have not included in this chapter measures of more interest to social scientists, such as user perception and acceptability. Most researchers now accept the "Receiver Operating Characteristic" (ROC) curve as the appropriate measure of the application-dependent technical performance of any biometric identification device. Further, we now agree that the error rates illustrated in the ROC must be normalized to be independent of the database size and other "accept/reject" decision parameters of the test. This chapter discusses the general approach to application-dependent, decision-policy independent testing and reporting of technical device performance and gives an example of one practical test. System performance prediction based on test results is also discussed.*

Keywords: *Biometric identification, testing, receiver operating characteristic curve.*

1. Introduction

We can say, somewhat imprecisely, that there are two distinct functions for biometric identification devices: 1) to prove you are who you say you are, and 2) to prove you are not who you say you are not. In the first function, the user of the system makes a "positive" claim of identity. In the second function, the user makes the "negative" claim that she is not anyone already known to the system.

Biometric systems attempt to use measures that are both distinctive between members of the population and repeatable over each member. To the extent that measures are not distinctive or not repeatable, errors can occur. In discussing system errors, the terms "false acceptance" and "false rejection" always refer to the claim of the user. So a user of a positive identification system, claiming to match an enrolled record, is "falsely accepted" if incorrectly matched to a truly non-matching biometric measure, and "falsely rejected" if incorrectly not matched to a truly matching biometric measure. In a negative identification system, the converse is true: "false rejection" occurring if two truly non-matching measures are matched, and "false acceptance" occurring if two truly matching measures are not matched. Most systems have a policy allowing use of multiple biometric samples to identify a user. The probability that a user is ultimately accepted or rejected depends upon the accuracy of the comparisons made and the accept/reject decision policy adopted by the system management. This decision policy is determined by the system manager to reflect the operational requirements of acceptable error rates and transaction times and, thus, is not a function of the biometric device itself.

Consequently, in this chapter we refer to "false matches" and "false non-matches" resulting from the comparison of single presented biometric measure to a single record previously enrolled. These measures can be translated into "false accept" and "false reject" under a variety of system decision policies.

In addition to the decision policy, the system "false rejection" and "false acceptance" rates are a function of five inter-related parameters: single comparison false match and false non-match rates, binning error rate, penetration coefficient, and transaction speed. In this chapter, we will focus on testing of these basic parameters and predicting system performance based on their resulting values and the system decision policy.

Regardless of system function, the system administrator ultimately has three questions: What will be the rate of occurrence of false rejections, requiring intervention by trained staff?; Will the probability of false acceptance be low enough to deter fraud?; Will the throughput rate of the system keep up with demand? The first question might further include an estimate of how many customers might be unable to enroll in or use the system. The focus of this chapter will be on developing predictive tools to allow "real-world" estimates of these numbers from small-scale tests.

2. Classifying Applications

Technology performance is highly application dependent. Both the repeatability and distinctiveness of any biometric measure will depend upon difficulty of the application environment. Consequently, we must test devices with a target application in mind. Although each application is clearly different, some striking similarities emerge when considered in general. All applications can be partitioned according to at least seven categories:

1. Cooperative versus Non-cooperative: Is the deceptive user attempting to cooperate with the system to appear to be someone she is not, or attempting not to cooperate to not appear to be someone known to the system?

2. Overt versus Covert: Is the user aware that the biometric measure is being taken?

3. Habituated versus Non-habituated: Is the user well acquainted with the system?

4. Attended versus Non-attended: Is the use of the biometric device observed and guided by system management?

5. Standard Environment: Is the application indoors or in an outdoor, or environmentally stressful, location?

6. Public versus Private: Will the users of the system be customers (public) or employees (private) of the system management?

7. Open versus Closed: Will the system be required, now or in the future, to exchange data with other biometric systems run by other management?

This list is incomplete, meaning that additional partitions might also be appropriate. We could also argue that not all possible partition permutations are equally likely or even permissible. A cooperative, overt, habituated, attended, private, application in a laboratory environment will generally produce lower error rates than outdoor applications on a non-habituated, unattended population.

3. The Generic Biometric System

Although biometric devices rely on widely different technologies, much can be said about them in general. Figure 17.1 shows a generic biometric identification system, divided into five sub-systems: data collection, transmission, signal processing, decision and data storage. The key subsystems are:

1. Data collection, which includes the imaging of a biometric pattern presented to the sensor.

2. Transmission, which may include signal compression and re-expansion and the inadvertent addition of noise.

3. Signal processing, in which the stable, yet distinctive, "features" are extracted from the received signal and compared to those previously stored.

4. Storage of "templates" derived from the "features" and possibly the raw signals received from the transmission subsystem.

5. Decision, which makes the decision to "accept" or "reject" based upon the system policy and the scores received from the signal processing system.

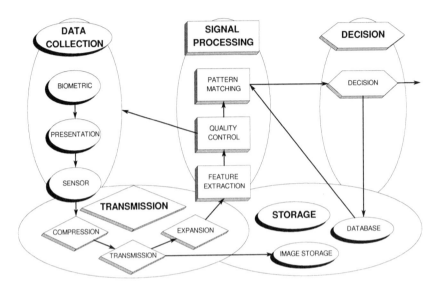

Figure 17.1 General biometrics system.

4. Application-Dependent Device Testing

We are now in a position to present a more mathematical development of the above ideas and to explain more precisely three major difficulties in biometric testing: the dependence of measured error rates on the application classification, the need for a large test population which adequately models the target population, and the necessity for a time delay between enrollment and testing. This section will present a mathematical development of the five basic system parameters: false match rate, false non-match rate, binning error rate, penetration coefficient and transaction speed. We will also discuss Receiver Operating Characteristic curves [1-5] and confidence interval estimations.

Features

The features extracted by the signal processing sub-system of Figure 17.1 are generally vectors in a real or complex [6] metric space, with components generally taking on integer values over a bounded domain. In some systems [7], the domain of each component is restricted to the binary values of {0,1}. Fingerprint systems are the primary exception to this rule, using features not in a vector space. In this chapter, we will suppose that the components are any real number.

If each feature vector, X, has J components, x_j, we can write

$$X = (x_j), \ j=1,2,..., \ J. \tag{17.1}$$

where

$$x_j = \mu_j + \varepsilon. \tag{17.2}$$

The components of the feature vector, X, consist of a fundamental biometric measure, μ_j, and an error term, ε, both assumed to be time-invariant and independent over all J. The error term, ε, has some distribution, $\xi(0,\sigma^2)$, not presumed to be normal. To simplify the development, we will assume that the distribution of the errors is identical for all components of X. We can say, therefore, that the components of X come from a distribution

$$x_j \sim \xi(\mu_j, \sigma^2). \tag{17.3}$$

Errors arise from the data collection sub-system of Figure 17.1, perhaps owing to random variations in the biometric pattern, pattern presentation or the sensor. Errors owing to the transmission or compression processes of the transmission sub-system of Figure 17.1 may also be important. Assuming the errors to be uncorrelated, we can write

$$\sigma^2 = \sigma^2_{biometric} + \sigma^2_{presentation} + \sigma^2_{sensor} + \sigma^2_{transmission}, \tag{17.4}$$

where subscripted terms on the right hand side are the error variances associated with changes in the biometric pattern, the presentation and the sensor, accordingly, along with the transmission error variance. In reality, compression errors, included in the transmission error term, may be a function of the sensor error, adding correlations to the error process.

We note the first major problem with error testing of biometric devices: the error variance of each of the terms in Eq. (17.4) is highly application dependent. There is currently no way to predict the error terms for all applications from measurements made in any one test environment. Consequently, test results are always dependent upon the test environment and will not reflect errors in dissimilar application environments of the "real world". Testing of the individual error variances as noted in Eq. (17.4) has not been done, so we will consider in this chapter only the composite variance σ^2.

Templates

At the time of enrollment, the user presents $M \geq 1$ samples of the biometric measure for the creation of a "template", $\overline{X}$, from the M feature vectors, X^i. The superscript, i=1,2,...,M, has been added to the feature vector, X, to indicate multiple samples from the same user. The template, $\overline{X}$, may be computed as the average of the M feature vectors, X^i, in which case we can write,

$$\overline{X} = (\overline{x}_j),$$ (17.5)

where

$$\overline{x}_j = \frac{1}{M} \sum_{i=1}^{M} x_j^i.$$ (17.6)

When computing the weighted sum of uncorrelated random variables, the following relationships hold [8]:

$$\mu = \sum_{i=0}^{M} c_i \mu_i$$ (17.7)

and

$$\sigma^2 = \sum_{i=0}^{M} c_i^2 \sigma_i^2,$$ (17.8)

where c_i i=1,2,...,M is the weighting vector.

Under some weighting vectors, such as uniform c_i, the distribution of the weighted sum of uncorrelated random variables will, by the Central Limit Theorem, tend toward normality as M increases.

Applying Eqs. (17.7) and (17.8) to Eq. (17.6), the components of the template $\overline{X}$ are seen to be distributed as

$$\overline{x}_j \sim \overline{\xi}(\mu_j, \frac{\sigma^2}{M}),$$ (17.9)

where $\overline{\xi}$ indicates a distribution tending toward normality. In many systems, however, M may be one or three, meaning that $\overline{\xi}(\mu_j, \frac{\sigma^2}{M})$ cannot generally be considered normal.

"Genuine" Distances

The feature vectors, X, vary across users, which we will express by adding a subscript, h=1,2,..N, to our notation, where N is the number of enrolled users. Our original assumption, that the components of the feature vector, X, are independently distributed random variables, is now expanded to include independence over users, as well. The sample data across the entire population of users is "non-stationary", meaning that multiple measures from a single user cannot be used to approximate the distribution for the entire population. This adds a second major complication to biometric testing, the requirement for a large test population that adequately represents the target population of the application.

For every user, a template is created from M samples of the biometric measure. Then for each user, the biometric feature vector is re-sampled and a distance measure, d_{hh}, is computed between the additional sample and the user's template.

$$d_{hh} = \left\| \overline{X}_h - X_h^{M+1} \right\| = \left\| \Delta X_{hh} \right\|, \qquad (17.10)$$

where the double brackets indicate a general distance measure and

$$\Delta X_{hh} = (\overline{x}_{hj} - x_{hj}^{M+1}) \quad \text{for } j=1,2\ldots,J. \qquad (17.11)$$

Applying equation (17.6), the components of ΔX_{hh} become

$$\Delta x_{hhj} = \frac{1}{M} \sum_{i=1}^{M} x_j^i - x_j^{M+1}. \qquad (17.12)$$

Referring to equation (17.2),

$$\Delta x_{hhj} = \mu_{hj} - \mu_{hj} + \varepsilon = \varepsilon, \qquad (17.13)$$

where, by equation (17.8), the error term, ε, is distributed as

$$\varepsilon \sim \overline{\xi}(0, \frac{(M+1)\sigma_h^2}{M}). \qquad (17.14)$$

Consequently, the distribution of the component, Δx_{hhj}, used to compute the genuine distance d_{hh}, does not tend toward normality with increasing M, but rather to $\xi(0, \sigma_h^2)$, the original distribution of the error terms for user h.

One of the tasks in testing will be to develop the probability distribution of these distance measures over the entire user population. We will call this density function $F'_{GEN}(d)$, where "GEN" indicates "genuine", indicating that samples are being compared to each user's own (genuine) template.

We have assumed, for simplicity, that both μ_{hj} and σ_h^2 are time-invariant. In reality, however, both may drift over time. The measurement means, μ_{hj}, may move as a "random walk". In general, biometric system identification errors increase with the passage of time after enrollment. This phenomenon is generally attributed to changes in the underlying biometric measures, μ_{hj}, and, consequently, is referred to as "template aging". Sensor and presentation changes over time may also be contributing factors. This represents a third major problem in the error testing of biometric devices: performance estimation may depend upon the time difference between enrollment and test samples.

Consider $\mu_{hj} = f(t)$, where t is time. Then the μ_{hj} of Eq. (17.13) are also functions of time, and can be given by

$$\Delta x_{hhj}(t) = \mu_{hj}(t_1) - \mu_{hj}(t_2) + \varepsilon,\qquad(17.15)$$

where, again,

$$\varepsilon \sim \overline{\xi}\left(0, \frac{(M+1)\sigma_h^2}{M}\right),\qquad(17.16)$$

and t_1 and t_2 are the times at enrollment and later sampling, respectively. To understand the effects of a time-varying mean, we compare the time invariant case of of Eq. (17.13) to the time varying case of Eq. (17.15). If our distance measure is Euclidean, then any variation over time in the μ_{hj} causes an increase in the expected distance values, $E(d_{hh})$, over the population, because

$$E\left(\left(\sum_{j=1}^{J}\left(\mu_{hj}(t_1) - \mu_{hj}(t_2) + \varepsilon\right)\right)^2\right)^{\frac{1}{2}} > E\left(\left(\sum_{j=1}^{J}\varepsilon^2\right)^{\frac{1}{2}}\right)\qquad(17.17)$$

if $\mu_{hj}(t_1)$ *and* $\mu_{hj}(t_2)$ are not always equal and the ε and μ terms are uncorrelated, as originally assumed.

Ideally, the time interval between enrollment and sampling in any test should be similar to the interval expected in the application. This is usually not possible to estimate or attain so, as a "rule of thumb", we would like the time interval to be at least on the order of the healing time of the body part involved. This would allow any temporary variations in the biometric measures to be considered in the computation of the template-to-sample distances. This requirement, of course, greatly increases test time and expense.

It has been commonly noted in practice [9] that users can be roughly divided into two groups depending upon distance measurements, d_{hh}: a large group, $\{N_1\}$ with small distance measures, called "sheep", and a smaller group, $\{N_2\}$, with high distance measures, called "goats" [10], where the total population $N = N_1 + N_2$. The preceding development leads us to believe that "sheep" and "goats" may be distinguished either by the value of σ^2_h, with users in $\{N_1\}$ having smaller error variance σ^2 then users in $\{N_2\}$, or by the time-variability of their fundamental measures, μ_{hj}.

More precisely, the terms "goats" and "sheep" have generally been applied to indicate the chronic classification of individuals, "goats" being users who consistently return large distance measures when samples are compared to stored templates. Multiple test samples over time from the same user do not return additional independent data for population estimates and may result in the mixing of "habituated" and "non-habituated" user interaction with the system. Consequently, previous tests have not followed users in time. From any single set of test distance samples, the large distances will represent both chronic "goats" and a few "sheep", who simply happen to return a large distance score at the tail of the "sheep" error distribution.

Histograms of "genuine" distances are noted in practice to be bi-modal, the distance measures from the "sheep" contributing to the primary mode, and the distance measures from the "goats" contributing to the secondary mode.

"Impostor" Distances

Using the same metric as used for establishing the "genuine" distances, a set of samples X_k^{M+1} could be compared to non-matching templates $\overline{X}_h$, $h \neq k$, to arrive at a non-matching distance, d_{hk}. We can rewrite Eqs. (17.10), (17.11) and (17.12) to get

$$\Delta x_{hkj} = \frac{1}{M} \sum_{i=1}^{M} x_{hj}^i - x_{kj}.$$ (17.18)

By Eq. (17.2), these components of the difference vector ΔX_{hk} can be written

$$\Delta x_{hkj} = \mu_{hj} - \mu_{kj} + \varepsilon,$$ (17.19)

where, by Eqs. (17.7) and (17.8)

$$\varepsilon \sim \overline{\xi}(0, \frac{\sigma_h^2}{M} + \sigma_k^2).$$ (17.20)

The distribution of the error term, ε, does not tend to normal with increasing M, but rather to the original, unspecified distribution $\xi(0, \sigma_k^2)$.

Over the entire population, these distance measures, d_{hk}, for $h \neq k$, have a density function $F'_{IMP}(d)$ where "IMP" means "impostor", so named because the density is of measures from an "impostor" sample to a non-matching template. Some researchers [11] have suggested the use of additional templates not matched by samples for the calculation of "impostor" distributions. This is sometimes called a "background" database. In our notation, this would create two groups of templates, $\overline{X}_h, h \in \{H_1\}$, those matched by test samples, and $\overline{X}_h, h \in \{H_2\}$, those not matched. The genuine distance distribution, based on distances whose components are distributed as Eq.

(17.14), considers only the σ_h^2 for the users, $h \in \{H_1\}$, with matching samples. By Eq. (17.20), however, the impostor distribution is impacted by the distribution of variances σ_h^2 for users in both matched and unmatched groups. Unless we are certain that the populations are the same, such that the distribution of the terms $\left(\mu_{hj} - \mu_{kj} \right)$ does not depend upon the group $\{H_1\}$ or $\{H_2\}$ from which the members come, and that the application environment is the same, such that σ_h^2 is also group independent, "background" databases only add uncertainty to the measurements.

"Inter-Template" Distances

Between each pair of templates, $\overline{X}_h$ and $\overline{X}_k$, $h \neq k$, a distance, δ_{hk}, can be computed using the same metric as was used to compute the genuine and impostor distances:

$$\delta_{hk} = \left\| \overline{X}_h - \overline{X}_k \right\| = \left\| \Delta \overline{X}_{hk} \right\|. \tag{17.21}$$

We use the Greek symbol, δ, to differentiate this "inter-template" distance from the "impostor" distance of the preceding section.

Because we are working in a metric space, the distances are symmetric such that $\delta_{hk} = \delta_{kh}$, and the distance of any vector from itself is zero, so $\delta_{kk} = 0$. Therefore, $N(N-1)/2$ non-independent distances can be computed between all templates. From Eq. (17.6), we have for the components of ΔX_{hk},

$$\Delta \overline{X}_{hkj} = \frac{1}{M} \sum_{i=1}^{M} x_{hj}^i - \frac{1}{M} \sum_{i=1}^{M} x_{kj}^i \quad \text{for } h \neq k. \tag{17.22}$$

By Eq. (17.2),

$$\Delta \overline{x}_{hkj} = \mu_{hj} - \mu_{kj} + \varepsilon, \tag{17.23}$$

where, by Eqs. (17.7) and (17.8),

$$\varepsilon \sim \overline{\xi}(0, \frac{\sigma_h^2 + \sigma_k^2}{M}). \tag{17.24}$$

For the inter-template distance, the error term is from a distribution tending toward normality as M increases. Further, in the limit, the variance of the error term goes to zero with increasing M. This indicates that the inter-template terms are not impacted by the measurement error for large M.

We denote the density function of δ_{hk} over the population as $F'_{IT}(\delta)$, where "IT" indicates "inter-template". Comparing Eq. 17.20 to Eq. 17.24, the distributions of the terms composing the "impostor" and "inter-template" distributions are equivalent only when M=1. For M>1,

$$\frac{\sigma_h^2}{M} + \sigma_k^2 > \frac{\sigma_h^2 + \sigma_k^2}{M}, \tag{17.25}$$

indicating that the variance in the error terms of components comprising the impostor distance vector will be larger than that of the terms comprising the inter-template distance vector. For uncorrelated means and errors, as assumed, and Euclidean distances, the expected values of the impostor distances will be greater than for the inter-template distances. Consequently, the impostor and inter-template distributions will only be equivalent for M=1. The inter-template distribution makes an increasingly poor proxy for the impostor distribution as M increases.

The three distributions, "genuine", "impostor" and "inter-template", are shown in Figure 17.2. Both the impostor and inter-template distributions lie generally to the right of the genuine distribution, which shows the second mode noted in all experimental data.

Decreasing the difficulty of the application category (changing from non-habituated, non-attended to habituated, attended, for instance) will effect the genuine distribution by making it easier for users to give repeatable samples, decreasing the value of σ_h^2, and thus moving the genuine curve to the left. Decreasing the measurement errors, σ_h^2 and σ_k^2, also causes movement in the impostor distribution to the left, but causes movement in the "inter-template" distribution only for small M.

Operational systems store templates and transaction distance measures, but rarely store the samples acquired during operations. Consequently, under the assumption that all users are "genuine", the genuine distribution can be constructed directly from the transaction distance measures. The "inter-template" distribution can be constructed by "off-line" comparison of the distances between templates. The "impostor" distribution, however, cannot be reconstructed without operational samples. Methods for convolving $F'_{GEN}(d)$ and $F'_{IT}(\delta)$ to determine $F'_{IMP}(d)$, under some simplifying assumptions, have been discussed in [12] and [13].

A decision policy commonly accepts as genuine any distance measure less than some threshold, τ. In non-cooperative applications, it is the goal of the deceptive user ("wolf") not to be identified. This can be accomplished by willful behavior to increase his/her personal σ_h^2, moving a personal genuine distribution to the right and increasing the probability of a score greater than the decision policy threshold, τ. We do not know for any non-cooperative system the extent to which "wolves" can willfully increase their error variances.

ROC Curves

Even though there is unit area under each of the three distributions, the curves themselves are not dimensionless, owing to their expression in terms of the dimensional distance. We will need a non-dimensional measure, if we are to compare two unrelated biometric systems using a common and basic technical performance measure.

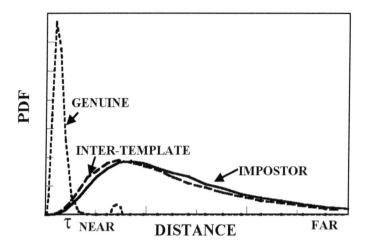

Figure 17.2 Distance distributions.

The false non-match rate, FNMR, at any τ is the percentage of the distribution $F'_{GEN}(d)$ greater than $d = \tau$ and can be given by

$$\text{FNMR}(\tau) = \int_{\tau}^{\infty} F'_{GEN}(d)\, d\, d = F_{GEN}(d)|_{\tau}^{\infty} = 1 - F_{GEN}(d)|_{0}^{\tau}. \qquad (17.26)$$

The false match rate, FMR, at any τ is the percentage of the distribution $F'_{IMP}(d)$ smaller than $d = \tau$ and can be given by

$$\text{FMR}(\tau) = \int_{0}^{\tau} F'_{IMP}(d)\, d\, d = F_{IMP}(d)|_{0}^{\tau}. \qquad (17.27)$$

The "Receiver Operating Characteristic" (ROC) curve is the two-dimensional curve represented parametrically in τ by the points $\left[F_{IMP}(d)|_{0}^{\tau}, F_{GEN}(d)|_{0}^{\tau} \right]$. We find the ROC curve to be more intuitive when displayed as the points $\left[\text{FMR}(\tau), \text{FNMR}(\tau) \right]$.

As previously noted, the probability densities $F'_{IMP}(d)$ *and* $F'_{IT}(\delta)$ are equivalent only when M=1. When M>1 and the distances are Euclidean with component means and error uncorrelated, the expected values of the impostor distances will be larger than the expected values of the inter-template distances. Consequently, in this case, if we compare the integrals of the distributions between 0 and some threshold, τ,

$$F_{IMP}(d)|_{0}^{\tau} = \int_{0}^{\tau} F'_{IMP}(d)\, d\, d \qquad (17.28)$$

and

$$F_{IT}(\delta)|_0^\tau = \int_0^\tau F'_{IT}(\delta)\,d\delta\,,\tag{17.29}$$

we find that

$$F_{IT}(\delta)|_0^\tau \geq F_{IMP}(d)|_0^\tau\,.\tag{17.30}$$

By this last equation, we can see that if the inter-template distribution, $F_{IT}(\delta)|_0^\tau$, is used to replace the impostor distribution under these conditions, the false match rate will be overestimated.

We note that the ROC curve is non-dimensional. Other non-dimensional measures have been suggested for use in biometric testing [14], such as "D-prime"[1,2] and "Kullback-Leibler" [15] values. These are single, scalar measures, however, and are not translatable to error rate prediction. The "equal error rate" (EER) is defined as the point on the ROC where the false match and false non-match rates are equivalent. The EER is non-dimensional, but not all biometric systems have meaningful EERs owing to the tendency of the genuine distribution to be bimodal. False match and false non-match error rates, as displayed in the ROC curve, are the only appropriate test measures for system error performance prediction.

Penetration Rate

In systems holding a large number, N, of templates in the database, search efficiencies can be achieved by partitioning them into smaller groups based both upon information contained within (endogenous to) the templates themselves and upon additional (exogenous) information, such as the customer's name, obtained at the time of enrollment. During operation, submitted samples are compared only to templates in appropriate partitions, limiting the required number of template-to-sample comparisons. Generally, a single template may be placed into multiple partitions if there is uncertainty regarding its classification. Some templates of extreme uncertainty are classified as "unknown" and placed in all of the partitions. In operation, samples are classified according to the same system as the templates, then compared to only those templates from the database which are in communicating partitions. The percentage of the total database to be scanned, on average, for each search is called the "penetration coefficient", P, which can be defined as

$$P = \frac{E(number\ of\ comparisons)}{N}\,,\tag{17.31}$$

where E(number of comparisons) is the expected number of comparisons required for a single input sample. In estimating the penetration coefficient, it is assumed that the search does not stop when a "match" is encountered, but continues through the entire partition. Of course, the smaller the penetration coefficient, the more efficient the system. Calculation of the penetration coefficient from the partition probabilities is discussed in [16,17].

The general procedure in testing is to calculate the penetration coefficient empirically from the partition assignments of both samples and templates. Suppose there are K partitions, C_i, for $i=1,2,...,K$ and there are L sets, S_l, $l=1,2,...,L$, indicating which partitions communicate. For instance, an "unknown" partition communicates with every other partition individually.

There are N_S samples, X_h, $h=1,2...,N_S$, and N_T templates, $\overline{X}_k$, $k=1,2...,N_T$. Each sample, X_h, can be given multiple partitions, C_{hi}, the precise number of which, I_h, will depend upon the sample. Similarly, each template, $\overline{X}_k$, can have multiple partitions $\overline{C}_{kj}$, $j=1,2,...,I_k$. For any sample-template pair, if any of the partitions are in a communicating set, the sample and template must be compared. However, they need to be compared at most only once, even if they each have been given multiple partitions in multiple communicating sets.

We define the "indicator" function,

$$1_l\left(C_{hi}, \overline{C}_{kj}\right) = \begin{cases} 1 \text{ if } C_{hi} \text{ AND } \overline{C}_{kj} \in \{S_l\} \\ 0 \text{ if } C_{hi} \text{ OR } \overline{C}_{kj} \notin \{S_l\} \end{cases}, \quad (17.32)$$

so that the function equals unity if the partitions C_h and $\overline{C}_k$ are both elements of the set S_l and zero otherwise. For each of the samples, h, and each of the templates, k, we must search all partitions, $i=1,2,...,I_h$, and $j=1,2,...,J_k$, against all L sets to determine if any C_{hi} and $\overline{C}_{kj}$ communicate. However, a single sample and single template never need be compared more than once. The penetration coefficient will be

$$P = \frac{\sum_{h=1}^{N_S} \sum_{k=1}^{N_T} H\left(\sum_{i=1}^{I_h} \sum_{j=1}^{J_k} \sum_{l=1}^{L} 1_l(C_{hi}, \overline{C}_{kj})\right)}{N_S N_T}, \quad (17.33)$$

where H(.) is the Heavyside unity function, defined as unity if the argument is greater than zero and zero otherwise.

There may be multiple, say B, independent, filtering and binning methods, P_i, $i=1,2,...,B$, used in any system. If the methods are truly independent, the total penetration coefficient for the system, P_{SYS}, using all B methods, can be written

$$P_{SYS} = \prod_{i=1}^{B} P_i. \quad (17.34)$$

If correlations exist between any of the partitioning schemes, Eq. (17.34) will under-estimate the true penetration coefficient.

Bin Error Rate

The bin error rate reflects the percentage of samples falsely not matched against their templates because of inconsistencies in the partitioning process. This error rate is determined by the percentage of samples not placed in a partition which communicates with its matching template. For each partitioning method employed, a single test can be designed to determine the bin error rate, e. Consider N matched sample-template pairs, X_h and $\overline{X}_h$. The percentage of the pairs for which each member is placed in a communicating partition is an estimate of the complement of the bin error rate. This percentage can be computed by

$$1-e = \frac{\sum_{h=1}^{N} H\left(\sum_{i=1}^{I_h} \sum_{j=1}^{J_h} \sum_{l=1}^{L} 1_l (C_{hi}, \overline{C}_{hj}) \right)}{N}. \qquad (17.35)$$

The bin error rate for the system, however, will increase as the number, B, of independent binning methods increase. If any one of the methods is inconsistent, a system binning error, ε_{SYS}, will result. Therefore, the probability of no system binning error over B binning methods is

$$1 - e_{SYS} = \prod_{i=1}^{B} (1 - e_i). \qquad (17.36)$$

Transaction Speed

The time required for a single transaction, $T_{transaction}$, is the sum of the data collection time, $T_{collect}$, and the computational time, $T_{compute}$.

$$T_{transaction} = T_{collect} + T_{compute}. \qquad (17.37)$$

For positive identification systems, only a very few comparisons between templates and submitted samples are required and generally $T_{collect} > T_{compute}$. The collection times are highly application dependent, varying from a very few seconds [18] to a couple of minutes [19]. Transaction times are best estimated by direct measurement of the system throughput, S, as given by

$$S = \frac{1}{T_{transaction}}. \qquad (17.38)$$

For large-scale, negative identification systems, the computational time can be much greater than the collection time. The challenge is to reduce the computational time so that the throughput is not limited by the computer hardware. The computational time can be estimated from the hardware processing rate, C, and the number of comparisons required for each transaction. If m is the number of biometric records

collected and searched from each user during a transaction and N is the total number
of records in the database, then

$$T_{compute} = \frac{m\, P_{SYS}\, N}{C}, \tag{17.39}$$

where P_{SYS} is again the system penetration coefficient. Methods for estimating
hardware processing speeds are given in texts such as [20].

Confidence Intervals

The concept of "confidence intervals" refers to the inherent uncertainty in test results
owing to small sample size. These intervals are *a posteriori* estimates on the
uncertainty in the results on the test population in the test environment. They do not
include the uncertainties caused by errors (mislabeled data, for example) in the test
process. Future tests can be expected to fall within these intervals only to the extent
that the distributions of μ_{hj} *and* σ_h^2, and the errors in the testing process, do not
change. The confidence intervals do not represent *a priori* estimates of performance
in different environments or with different populations. Because of the inherent
differences between test and application populations and environments, confidence
intervals have not been widely used in reported test results and are of limited value.

The method of establishing confidence intervals on the ROC is not well
understood. Traditionally, as in [14], they have been found through a summation of
the binomial distribution. The confidence, β, given probability p, of K distances, or
fewer, out of N independent distances being on one side or the other of some
threshold, τ, would be

$$1 - \beta = \Pr\{i \le K\} = \sum_{i=0}^{K} \binom{N}{i} \frac{N!}{i!(N-i)!} p^i (1-p)^{N-i}. \tag{17.40}$$

When computing the confidence interval on the false non-match rate, for instance,
K would be the number of the N independent, genuine distance measures greater than
the threshold τ. The best "point estimate" of the false non-match rate would be K/N.

The probability, p, calculated by inverting Eq. (17.40), would be the upper bound
on the confidence interval. The lower bound could be calculated from the related
equation for $\Pr\{i \ge K\}$. In practice, values of N and K are too large to allow equation
(17.40) to be computed directly and p may be too small to allow use of normal
distribution approximations. The general procedure is to use the "incomplete Beta
function" [21,22]

$$I_p(K+1, N-K) = \sum_{i=K+1}^{N} \binom{N}{i} \frac{N!}{i!(N-i)!} p^i (1-p)^{N-i} = \beta \tag{17.41}$$

and numerically invert to find p for a given N, K, and β.

One interesting question to ask is "What is the lowest error rate that can be statistically established with a given number of independent comparisons?". We want to find the value of p such that the probability of no errors in N trials, purely by chance, is less than 5%. This gives us the 95% confidence level, β. We apply Eq. (17.40) using K=0,

$$0.05 > \Pr(K = 0) = \sum_{i=0}^{0} \frac{N!}{i!(i-N)!}\, p^i\, (1-p)^{N-i} = (1-p)^N .$$ (17.42)

This reduces to

$$\ln(0.05) > N\, \ln(1-p) .$$ (17.43)

For small p, ln (1-p) ≈ -p and, further, ln (0.05) ≈ -3. Therefore, we can write

$$N > 3/p .$$ (17.44)

Recent work indicates that while this approach is satisfactory for error bounds on the false non-match rate, where distance measures are generally calculated over N independent template-sample pairs, it cannot be applied for computing confidence intervals on false match results where cross-comparisons are used. Bickel [23] has given the confidence intervals for the false match rate when cross-comparisons are used and the templates are created from a single sample, such that M=1. For N samples, there are N(N-1) non-independent cross-comparisons. We will denote a cross-comparison distance less than or equal to the threshold, τ, by

$$r(h,k) = 1\!\left(d_{hk} \le \tau\right),$$ (17.45)

where 1(.) is again the indicator function. So the best estimate of the probability, FMR(τ), of a cross-comparison being τ, or less, would be the number of such cross-comparisons divided by the total number available,

$$\hat{FMR}(\tau) = \frac{1}{N(N-1)} \sum_{h=1}^{N} \sum_{k=1}^{N} r(h,k) \quad for\ h \ne k .$$ (17.46)

The (1- α)% confidence bounds are

$$\hat{FMR}(\tau) \pm z_{\left(1-\frac{\alpha}{2}\right)} * \left(\frac{\hat{\sigma}(\tau)}{\sqrt{N}}\right),$$ (17.47)

where

$$\hat{\sigma}^2 = \frac{1}{N(N-1)^2} \sum_{h=1}^{N} \left(\sum_{k \ne h} r(h,k) + \sum_{k \ne h} r(k,h) \right)^2 - 4*\left(\hat{FMR}\right)^2$$ (17.48)

and $Z_{\left(1-\frac{\alpha}{2}\right)}$ indicates the number of standard deviations from the origin required to

encompass $\left(1-\frac{\alpha}{2}\right)\%$ of the area under the standard normal distribution. For $\alpha=5\%$, this value is 1.96. The explicit dependency on τ of all quantities in Eq. (17.48) has been dropped for notational simplicity.

In practice, time and financial budgets, not desired confidence intervals, always control the amount of data that is collected for the test. From the test results, we can calculate the upper bound on the confidence interval, "guess-timate" the potential effect of differences between test and operational populations and environments, then over-design our system decision policy to account for the uncertainty.

5. Testing Protocols

The general test protocol is to collect one template from each of N users in an environment that closely approximates that of the proposed application, ideally within the same application partitions as described in Section 17.2. The value of N should be as large as time and financial budget allow and the sample population should approximate the target population as closely as practicable. Some time later, on the order of weeks or months if possible, one sample from each of the same N users is collected. Then, in "off-line" processing, the N samples are compared to the N previously stored templates to establish N^2 non-independent distance measures. For all distance thresholds, τ, point estimates of the false match and false non-match error rates are given by

$$\hat{FMR}(\tau) = \frac{\sum_{h=1}^{N}\sum_{k\neq h}^{N} 1\left(d_{hk} < \tau\right)}{N(N-1)} \qquad (17.49)$$

and

$$\hat{FNMR}(\tau) = 1 - \frac{\sum_{h=1}^{N} 1\left(d_{hh} < \tau\right)}{N}, \qquad (17.50)$$

where $1(.)$ is the indicator function, equal to unity if the argument is true and zero otherwise, and the hat indicates the estimation. When testing from operational data, substitution of the inter-template distances, δ_{hk}, for the impostor distances, d_{hk}, in Eq. (17.49) will generally result in overestimation of the false match rate.

For systems employing binning, estimates of penetration coefficient and binning error rate are estimated from Eqs. (17.33) and (17.35) by comparing partition assignments of the templates to those of the samples. Results from one test [24] on four Automatic Fingerprint Identification System (AFIS) vendors are given in Figures 17.3 and 17.4.

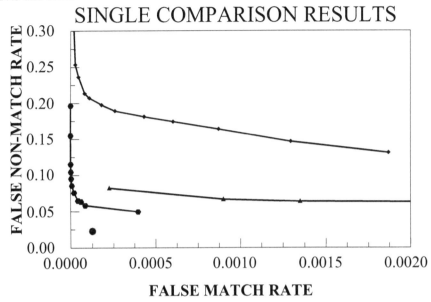

Figure 17.3 ROC curve.

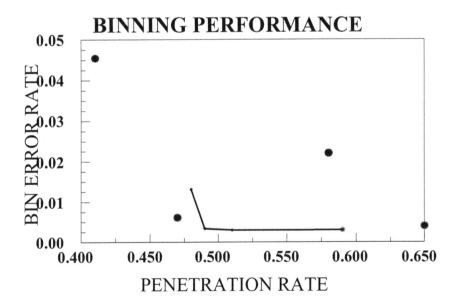

Figure 17.4 Binning error and penetration rates.

6. System Performance Prediction from Test Results

The five basic system performance parameters, false match rate, false non-match rate, penetration coefficient, bin error rate and transaction speed, can be used to predict system "false accept/false reject" rates and throughput under a wide variety of system decision policies [16]. Recall that the concern of every system manager is three-fold: the number of false rejections requiring human intervention, including the percentage of the population who are unable to enroll; the deterrence value of the false acceptance rate; and the ability of the system throughput rate to meet the input demand.   In this section, we will consider the single example of a negative identification system using two independent biometric measures and a system policy that declares a "rejection" if both of the measures are found to match both measures of any previously enrolled individual. All calculations will assume statistical independence.

If the penetration coefficient is found to be 0.5 on each measure, then by Eq. (17.34), the system penetration coefficient will be P_{SYS} = 0.5*0.5=0.25. If the bin error rate is 0.01 for each measure, by Eq. (17.36) the system bin error rate will be e_{SYS} =1-(1-0.01)*(1-0.01)=0.02.

A false rejection occurs if both submitted samples from a single user are found to falsely match both templates of one of the previously enrolled N individuals. Assuming statistical independence of error rates, if the first sample pair is compared to the two stored templates from just one enrolled user, the chance of a false rejection, FRR, occurring is

$$FRR = FMR^2 . \tag{17.51}$$

For notational simplicity, we have not indicated the dependence of FMR on the threshold, τ. The probability of not getting a false rejection over of P_{SYS}*N searched template pairs is given by

$$1 - FRR = (1 - FMR^2)^{N*P_{SYS}} . \tag{17.52}$$

Suppose that our working estimate of the false match rate, based on testing, is 10^{-6} and that the system will be designed for N=4x10^6 users.   Then, FRR $\approx 10^{-6}$.   The expected number of users falsely rejected during enrollment of the entire population will be $N*FRR < 4$, thereby requiring limited human intervention for exception handling over the course of enrollment of the population.

Assuming that the database is "clean", meaning only one template set for any single user, a false acceptance will not occur if both samples are matched to the enrolled templates and no binning error occurs. Therefore, the complement of the false acceptance rate, FAR, can be given as

$$1 - FAR = (1 - e_{SYS})(1 - FNMR)^2 . \tag{17.53}$$

If our working estimate of the false non-match rate is 7%, then FAR=15%. The number of fraudsters, F, in the system will be

$$F = N*FR*FAR , \qquad\qquad (17.54)$$

where FR is the fraud rate, or percentage of the population that is attempting to defraud the system. The fraud rate depends not only on the inherent honesty of the population, but also on the perceived chance of getting caught. The true chance of getting caught, of course, is 1-FAR, or 85% in this example, but the perceived rate may be different. Consequently, estimation of the fraud rate is best left to social scientists. We hypothesize that a FAR of 15% is more than adequate for most real systems.

Usually, in large-scale systems, a throughput rate is specified as a system requirement and the throughput equations (17.38) and (17.39) are used to determine the necessary hardware processing speed. If our system is designed for 4×10^6 users, we may want to enroll them over a four-year period, about 1,000 days. We might design the system for a maximum capacity of 6,000 enrollments per day when the last of the users are being enrolled. We assume transaction time is controlled by the hardware processing rate. In our system, the number of samples, m, used for each individual is 2. Therefore, the processing rate, as calculated using Eqs. (17.38) and (17.39), must be 1.2×10^{10} computations per day, if no daily backlog is acceptable. Assuming 20 hour per day availability of the processing system, the required rate will be about 170,000 comparisons per second.

7. Available Test Results

Results of some excellent tests are publicly available. The most sophisticated work has been done on speaker verification systems. Much of this work is extremely mature, focusing on both the repeatability of sounds from a single speaker and the variation between speakers [25-31]. The scientific community has adopted general standards for speech algorithm testing and reporting using pre-recorded data from a standardized "corpus" (set of recorded speech sounds), although no satisfactory corpus for speaker verification systems currently exists. Development of a standardized database is possible for speaker recognition because of the existence of general standards regarding speech sampling rates and dynamic range. The testing done on speech-based algorithms and devices has served as a prototype for scientific testing and reporting of biometric devices in general.

In 1991, the Sandia National Laboratories released an excellent and widely available comparative study on voice, signature, fingerprint, retinal and hand geometry systems [32]. This study was of data acquired in a laboratory setting from professional people well-acquainted with the devices. Error rates as a function of a variable threshold were reported, as were results of a user acceptability survey. In April, 1996, Sandia released an evaluation of the IriScan prototype [33] in an access-control environment.

A major study of both fingerprinting and retinal scanning, using people unacquainted with the devices and in a non-laboratory setting, was conducted by the California Department of Motor Vehicles and the Orkand Corporation in 1990 [19]. This report measured the percentage of acceptance and rejection errors against a database of fixed size, using device-specific decision policies, data collection times, and system response times. Error results cannot be generalized beyond this test. The report includes a survey of user and management acceptance of the biometric methods and systems.

In 1996, an excellent comparative study on facial recognition systems was published by the U.S. Army Research Laboratory [34]. This study used as data facial images collected in a laboratory setting and compared the performance of four different algorithms using this data. Both test and enrollment images were collected in the same session and false match and false non-match rates are reported as a type of "rank order" statistic, meaning that the results are dependent on the size of the test database and cannot be used for general performance prediction. Earlier reports from this same project included a look at infrared imagery as well [35].

In 1998, San Jose State University released the final report to the Federal Highway Administration [24] on the development of biometric standards for the identification of commercial drivers. This report includes the results of an international automatic fingerprint identification benchmark test.

The existence of a dozen annual industry conferences, including the U.S. Biometric Consortium and the European Association for Biometrics meetings and the CardTech/SecurTech conferences, in addition to other factors such as the general growth of the industry, has encouraged increased informal reporting of test results

8. Conclusions

The science of biometric device analysis and testing is progressing extremely rapidly. Just as aeronautical engineering took decades to catch up with the Wright brothers, we hope to eventually catch up with the thousands of system users who are successfully using these devices in a wide variety of applications. The goal of the scientific community is to provide tools and test results to aid current and prospective users in selecting and employing biometric technologies in a secure, user-friendly, and cost-effective manner.

Acknowledgements

The framework for a scientific approach to biometric testing was established several years ago in a series of questions posed to the biometric identification community by Joseph P. Campbell, Jr. This paper was created as a response to those questions. The mathematical notation and approach was suggested by Peter Bickel. Thoughtful input by Kang and Barry James of the University of Minnesota, Deluth, and assistance from Jim Maar and MAJ. John Colombi, USAF, was most helpful and appreciated. The author, however, claims sole credit for all errors and omissions.

References

[1] W. W. Peterson, and T. G. Birdsall, *The Theory of Signal Detectability*, Electronic Defense Group, University of Michigan, Technical Report 13, 1954.

[2] W. P. Tanner, and J. A. Swets, "A Decision-Making Theory of Visual Detection," *Psychological Review*, Vol. 61, pp. 401-409, 1954.

[3] D. M. Green and J. A. Swets, *Signal Detection Theory and Psychophysics*, Wiley, 1966.

[4] J. A. Swets (editor), *Signal Detection and Recognition by Human Observers*, Wiley, 1964.

[5] J. P. Egan, *Signal Detection Theory and ROC Analysis*, Academic Press,1975.

[6] J.L. Wayman, "A Dual Channel Approach to Speaker Verification," *Proceedings Speech Research Symposium XII*, Rutgers University, June 1992.

[7] J. G. Daugman, "Biometric personal identification system based on iris analysis," *U.S. Patent 5291560*, 1994.

[8] S.S. Wilks, *Mathematical Statistics*, Wiley and Sons, New York, pp. 83, 1962.

[9] J.P. Campbell, address to the 9th Biometric Consortium meeting, Crystal City, VA, April 1997.

[10] J. Williams, *Glossary of Biometric Terms*, Security Industry Association, 1995.

[11] W. Shen, M. Surette, and R. Khanna, "Evaluation of Automated Biometrics-Based Identification and Verification Systems," *Proceedings IEEE*, Vol. 85, pp. 1464-1479, September 1997.

[12] P. Bickel, response to NSA-MSP Problem #97-21, University of California, Berkeley, Department of Statistics, 1998.

[13] C. Franzen, "Convolution Methods for Mathematical Problems in Biometrics," U.S. Naval Postgraduate School Technical Report, 1998.

[14] J. Williams, "Proposed Standard for Biometric Decidability," *Proceedings CTST'96*, pp. 223-234 , 1996.

[15] S. Kullback and R. Leibler, "On Information and Sufficiency," *Annals of Mathematical Statistics*, Vol.22, pp. 79-86, 1951.

[16] J.L. Wayman, "Error Rate Equations for the General Biometric System," *Automation and Robotics Magazine*, Special Issue on Biometric Identification, January 1999 (To appear).

[17] K. James and B. James, *NSA SAG Problem 97-25*, Dept. of Statistics, UC Berkeley, Monograph, 1998

[18] D. Welsh and K. Sweitzer (of Ride and Show Engineering at Walt Disney World), Presentation, *CardTech/SecurTech '97* , May 21, 1997.

[19] Orkand Corporation, "Personal Identifier Project: Final Report," State of California Department of Motor Vehicles report DMV88-89, April 1990 (reprinted by the U.S. National Biometric Test Center).

[20] J.L. Hennessy and D.A. Patterson, *Computer Architecture: A Quantitative Approach*, 2nd ed., Morgan Kaufman, San Francisco, 1996.

[21] M. Abromowitz and I. Stegun, *Handbook of Mathematical Functions with Formulas, Graphs, and Mathematical Tables*, John Wiley and Sons, New York, 1972.

[22] W. H. Press, B. P. Flannery, S. A. Teukolsky, W. H. Vetterling, *Numerical Recipes*, 2nd ed., Cambridge University Press, Cambridge, 1988.

[23] P. Bickel, personal correspondence in response to NSA-SAG Problem #97-23, March 1998.

[24] J. L. Wayman, *Biometric Identifier Standards Research Final Report*, College of Engineering, San Jose State University (sponsored by the Federal Highway Adminstration), October 1997.

[25] B. Atal, "Automatic Recognition of Speakers from Their Voices," *Proceedings IEEE*, Vol. 64, pp. 460-475, 1976.

[26] A. Rosenberg, "Automatic Speaker Verification," *Proceedings IEEE*, 64, pp. 475-487, 1976.

[27] N. Dixon and T. Martin, *Automatic Speech and Speaker Recognition*, IEEE Press, NY, 1979.

[28] G. Doddington, "Speaker Recognition: Identifying People by Their Voices", *Proceedings IEEE*, Vol. 73, pp. 1651-1664, 1985.

[29] A. Rosenberg and F. Soong, "Recent Research in Automatic Speaker Recognition," *Advances in Speech Signal Processing* (Eds. S. Furui and M. Sondhi), Marcel Dekker, 1991.

[30] J. Naik, "Speaker Verification: A Tutorial," *IEEE Communications Magazine*, pp. 42-48, 1990.

[31] J.P.Campbell, Jr., "Speaker Recognition: A Tutorial," *Proceedings IEEE*, Vol. 85, pp. 1437-1463, September 1997.

[32] J. P. Holmes, et al., "A Performance Evaluation of Biometric Identification Devices," Sandia National Laboratories, SAND91-0276, June 1991.

[33] F. Bouchier, J. Ahrens, and G. Wells, "Laboratory Evaluation of the IriScan Prototype Biometric Identifier," Sandia National Laboratories, SAND96-1033, April 1996.

[34] P.J. Phillips, et al., "FERET (Face-Recognition Technology) Recognition Algorithm Development and Test Results," Army Research Laboratory, ARL-TR-995, October 1996.

[35] P.J. Rauss, et al., "FERET (Face-Recognition Technology) Recognition Algorithms," *Proceedings of ATRWG Science and Technology Conference,* July 1996.

18 SMARTCARD BASED AUTHENTICATION

Nalini K. Ratha and Ruud Bolle
IBM T. J. Watson Research Center
Yorktown Heights, NY
{ratha,bolle}@us.ibm.com

Abstract *In the modern electronic world, authentication of a person is an important task in many areas of day-to-day life. Using a biometrics to authenticate a person's identity has several advantages over the present practices of passwords and/or authentication cards using magnetic stripes or bar codes. However, with the use of a biometrics there is an open issue of misuse of the biometrics for purposes that the owner of the biometrics may not be aware of. In this chapter, we propose a new method of remote authentication that combines the security of a smartcard with the accuracy and convenience of biometrics to authenticate the identity of a person. With this approach, the need to access a large biometrics database is eliminated. The proposed method can be used in many application areas including system security, electronic commerce and access control.*

Keywords: *authentication, smartcard, encryption, biometrics.*

1. Introduction

Identification of a person is a basic task in day-to-day life. We identify our friends, family members and business associates effortlessly. In a small village or in a small community, in olden days business was run on personal identification methods without computers. For example, a village banker approved a loan application based on the personal knowledge of the background of the applicant. There were no credit checks or credit rating bureaus either. With the growth in transportation and communication, now the concept of a village or community where every one knew one another is becoming extinct. Often, business needs demand that a person moves to a new location for a brief period. To cater to such needs in a cashless society, automatic methods of identification are required moving beyond the old methods of family and personal trust.

In a complex and fast-moving world, one has to prove several times a day who they claim to be. Examples range from need for identification in a child care facility to identification at a port of entry in a country. Currently, several methods are in vogue depending on the seriousness of the task. In order to get money from a bank ATM, an ATM card and knowledge of the PIN is required. For entry into a bar in the US, proof of age is required by way of a driver's license or a government agency issued identity card. Many countries require possession of a citizen card to avail some facilities. Most of these methods have been known to have problems defeating the whole purpose of identification. ATM related fraud runs into billions of dollars every year. Similarly, production of fake ID cards is a common practice. There are more than 12 million point-of-sale terminals and close to half a million ATM machines [1]. The most common method of identification based on a magnetic stripe card with a signature can be easily fooled, leading to billions of dollars of fraud all over the world. Yet another example of identification needs arise in providing access to computing systems. In large organizations, it is common to have several user accounts for different purposes. Though the systems require a password and user ID, often the passwords are easily accessible to colleagues and even occasional visitors as the passwords are prominently on display on the machines. Though we have described these scenarios as identification scenarios, technically speaking these are *authentication* scenarios. By authentication, we mean methods to validate claims by persons who they claim to be. The system either agrees with the claim or rejects the claim. The need for a secure method of identification has always been a challenge to system designers. In a highly connected networked electronic world, a secure and automatic method of identification can change the way the businesses run by saving billions of dollars of fraud. With the unbounded growth of Internet-based electronic commerce, the need to securely and accurately authenticate a person has become very important to facilitate remote transactions.

In summary, the following observations can be made.

- Today's societies are highly networked and mobile.

- Prevailing methods of identification and authentication are insufficient.

- New ways of electronic economy require different secure authentication methods.

In this chapter, we present integration of two technologies, namely biometrics and smartcard to meet some of the technical challenges posed in a network-based authentication system. Biometrics provide the accuracy needed by these systems and smartcards provide the security far beyond the magnetic stripe cards. By combining the two, the overall system requirements are better met than each of them individually.

The chapter is organized as follows. A quick review of online transaction processing in a credit card-based system is provided in the next section to motivate the need for a remote authentication paradigm. Smartcards are introduced in Section 3. Programming a smartcard needs understanding of the smartcard interface and the security features available on a smartcard. Details about how to program smartcards and the security facilities are also described in Section 3. Two solution paradigms covering integration of biometrics and smartcard are presented in Section 4. Several applications are described in Section 5 with emphasis on fingerprints as the biometrics.

2. Online Credit Card Transaction Processing

In order to describe the importance of our model of network-based authentication and its benefits, we chose the credit card transaction processing as an example. The model can be extended to other applications. Several events take place from the time a consumer presents a credit card to the time the transaction gets approved and the merchant gets the money transferred to his account. A model presented in [2] is described here. Similar models have been explained in [3]. In order to provide immediate gratification of consumer needs, credit cards allow the purchase of goods and services on credit wherein the credit card along with the signature is used as a token of authentication. Let's look at the various steps that a transaction initiated at a retail store point of sale terminal goes through as shown in Figure 18.1. The merchant obtains an authorization from the card-issuing bank's authorization system. This authorization assures the merchants their payments. The authorization system verifies the credit limit and authenticity of the card before granting the authorization to the merchant. Periodically, the merchant submits a collection of authorized transactions to the card clearing center. In this model, the merchant validates the payer's identity by matching the signature on the back of the card against the one on the charge slip. Integrity of the transaction is protected by handing over a copy of the transaction to the card owner. The credit card number is used to identify the account of the consumer. A transaction authorization and communication between the merchant and the authorization center as well as the card clearing/acquiring center takes place over a private network. Often, these messages are coded using encryption techniques.

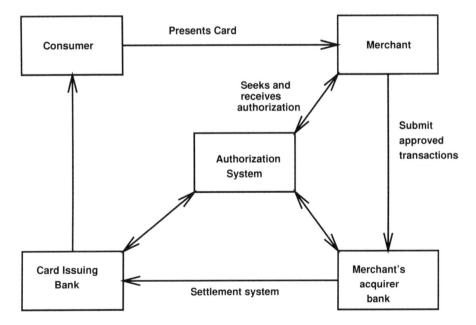

Figure 18.1 Steps in online transaction processing.

This model is   prone to various sources of fraud, including frauds that may be caused by orders received by merchants over telephone or mail where the card owner's signature is not available to them. The first source is the manual verification of a signature which is quite often ignored. High-tech snoopers on the public network might break into the communication channel to "steal" card numbers. The main problem with this model is that it requires an online communication facility to obtain a real-time authorization for every transaction. Often, there are direct and indirect charges associated with every transaction authorization.

In contrast to this model, if the credit card possessed intelligence and security to approve transactions and debit money from consumer's bank while verifying the identity of the owner of the card, many potential problems that lead to fraud can be prevented. Such a model adapted from [4] is shown in Figure 18.2. Smartcards provide support for such a model.

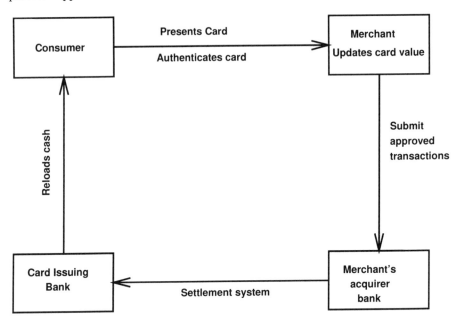

Figure 18.2 Steps in online transaction processing with an intelligent card.

3. Smartcards

A smartcard resembles a credit card in terms of physical look and size with one or more semiconductor devices attached to a module embedded in the card. More specifically, the smartcard is a portable, very secure, low cost, intelligent device, capable of manipulating and storing data. Its intelligence is due to a microprocessor that is suitable for use in a wide range of applications. Smartcards manufactured by different vendors differ significantly from each other even though they follow a standard specification such as ISO 7816 or CEN 726. An ISO 7816 based IBM smartcard is shown in Figure 18.3. Conventional smartcards are based on an invention made in 1974 by Roland Moreno in 1974 [5]. Over the last 25 years, smartcards have gone through several phases of development resulting in today's credit-card-sized cards with embedded processor and memory to store and process data. A taxonomy of smartcards is shown in Figure 18.4. Memory cards are primarily information storage cards that can store information about the amount of money as in vending machine or cash cards and phone time units as in phone cards. These cards typically provide more storage than a magnetic stripe card and are more secure than a magnetic stripe card. The smartcards with a processor and memory are more interesting for their intelligence. The intelligent smartcards can be either powered through contacts or by a RF coil as in contactless smartcards. There are different standards available for contactless smartcards [6]. The hybrid cards are combinations of both contact and contactless technologies. For example, a corporate smartcard can be used for contactless access to the buildings while sign on for the corporate computer system can be through the contact card, available on the hybrid card. The contact-based smartcards are more commonly used in many applications. Contact cards can also be further classified on the basis of presence or absence of a crypto coprocessor. The need for a special coprocessor to support cryptography is explained later in this section.

Figure 18.3 An IBM smartcard.

Physical Components in a Smartcard

For wide acceptance and usage of smartcards in many applications such as telephony, travel, health care and retail industries, standards provide interoperability on different smartcard readers. A contact type smartcard has to meet physical sizes specified in ISO 7816 standards. In addition to the ISO 7816 standard, there are several other standards for smartcards.

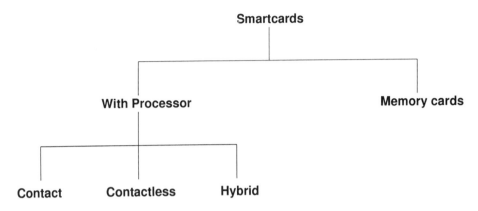

Figure 18.4 A taxonomy of smartcards.

- ISO 7816: The International Standards Organization (ISO) began standardization of chip cards since early 1983. The physical characteristics of identification cards described in ISO 7810, ISO 7811, ISO 7812, and ISO 7813 form the basis. All smartcards today follow ISO 7816 specifications for physical, electrical characteristics of smartcard interface, formats and protocols for information exchange with smartcards.   A smartcard with its mechanical dimensions as specified in ISO 7816 is shown in Figure 18.5. More details about the ISO 7816 standards are available in [6].

- CEN 726 in Europe: Smartcards had an early start in Europe, consequently, Europians are leaders in this industry. For interoperability of smartcards in different telecommunication organizations in Europe, CEN 726 standard was developed. The ISO 7816 standard has been influenced by several features from the CEN 726.

- EMV: In addition to the above two standards, there are industry-based standards emerging.   Europay, Mastercard and Visa (EMV) have cooperated to create global specifications for the payment industry. The application of smartcards in financial industry depends on this standard.

In addition to these standards, Microsoft's PC/SC standard [7], OpenCard standard from a consortium of vendors [8] and JavaCard [9] are the other standards.

Smartcard Architecture

There are three basic components in a smartcard as described below.
1. A processor to manipulate and interpret data

2. Memory

3. Input/Output handler

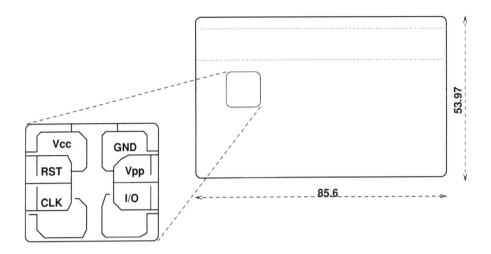

Figure 18.5 Physical dimensions of an ISO 7816 smartcard.

Architectural components of a typical smartcard are shown in Figure 18.6. The smartcard processor is usually a 8-bit microcontroller-based, though, there are several efforts to upgrade it to a 16- or 32-bit processor in more recent smartcards. The memory in a smartcard consists of three different memory types: (i) ROM, (ii) RAM and (iii) EEPROM. The ROM is used for the smartcard operating system and is usually embedded during manufacture. The RAM memory is used by the operating system as temporary storage area. The user available data segments are allocated in the EEPROM memory segments. The first two types of memory are not available for user access. Several levels of access security are supported in the EEPROM. The methods of assigning access security can be controlled through use of a password or a biometrics or using cryptography. The security features on a smartcard are described later in this section.

Software for Smartcard

From the time of smartcard manufacture to the end of loading application and usage by consumers, different kinds of software are used to handle smartcards. The software components can be classified into three categories, namely

1. Operating system;

2. Initialization and personalization;

3. Application interface.

The operating system is a vendor dependent component of the software. The operating system supports a file system on the EEPROM storage, command interpretation and security options for the data stored on the smartcard. During initialization and personalization, application specific data structures are loaded. During the usage of the smartcard, the smartcard interacts with the application through the application interface component.

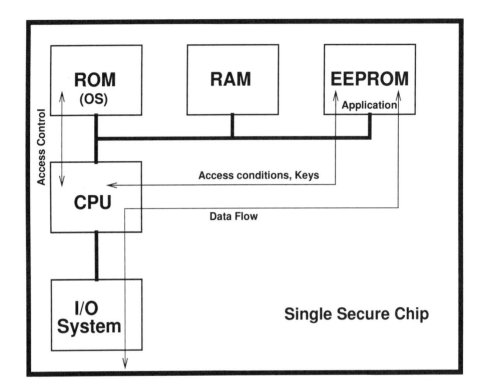

Figure 18.6 Architecture of a smartcard.

File System on Smartcard

In this subsection, we describe the file system supported on IBM smartcards. The file system on an IBM smartcard has a hierarchical structure as in a hard disk drive. The file system is characterized by

- a Master file or root of the directory (MF),

- Dedicated files (DF) also known as application directories containing data and executables for an application,

- Elementary file (EF) also known as data files containing actual data structures and data elements of an application.

The file system is shown in Figure 18.7. There are five types of elementary files as shown in Figure 18.8.

- Transparent

- Linear files with fixed length records

- Formatted files with variable length records

- Application specific command files

- Formatted cyclic fixed length files

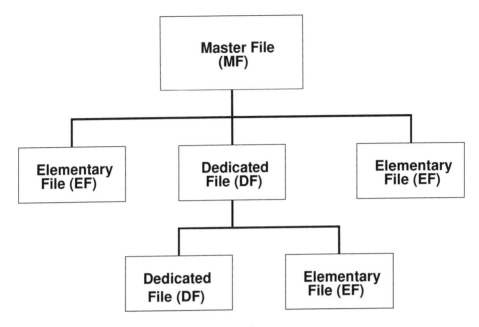

Figure 18.7 File system on an IBM smartcard.

At the time of file creation, access conditions can also be defined. Following access conditions are allowed.

- Read: read and seek on the records of the file allowed.

- Update: Updating a record such as decrementing a value or modifying a field allowed.

- Administer: create, delete, validate and rehabilitate files.

A file on the smartcard can have one or more of the following protection modes.

- always: not protected and available always.

- secret password based: A password (set during the file creation) is required to access the file.

- a second password based: A second password-based access in case the first one is lost.

- external authentication: access is allowed if an external cryptographic authentication is successful.

- protected: The file is protected using a message authentication code.

- encrypted: update of sensitive files requiring encryption.

- never: Data never accessible

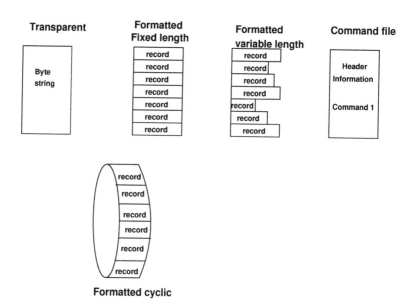

Figure 18.8 Supported elementary file types on a smartcard.

Commands to Smartcard

The commands to which a smartcard responds can be classified into five classes.

- application data commands

- security commands

- card management functions

- electronic purse commands

- miscellaneous commands

Additional commands can be developed and loaded during the initialization phase of the card.

Application Interface

Different software components in the IBM smartcard are shown in Figure 18.9. The designer of a smartcard application describes the file structure on a smartcard in a high-level language. This description is compiled using a layout compiler that produces an initialization and personalization bit stream for the card. After the card is initialized, the application program can access the files through the smartcard agents defined in the IBM smartcard toolkit [10] or through the Java interface supported by Opencard forum [11].

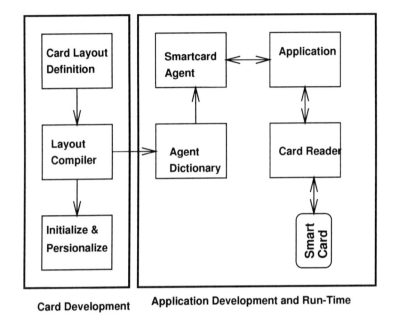

Figure 18.9 Software components of IBM smartcard.

Security Issues in a Smartcard

The most attractive feature of a smartcard is the variety of security features that it can support. At the card level, it can be protected by several passwords. At the file level, we described different kinds of access protection methods in the previous section. The data contents in the card get reset in case of possible hardware attacks on the card. In addition to these, cryptographic authentication can also be supported on many smartcards. In this method, access to data on the smartcard is permitted if a challenge presented by the smartcard is decoded correctly by the external world. Often, it is also desired to verify the authenticity of the card. For this purpose, the external world poses a challenge to the card. The processor in the card decodes the message in the challenge and presents it to the external world. The card is considered genuine if the decoded message matches with the original message. Before we describe other encryption methods, we present a short analysis of encryption methods. Detailed accounts of encryption are available in [7, 11].

Encryption Methods
Cryptographic algorithms are used to encode and decode messages. There are two kinds of cryptographic algorithms.

- symmetric secret key algorithms

- asymmetric secret key algorithms

In the symmetric algorithms the same secret key is used in encryption and decryption. Hence, it is required that the secret key be known to both sender and receiver. In the asymmetric method, the message is encoded using public key and decrypted using a private key. The public key is different from the private key. As the name indicates, the public key can be known to many parties. The private key is only known to the decryption module which can decrypt the message. Figure 10 shows the two methods. Public key algorithms are based on corresponding key pairs consisting of a secret private key and a public key. The private secret key is managed by the owner whereas the publick key is known to everyone. This feature of public key algorithms enables them to work with digital signatures. Cryptographic information created with the secret key of the sender is called a "digital signature". The public key can be used to verify the signature associated with the message without the need to know the secret key. Furthermore, as the secret key is unique, any message with the signature generated by the key uniquely refers back to the owner.

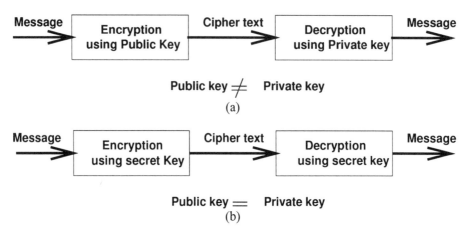

Figure 18.10 Encryption methods. (a) Asymmetric; (b) Symmetric.

Crypto Options in a Smartcard
Public key algorithms based on asymmetric key crypto algorithms are used in special smartcards to protect data. These smartcards require larger RAM memory and more powerful processors. Hence, there are special smartcards with crypto coprocessors. Data integrity during communication between smartcard and external world is ensured by associating access conditions described earlier. Message authentication methods are also used. A message authentication code is computed based on a secret key, a random number and the data which is to be exchanged. The code is appended to the message and transmitted. The authentication code is regenerated at the receiving end from the message and the key and compared with original code with the message. A digital signature can also be used to ensure data integrity.

4. Biometrics and Smartcard

Biometrics can play an important role in making smartcards more secure and smartcards can make biometrics more pervasive and useful. By biometrics, we refer to the science of identifying and authenticating by using personal characteristics [12]. An automated biometrics-based system can be used to authenticate a person. Adding a smartcard option in the system can provide many advantages. In this section, we present two generic solution models where biometrics and smartcard can be integrated.

Crypto Key Management with Biometrics

In the previous section, it was stated that the secret key management is the owner's responsibility. Often, the keys are long strings of numbers which can not be remembered easily, resulting in the key being kept in a file. This can be a potential source of insecurity. One way to avoid the problem is to release the secret key to the software subsystem after verifying the owner by a biometrics. This is very important in case the card contains a large amount of electronic cash or an important document. Our proposed model is shown in Figure 18.11.

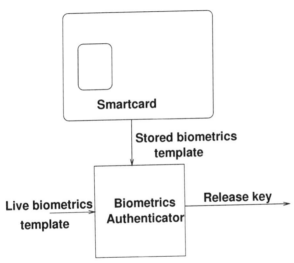

Figure 18.11 Biometrics in smartcard security.

Remote Authentication with Smartcard and Biometrics

Smartcards can play an important role in biometrics too. In an identification system, the biometrics templates are often stored in a central database. With the central storage of a biometrics, there is an open issue of misuse of the same for purposes that the owner of the biometrics may not be aware of. Large collection of biometrics data can be sold or given away to unsolicited parties who can misuse the information. We can decentralize the database storage part into millions of smartcards and give it to the owners. The system does not store the biometrics on its central database anymore. The biometrics data can be stored along with other information such as message authentication code to maintain its integrity. Other textual information can also be a part of the data. This model is more attractive in contrast to storing the whole database on a central system. For authentication of the user, the smartcard and the live biometrics are presented to the system. The system generates the matching template from the biometrics and decodes the stored template from the smartcard and examines the similarity between the two templates. The two steps in this proposed method are shown in Figure 18.12. The owner must be enrolled into the system. During the enrollment, the user's biometrics is collected and the features are extracted from the biometrics signal after ensuring the quality of the input signal. The template is then stored in the smartcard along with other information such as user ID and other user dependent parameters. For example, in a fingerprint-based system [18], the fingerprint features along with, say, the threshold to be used during matching can be stored. Similarly, in a face recognition-based system [19], the face template and the parameters used in the face matching system can be stored. For speech-based systems [20], this method can provide extra benefit, the parameters used in speech modeling can also be stored on the smartcard. These parameters can be restored during the matching stage and used by the matcher, thus resulting in better performance.

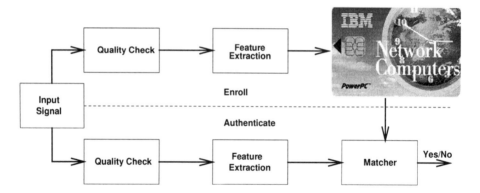

Figure 18.12 System architecture using biometrics and smartcard.

5. Applications

In this section, we describe a few example applications that can benefit from the combination of biometrics and smartcard.

- Network Computer secure login: Network computers are intelligent terminals with no local storage. They connect to a remote server. The server must authenticate the users every time during login. We have implemented a secure login scheme using fingerprint and smartcard for an IBM Network computer. The NCs are equipped with smartcard readers. A registered user of the system is issued a smartcard that contains his fingerprint template and user ID and access rights on the server. Other personalization information can also be stored on the smartcard. The NC locally matches the fingerprint of the user against the template stored on the smartcard and grants access to the server only if the two templates match. This reduces network traffic and load on the sever to validate users and it becomes a local function on each NC. This model can be extended further to single sign on cards. The smartcard can contain details of other systems or network passwords. Once the user has been authenticated at the NC by matching the fingerprint, the NC can release the appropriate password from the smartcard.

- NT Secure logon/Screen saver: An extension of the above scheme is to provide a smartcard and fingerprint based secure logon for NT. Even screen savers can be designed with a biometrics.

- Home banking: With the growth of Internet and availability of PCs at home, banks are already providing many banking functions on home PCs. Most of these functions are limited to accessing the account information or transferring money between accounts. With smartcards acting as 'cash cards', the banks can soon provide functions to download money from their accounts at home. By use of a biometrics to authenticate the person, it is a simple extension from our model. The card can be loaded with extra money if the biometrics is successfully matched.

- Internet commerce: Over the last couple of years, commerce on the Internet is growing exponentially. Coupled with a web browser, many purchases are made from home. However, the security of the credit card information being exchanged is always in question. Often a smartcard is used to enhance the security in these applications. By adding a biometrics, the system security is further enhanced.

- Automotive: Future cars and automobiles are aiming at more personalization and security with more compute power being made available in a vehicle. Concept cars with voice recognition and keyless ignition facilities are being demonstrated. A personalized security system can be achieved with the combination of smartcard and biometrics.

- Health care: Privacy of health care information is very important. Often health care information is handled with a minimal care and can be easily bought from health care providers. This information can be easily protected by use of a smartcard and biometrics. The biometrics authenticates the rightful owner of the information and can be presented to the health care providers depending on their access level. For example, a doctor can have access to all the test results whereas an accounting personnel should have access to only the billing information and insurance information.

6. Conclusions

Smartcards are a model of very secure storage and biometrics is the ultimate technology for authentication. The two can be combined in many applications to enhance both the security and authentication. With the cost of biometrics sensors, in particular, fingerprint sensors on the decline, the day isn't far off when every person will have his personal smartcard with a built-in fingerprint sensor as the authentication tool. That will be the most secure way of carrying out authentication in remote applications such as home banking, access control and electronic commerce.

References

[1]   J. J. Farrell, "Smartcards become an international technology," *in Proceedings of 13th TRON Project International Symposium, Tokyo, Japan*, Dec. 1996, pp. 134-140.

[2]   M. A. Sirbu, "Credits and debits on the Internet," *IEEE Spectrum*, Vol. 34, No. 2, pp. 23-29, Feb. 1997.

[3]   G. Lisimaque, "Almost everything you need to know to get started with smart cards," in *Proceedings CardTech/SecureTech 97*, pp. 305-319, May 1997.

[4]   S. Elliot and C. Loebbecke, "Smart card based electronic commerce: characterisitics and roles," in *Proceedings 31st Annual Hawaii International Conference on System Sciences, Hawaii*, pp. 242-250, January, 1998.

[5]   J. L. Zoreda and J.M. Oton, *Smart cards*, Artech House, London, 1994.

[6]   C. H. Fancher, "In you pocket: smartcards," *IEEE Spectrum*, Vol. 34, No. 2, pp. 47-53, Feb. 1997.

[7]   H. Dreifus and J. T. Monk, *Smart cards*, John Wiley and Sons, New York, 1997.

[8]   R. Hermann and D. Husemann, "Opencard framework 1.0 white paper," in *Opencard forum web site URL: http://www.opencard.org*, 1998.

[9]   S. B. Guthery, "Java card: Internet computing on a smart card," *IEEE Internet Computing*, pp. 57-59, Jan-Feb. 1997.

[10] IBM Germany Development Laboratory, Schoenaicher Strasse 220, 71032 Boeblingen, Germany, *Smart card toolkit programmer's reference*, August 1997.

[11] R. W. Baldwin and C. V. Chang, "Locking the e-safe," *IEEE Spectrum*, vol. 34, no. 2, pp. 40-46, February 1997.

[12] B. Miller, "Vital signs of identity," *IEEE Spectrum*, vol. 31, no. 2, pp. 22-30, February 1994.

[13] E. Turban and D. Mcelroy, "Using smart card in electronic commerce," in *Proceedings 31st Annual Hawaii International Conference on System Sciences, Hawaii*, pp. 62-69, January 1998.

[14] C. H. Fancher, "Smart cards," *Scientific American*, pp. 40-45, August 1997.

[15] J.P. Thomasson, "Smartcards: Portable security," in *Proceedings Second Annual IEEE Int. Conference on Innovative Systems in Sillicon*, pp. 259-265, 1997.

[16] N. Asokan, P. A. Janson, M. Steiner, and M. Waidner, "The state of the art in electronic payment systems," *IEEE Computer*, Vol. 30, No. 9, pp. 28-35, September 1997.

[17] D. A. Cunnigham, "The state of the smart card industry," in *Proceedings CardTech/SecureTech 97*, pp. 285-304, May 1997.

[18] A. K. Jain, L. Hong, and R. Bolle, "On-line fingerprint verification," *IEEE Transactions on Pattern Analysis and Machine Intelligence*, Vol. 19, No. 4, pp. 902-914, April 1997.

[19] K-M. Lam and H. Yan, "An analytic-to-holistic approach to face recognition based on single frontal view," *IEEE Transactions on Pattern Analysis and Machine Intelligence*, Vol. 20, No. 7, pp. 673-686, July 1998.

[20] J. P. Campbell, "Speaker Recognition: A tutorial," *Proceedings of IEEE*, Vol. 85, No. 9, pp. 1437-1462, September 1997.

19 BIOMETRICS: IDENTIFYING LAW & POLICY CONCERNS
John D. Woodward, Jr.

Abstract *Today's "new technological realities" force us to examine, from the law and policy perspectives, what is required to safeguard the public interest and to ensure optimal results for society. Biometrics is one such new technology reality. While not enjoying the media stature and public controversy associated with high tech issues like genetic cloning and cyberspace, biometrics -- which seeks a fast, foolproof answer to the questions, "Who are you?" or "Are you the person whom you claim to be?" -- will cause the law to take notice as it becomes more extensively used in the public and private sectors. Businesses, numerous government agencies, law enforcement and other private and public concerns are making increasing use of biometric scanning systems. As computer technology continues to advance and economies of scale reduce costs, biometrics will become an even more effective and efficient means for identification and verification. After briefly discussing biometric technologies and biometric applications, this chapter defines privacy in the context of biometrics and discusses which specific privacy concerns biometrics implicates. This chapter concludes that biometrics is privacy's friend because it can be used to help protect information integrity. The author also contends that any legitimate privacy concerns posed by biometrics, such as the possibility of a secondary market in individual biometric identification information, can be best handled by the existing law and policy framework. The author next considers the future of biometrics, and contends that "biometric balkanization," or the use of multiple biometric technologies deployed for multiple applications, provides greater privacy protections than does biometric centralization, or the use of one dominant biometric technology for multiple applications.*

Keywords: *Privacy, biometric law, public policy, government, information policy, constitutional law.*

1. Introduction[1]

On May 18, 1997, in his commencement address at Morgan State University, President William J. Clinton stated:

> The right to privacy is one of our most cherished freedoms. As society has grown more complex and people have become more interconnected in every way, we have had to work even harder to respect privacy, the dignity, the autonomy of each individual . . . [w]e must develop new protections for privacy in the face of new technological reality [1].

While it is doubtful that President Clinton had biometrics in mind during that Sunday speech, biometrics is clearly emerging as one such "new technological reality." From activities as diverse as the Winter Olympics in Nagano, Japan to the prisons of Cook County, Illinois, both the public and private sectors are making extensive use of biometrics. This new technological reality relies on "the body as password" for human recognition purposes to provide better security, increased efficiency and improved service [2,3,53,54]. As the technology becomes more economically viable, technically perfected and widely deployed, biometrics could become the passwords and PINs of the twenty-first century. In the process, biometrics could refocus the way Americans look at the brave new world of personal information [4].

Understanding biometrics is thus essential for elected officials and policymakers charged with determining how this new technology will be used and what role, if any, government should play in its regulation. Familiarity with biometrics is also important for the legal, business and policy advocacy communities so that they can meaningfully participate in the public debate related to biometrics.

Similarly, understanding the law and policy concerns of biometrics is necessary for the engineers and scientists who have brought about this new technological reality. History teaches us that new technologies, created by engineers and scientists, spark new law and cause old legal doctrines to be rethought, rekindled and reapplied by the nation's law and policy makers.[2]

New technology can cause a creative reshaping of existing legal doctrine when, for example, the judiciary has embraced a technology more quickly than the legislature, the executive branch or even the actual marketplace for the technology. To consider a well-known example from the legal casebooks, in 1928, there was no law or regulation requiring coastwise seagoing carriers to equip their tugboats with radio receiver sets. Moreover, no such custom or practice existed in the maritime industry, despite the fact that such sets could easily be used by tugs at sea to receive storm weather warnings. In a landmark legal case, Federal Circuit Judge Learned Hand, one of the great American jurists of this century, deemed that tugboats without radio receiver sets were unseaworthy because "a whole calling may have unduly lagged in the adoption of new and available devices" [6]. By accepting a new technology -- in this case, wireless communications -- more quickly than the legislative and executive branches or even the affected industry, Judge Hand, in effect, creatively reshaped the law. No longer would strict adherence to local custom and industry practice offer a

---

[1] This chapter is largely based on a previously-published article by the same author: John D. Woodward, "Biometrics: Privacy's Foe or Privacy's Friend?" in *Proceedings of the IEEE*, Sept. 1997.

[2] For an excellent examination of the intersection of science and technology with law and policy, see [5].

guaranteed defense against charges of negligence when a readily-available technology could result in greater utility to society.

Similarly, today's new technological reality of biometrics should force us to explore from the law and policy perspectives what is required to safeguard the public interest and to ensure optimal results for society. Engineers and scientists should not be excluded from this law and policy examination. Indeed, the law and policy concerns raised by biometrics are far too important to be left solely to politicians and lawyers.

In examining these law and policy concerns, this chapter focuses on privacy. After briefly discussing biometric technologies in sections 2 and 3, the author, in section 4, defines privacy in the context of biometrics and examines which specific privacy concerns biometrics implicates. The author then analyzes the various arguments often made that biometrics poses a threat to privacy. The author concludes that, to the contrary, biometrics is privacy's friend. Biometrics is privacy's friend because it can be used to help protect information integrity and to deter identity theft. Nonetheless, the author suggests that government can play a positive role in regulating and thereby promoting public acceptance of this new technology. Section 5 examines the biometric future, and contends that "biometric balkanization," or the use of multiple biometric technologies deployed for multiple applications, provides greater privacy protections than does biometric centralization, or the use of one dominant biometric technology for multiple applications.

2. What is Biometrics?

Definition of Biometrics & Biometric Scanning

While the word, "biometrics," sounds very new and "high tech," it stands for a very old and simple concept -- human recognition. In technical terms, biometrics is the automated technique of measuring a physical characteristic or personal trait of an individual and comparing that characteristic or trait to a database for purposes of recognizing that individual [7,56].

Biometrics uses physical characteristics, defined as the things we are, and personal traits, defined as the things we do, to include:

Physical Characteristics
Chemical composition of body odor
Facial features & thermal emissions
Features of the eye: retina & iris
Fingerprints
Hand geometry
Skin pores
Wrist/hand veins

Personal Traits
Handwritten signature
Keystrokes or typing
Voiceprint

Of these, only three of the physical characteristics and personal traits currently used for biometrics are considered truly consistent and unique: the retina, the iris and fingerprints [8,55]. As such, these three physical characteristics provide the greatest reliability and accuracy for biometrics.

Biometric scanning is the process whereby biometric measurements are collected and integrated into a computer system, which can then be used to automatically recognize a person. Biometric scanning is used for two major purposes: Identification and verification. Identification is defined as the ability to identify a person from among all those enrolled, i.e., all those whose biometric measurements have been collected in the database. Identification seeks to answer the question: "Do I know who you are?" and involves a one-compared-to-many match (or what is referred to as a "cold search").

Second, biometric scanning is used for verification, which involves the authentication of a person's claimed identity from his previously enrolled pattern. Verification seeks to answer the question: "Are you who you claim to be?" and involves a one-to-one match.

Advantages of a Biometric Scanning System

Biometric scanning can be used for almost any situation calling for a quick, correct answer to the question, "Who are you?" The great advantage of biometric scanning is that it bases recognition on an intrinsic aspect of a human being. Recognition systems that are based on something other than an intrinsic aspect of a human being are not always secure. For example, keys, badges, tokens and access cards (or things that you must physically possess) can be lost, duplicated, stolen or forgotten at home. Passwords, secret codes and personal identification numbers (PINs) (or things that you must know) can be easily forgotten, compromised, shared or observed [9].

Biometric technologies, on the other hand, are not susceptible to these particular problems because biometrics relies on things that you are. For example, one industry representative has summed up the inherent strengths of the biometric his company promotes in the following humorous way: "Your iris: You can't leave home without it."

Depending on the exact use for which the technology is envisioned, an ideal biometric technology would include a system based on: (i) a consistent and unique biometric characteristic, (ii) non-intrusive data collection, (iii) no or minimal contact between the person being scanned and the equipment doing the scanning, (iv) an automated system, i.e., no human decision maker in the decision loop, (v) very high accuracy, and (vi) high speed.

According to Dr. Joseph P. Campbell, Jr., a National Security Agency (NSA) researcher and the former Chairman of the Biometric Consortium (BC), the U.S. Government's focal point for biometric research; no one technology has emerged as the "'perfect biometric,' suitable for any application" [10]. While there is no "perfect biometric," a good biometric scanning system is fast, accurate, dependable, user-friendly and low-cost.

3. How are Biometrics Used?

Biometric applications are increasingly broad-based, expanding and international; as one industry expert has stated, "The influence of biometric technology has spread to all continents on the globe" [11]. In concrete terms, this influence translates into about $1 billion worth of computer systems that include biometric devices which were estimated to be installed worldwide during 1997 [12].

While biometric devices are deployed in many computer systems, the overall size of the biometric industry remains relatively small though rapidly growing. For example, in 1992, revenue from biometric devices was estimated at $8.3 million with 1,998 units being sold. By 1999, revenue is projected at $50 million with 50,000 units being sold. Accordingly, biometric scanning is likely to have a substantial impact on the way we interact and conduct our affairs in the foreseeable future.

While a detailed discussion of biometric applications is beyond the scope of this chapter, the following three major categories of biometric applications -- High Government Use, Lesser Government Use, and Private Sector Use -- highlight how biometric scanning is beginning to touch our lives:

High Government Use
Law Enforcement
Prison Management
Military & National Security Community

Lesser Government Use
Border Control & Immigration Checks
Entitlement Programs
Licensing
National Identity Card & Voter Registration

Private Sector Use
Banking and Financial Services
Personnel Management
Access Control
Information System Management

4. What is Privacy in the Context of Biometrics?

Working Definition

The issue of privacy is central to biometrics. Critics complain that the use of biometrics poses a substantial risk to privacy rights. Evaluating this argument requires, in the first instance, an understanding of what privacy rights entail. The word "privacy" (like the word "biometrics") is nowhere to be found in the text of the United States Constitution, America's highest law. Perhaps the absence of any explicit textual reference to privacy or right of privacy, combined with the word's apparent flexibility of meaning, make it all the more difficult to define what privacy is and to explain what the right of privacy should be.

Most importantly from the standpoint of biometrics, privacy includes an aspect of autonomy – "control we have over information about ourselves" [12], "[c]ontrol over who can sense us" [13], "...control over the intimacies of personal identity" [14], or as a federal appeals court has phrased it, "control over knowledge about oneself. But it is not simply control over the quantity of information abroad; there are modulations in the quality of knowledge as well" [15].

In the context of biometrics, this control over information about ourselves, or information privacy, lies at the very heart of the privacy concerns raised by this new technology. Individuals have an interest in determining how, when, why and to whom information about themselves, in the form of a biometric identifier, would be disclosed.

In the American legal experience, privacy protections have followed two traditional pathways depending on whether the source of the privacy intrusion is a governmental or private sector activity. While privacy is not explicitly cited in its text, the Constitution, through the Bill of Rights, protects the individual from government's intrusion into the individual's privacy. For example, the Bill of Rights contains privacy protections in the First Amendment rights of freedom of speech, press and association; the Third Amendment prohibition against the quartering of soldier's in one's home; the Fourth Amendment right to be free from unreasonable searches and seizures; the Fifth Amendment right against self-incrimination; and the Ninth Amendment's provision that "[t]he enumeration in the Constitution, of certain rights, shall not be construed to deny or disparage others retained by the people" [16].

With respect to private sector actions, the Constitution traditionally embodies what is essentially a *laissez-faire* spirit. As constitutional law scholar Laurence Tribe has noted, "[T]he Constitution, with the sole exception of the Thirteenth Amendment prohibiting slavery, regulates action by the government rather than the conduct of private individuals and groups [66]." With respect to the conduct of private individuals, the Supreme Court has been reluctant to find a privacy right in personal information given voluntarily by an individual to private parties [17,57].

For private sector intrusions into privacy, the common-law, through its doctrines of contract, tort and property, has, in varying degrees, attempted to provide certain protections for the individual (e.g., [18])[19]. However, the law has not used these doctrines to protect individual information in private sector databases. Generally, as a matter of law, an individual in possession of information has the right to disclose it.

Accordingly, the private sector enjoys great leeway as far as what it can do with an individual's information. "Except in isolated categories of data, an individual has nothing to say about the use of information that he has given about himself or that has

---

[3]As early as 1879, Judge Thomas M. Cooley, in his treatise on torts, included "the right to be let alone" as a class of tort rights, contending that "the right to one's person may be said to be a right of complete immunity." Echoing and popularizing Cooley's phrase, Warren and Brandeis, in their classic article written over one hundred years ago, articulated their view of privacy as a "right to be let alone" which would enable society to "achieve control of press invasions of privacy" [58,59].

been collected about him. In particular, an organization can acquire information for one purpose and use it for another . . . generally the private sector is not legislatively-constrained" [19].

While the Supreme Court has never explicitly recognized a constitutional right to privacy (and has never dealt with biometrics), America's highest court has grappled with information privacy issues. In *Whalen v. Roe*, an influential case decided twenty-one years ago, the Court decided the constitutional issue of whether the State of New York could record, in a centralized database, the names and addresses of all individuals who obtained certain drugs, pursuant to a doctor's prescription. Rejecting the information privacy claim, the Court ruled that a government database, containing massive amounts of sensitive medical information, passed constitutional muster because of the security safeguards in place. The Court's opinion, however, concluded with a cautionary note that still echoes loudly today: "We are not unaware of the threat to privacy implicit in the accumulation of vast amounts of personal information in computerized data banks or other massive government files" [20].

What Privacy Concerns Are Implicated?

The Individual Gives Up a Biometric Identifier
To determine the specific privacy concerns implicated by biometrics, we must first focus on what exactly is disclosed when biometric scanning is used. Regardless of whether an individual voluntarily provides a biometric identifier or is forced to surrender it as part of a state action or government-required scheme, he is giving up information about himself. When biometrics like finger imaging, iris recognition or retinal scanning are used, he discloses consistent and unique information about his identity. When the other biometrics are used, at a minimum, he discloses accurate information about who he is.

Invasive Aspects of the Information
Beyond this fundamental disclosure, there also might be invasive implications related to privacy concerns which stem from the biometric identification information disclosed. These invasive implications for privacy are essentially two-fold: 1) the invasive effects of a secondary market defined as disclosure of the biometric identification information to third parties and 2) any invasive information which might be additionally obtained as part of the biometric identifier.

Invasive Secondary Market Effects
Once a biometric identifier is captured from an individual in the primary market, and even if it is captured only once, the biometric identifier could easily be replicated, copied and otherwise shared among countless public and private sector databases. This sharing in a secondary market could conceivably take place without the individual's knowledge or consent. Indeed, biometric identifiers could be bought and sold in a secondary market much the way names and addresses on mailing lists are currently bought and sold by data merchants.

Particularly with respect to the private sphere, where the conduct of private actors has traditionally been given a degree of freedom of action from government interference, there are few current legal limits on the use of biometric information held by private actors. This observation is not meant to suggest that the federal or state governments would not be able to regulate the use of biometric information held by private actors; rather, it emphasizes what the present regulatory baseline is with respect to the regulation of biometric information: Until affirmative action has been taken by government, the use of biometrics is left to the market.

In other similar contexts where an individual has surrendered personal information to private actors, the Supreme Court has not found a right to privacy in the information surrendered. For example, in *Smith v. Maryland*, the defendant claimed that information in the form of telephone numbers he dialed from his home telephone (what is known as a pen register), could not be turned over to the police absent a search warrant [21]. Rejecting this argument, the Court noted that it "consistently has held that a person has no legitimate expectation of privacy in information he voluntarily turns over to third parties" [22].

In *United States v. Miller*, a case involving a bootlegger's private financial records which were turned over to U.S. Treasury agents pursuant to a grand jury subpoena, the bootlegger's attempt to have the evidence excluded was unsuccessful [23]. The Court found that Miller had no expectation of privacy in the records, reasoning that: "The depositor takes the risk, in revealing his affairs to another, that the information will be conveyed by that person to the Government." Moreover, these records could not therefore be considered confidential communications because they had been voluntarily conveyed to the bank in the "ordinary course of business" [24].

Technology is fast and the law is slow [25]. Thus, biometrics is still relatively too new for the Congress or the various state legislatures to have acted from the standpoint of adopting privacy protections aimed specifically at this technology. In an impressive first step toward understanding private sector applications of biometrics, on May 20, 1998, Congress held hearings on the topic of "Biometrics and the Future of Money." These hearings, before the Subcommittee on Domestic and International Monetary Policy of the Committee on Banking and Financial Services of the U.S. House of Representatives, chaired by Michael N. Castle, featured panels of leading technologists as well as policy experts. In terms of activities in the state legislatures, California has been the most pro-active. There, the legislature has been considering biometric privacy legislation as part of its identity theft reforms.

Currently, private actors possessing biometric identification information generally follow a nondisclosure policy -- they do not disclose this information to third parties -- as part of a strategy of building public acceptance for the technology. However, such nondisclosure policies are completely voluntary. Critics contend that biometric identifiers, like other personal information such as names and addresses for mailing lists, might eventually be "considered to be in the public domain" [26]. The fear is that the individual will lose ultimate control over all aspects of her biometric identifier.

Invasive Information Is Obtained
In addition to the identification information associated with the biometric, invasive information threatening privacy could conceivably include three other types of concerns. First, biometric identifiers could be used extensively for law enforcement purposes. Fingerprints have long been used by law enforcement and finger images -- or what are in effect the next generation of fingerprints -- are presently being used by

various law enforcement agencies as part of their databases. For example, the Federal Bureau of Investigation (FBI) has embarked on a bold finger imaging project for its Integrated Automated Fingerprint Identification System (IAFIS). IAFIS would replace the present paper and ink based system with electronic finger images.

Secondly, it is possible (and the point needs to be stressed, only *possible*) that some biometrics might capture more than just mere identification information. Information about a person's health and medical history might also be incidentally obtained. Recent scientific research suggests that fingerprints and finger imaging might disclose medical information about a person [27,28]. For example, Dr. Howard Chen, in his work on dermatoglyphics, or the study of the patterns of the ridges of the skin on parts of the hands and feet, notes that "[c]ertain chromosomal disorders are known to be associated with characteristic dermatoglyphic abnormalities," specifically citing Down syndrome, Turner syndrome and Klinefelter syndrome as chromosomal disorders which cause unusual fingerprint patterns in a person [28]. Certain non-chromosomal disorders, such as chronic, intestinal pseudo-obstruction (CIP) (described below), leukemia, breast cancer and Rubella syndrome, have also been implicated by certain unusual fingerprint patterns.

For example, Dr. Marvin M. Schuster, director of the division of digestive diseases at Johns Hopkins Bayview Medical Center, has discovered a "mysterious relationship" between an uncommon fingerprint pattern, known as a digital arch, and a medical disorder called CIP which affects 50,000 people nationwide. Based on the results of a seven year study, Dr. Schuster found that 54 percent of CIP patients have this rare digital arch fingerprint pattern. Schuster's discovery suggests a genetic basis to the disease in that the more digital arches in the fingerprint, the stronger the correlation to CIP [29].

While still controversial within the scientific communities, several researchers report a link between fingerprints and homosexuality [30,60,61,62]. For example, psychologists at the University of Western Ontario report that homosexual males are more likely than their heterosexual counterparts to show asymmetry in their fingerprints. While this research is far from conclusive, the availability of such information with its possible links to medical information and lifestyle preferences again raises concern about the need to protect the privacy of the information.

From examining the retina or iris, an expert can determine that a patient may be suffering from common afflictions like diabetes, arteriosclerosis and hypertension; furthermore, unique diseases of the iris and the retina can also be detected by a medical professional [31,63]. While both the iris and retina contain medical information, it is by no means obvious that biometric scanning of the iris or retina automatically implicates privacy concerns related to the disclosure of medical information. A necessary area of further technical inquiry is whether the computerized byte code taken of the iris or retina actually contains this medical information or if the information captured is sufficient to be used for any type of diagnostic purpose.

While much research remains to be done, the availability of such information with its possible links to medical information raises important questions about the privacy aspects of the information disclosed as well as public perception concerns.

Biometrics as Privacy's Foe: Criticisms of Biometrics

The Loss of Anonymity; the Loss of Autonomy
A basic criticism of biometrics from the standpoint of privacy is that we, as individuals, lose our anonymity whenever biometric scanning systems are deployed. Part of controlling information about ourselves includes our ability to keep other parties from knowing who we are. While we all know that at some level, a determined party -- whether the state or a private actor -- can learn our identity (and much more about us), biometric scanning makes it plain that our identity is now fully established within seconds. As Professor Clarke explains, "The need to identify oneself may be intrinsically distasteful to some people ... they may regard it as demeaning, or implicit recognition that the organisation [sic.] with whom they are dealing exercises power over them" [32]. Privacy advocate Robert Ellis Smith agrees, noting that, "In most cases, biometric technology is impersonal." At the same time, if the technology meets with widespread success, individuals may find that they are required to provide a biometric identifier in unexpected, unwelcome or unforeseen future circumstances. Moreover, you cannot make up a biometric as easily as you can an address and phone number. In this sense, perhaps, the loss of anonymity leads to an inevitable loss of individual autonomy.

To the extent there is less individual anonymity today than in decades or centuries past, biometrics is not to blame. Rather, far larger economic, political and technological forces were at work. America's transformation from an agrarian to industrial to post-industrial, service economy combined with the massive growth of government since the New Deal of the 1930s have put a greater premium on the need for information about individuals and organizations. At the same time, technical advances have made it much easier and more convenient to keep extensive information on individuals. Summarizing this trend, one scholar has noted, "[I]n the present service economy, information has become an increasingly valuable commodity . . . The computer has exacerbated this problem through its capacity to disclose a large amount of personal information to a large number of unrelated individuals in a very short amount of time [18]."

While a biometric identifier is a very accurate identifier, it is not the first nor is it the only identifier used to match or locate information about a person. Names and numerical identifiers such as social security numbers, account numbers and military service numbers have long been used to access files with personal information. Moreover, the impressive search capabilities of computer systems with their abilities to search, for example, the full text of stored documents, make identifiers far less important for locating information about an individual.

Moreover, there is usually a good reason why recognition in the form of identification or verification is needed. Balancing the equities involved and depending on the case, the benefits of establishing a person's identity generally outweigh the costs of losing anonymity. For example, given the massive problem of missing and abused children, we eagerly support the idea of day care providers using biometrics to make certain that our children get released at the end of the day to a parent or guardian whose identity has been verified.

Similarly, to consider a "pocketbook" example, the world's financial community has long been concerned about growing problems of ATM fraud and unauthorized account access, estimated to cost $400 million a year [33,64]. Credit card fraud is estimated at $2 billion per year. The financial services industry believes that a significant percentage of these losses could be eliminated by biometric scanning.

Critics give too much credit to biometrics' alleged ability to erode anonymity without giving enough attention to the market's ability to protect privacy in response. It is not obvious that more anonymity will be lost when biometrics are used. Public and private sector organizations already have the ability to gather substantial amounts of information about individuals by tracking, for example, credit card use, consumer spending and demographic factors.

Drawing a parallel to the financial services industry, despite the existence of many comprehensive payment systems, like credit cards, which combine ease of service with extensive record-keeping, many Americans still prefer to use cash for transactions -- a form of payment that leaves virtually no record. An individual who wants anonymity might have to go to greater lengths to get it in the biometric world but the ability of the marketplace to accommodate a person's desire for anonymity should not be so readily dismissed. Moreover, as explained below, the ability of biometrics to serve as privacy enhancing technologies should not be discounted.

The Biometric-Based Big Brother Scenario

Aside from the alliterative qualities the phrase possesses, critics of biometrics seem to inevitably link the technology to Big Brother. Biometrics, in combination with impressive advancements in computer and related technologies, would, its critics argue, enable the State to monitor the actions and behavior of its citizenry. In this vein, concern has been expressed that biometric identifiers will be used routinely against citizens by law enforcement agencies. As Marc Rotenberg of the Electronic Privacy Information Center has succinctly explained, "Take someone's fingerprint and you have the ability to determine if you have a match for forensic purposes" [34].

This Big Brother concern, however, goes beyond normal police work. Every time an individual used her biometric identifier to conduct a transaction, a record would be made in a database which the government, using computer technology, could then match and use against the citizen -- even in ways that are not authorized or meet with our disapproval. To borrow the reasoning of a 1973 report on national identity card proposals, the biometric identifier, in ways far more effective than a numerical identifier, "could serve as the skeleton for a national dossier system to maintain information on every citizen from cradle to grave" [35]. Professor Clarke has perhaps offered the best worst-case *1984*-like scenario:

> Any high-integrity identifier [such as biometric scanning] represents a threat to civil liberties, because it represents the basis for a ubiquitous identification scheme, and such a scheme provides enormous power over the populace. All human behavior would become transparent to the State, and the scope for non-conformism and dissent would be muted to the point envisaged by the anti-utopian novelists.

There is at least one example from United States history where supposedly confidential records were used in ways never likely intended. In November 1941, almost two weeks before the Japanese attack on Pearl Harbor, President Franklin D. Roosevelt ordered a comprehensive list made, to include the names and addresses, of all foreign-born and American-born Japanese living in the United States. To compile the list, staffers used 1930 and 1940 census data. Working without the benefit of computers, staffers compiled the list in one week [36]. By the Spring of 1942, the United States Government forced persons of Japanese descent, including United States citizens, to relocate from their homes on the West Coast and report to "Relocation Centers" [37].

Function Creep
The biometric-based Big Brother scenario would not happen instantly. Rather, when first deployed, biometrics would be used for very limited, clearly specified, sensible purposes -- to combat fraud, to improve airport security, to protect our children, etc. But as Justice Brandeis warned in his famous *Olmstead* dissent:

> Experience should teach us to be most on our guard to protect liberty when the Government's purposes are beneficent. Men born to freedom are naturally alert to repel invasion of their liberty by evil-minded rulers. The greatest dangers to liberty lurk in insidious encroachment by men of zeal, well-meaning but without understanding [38].

What would inevitably happen over time, according to civil libertarians, is a phenomenon known as "function creep": identification systems incorporating biometric scanning would gradually spread to additional purposes not announced or not even intended when the identification systems were originally implemented.

The classic example of function creep is the use of the Social Security Number (SSN) in the United States. Originated in 1936, the SSN's sole purpose was to facilitate recordkeeping for determining the amount of Social Security taxes to credit to each contributor's account [39]. In fact, the original Social Security cards containing the SSN bore the legend, "Not for Identification." By 1961, the Internal Revenue Service (IRS) began using the SSN for tax identification purposes. By 1997, "[e]verything from credit to employment to insurance to many states' drivers licenses requires a Social Security Number." From "Not for Identification," the SSN has become virtual mandatory identification.

Moreover, given the consequences of function creep, the size, power and scope of government will expand as all citizens get their biometric identifiers thrown into massive government databases by the "men [and women] of zeal, well-meaning but without understanding" about whom Justice Brandeis warned. In effect, a Russian proverb aptly identifies the danger of biometrics for freedom-loving Americans, "If you are a mushroom, into the basket you must go."

By Using Biometrics, Government Reduces the Individual's Reasonable Expectation of Privacy
Just as function creep implies that biometrics will gradually (and innocently) grow to be used by zealous, well-meaning bureaucrats in numerous, creative ways in multiple fora, function creep will also enable the Government to use the new technology of biometrics to reduce further over time the citizenry's reasonable expectations of their privacy.

Analogies can be drawn from previous cases where the Government has used cutting-edge technology to intrude in an area where the private actor had manifested a subjective expectation of privacy. For example, the Environmental Protection Agency (EPA), in an effort to investigate industrial pollution, used "the finest precision aerial camera available" mounted in an airplane flying in lawful airspace to take photographs of Dow Chemical Company's 2,000 acre Midland, Michigan facilities [40,65]. Fearful that industrial competitors might try to steal its trade secrets, Dow took elaborate precautions at its facility. Despite the elaborate precautions the company took to ensure its privacy, the Supreme Court, in a 5-4 vote, found that Dow had no reasonable, legitimate and objective expectation of privacy in the area photographed. The dissent noted that, by basing its decision on the method of surveillance used by the Government, as opposed to the company's reasonable expectation of privacy, the Court ensured that "privacy rights would be seriously at risk as technological advances become generally disseminated and available to society" [41].

Biometrics is the kind of technological advance the *Dow* dissenters warned about. Citizens no longer would have a reasonable expectation of privacy any time they use a biometric identifier because the Government's use of biometrics and computer matching would be merely utilizing commercially available technologies.

The Case for Biometrics
While biometrics is an important technological achievement, its use should be kept in a law and policy perspective: Big Brother concerns implicate far more than biometrics. The broader underlying issue is not controlling biometrics but rather the challenge of how law and policy should control contemporary information systems. Computers and the matching they perform permit "various fragments of information about an individual to be combined and compiled to form a much more complete profile. These profiles can be collected, maintained and disclosed to organizations with which the individual has no direct contact or to which the individual would prefer to prevent disclosure[18]." Biometrics should be viewed as an appendage to this enormous challenge.

Critics also overlook the many legitimate reasons why the government needs to use biometric applications. Biometric applications related to national security and prison management are easy to grasp; all of us want solid guarantees that only the correct military personnel can access nuclear materials and that serial killers do not slip out of prison by masquerading as someone else. These same concerns related to the use of false identity really apply across the board; for example, the government has a legitimate purpose in preventing fraud in the programs it administers.

Fraud is a significant issue in public sector programs. A persistent problem of state welfare entitlement programs is fraud perpetrated by double-dippers -- individuals who illegally register more than one time for benefits using an alias or otherwise false information about themselves. Many experts believe that fraud in entitlement programs, like welfare, can be as high as ten percent, which translates in dollar terms to over $40 billion a year in potential savings.

Biometrics is being used to help stop this fraud. Bob Rasor, a senior U.S. Secret Service official, commented that, "Biometrics would put a sudden and complete stop to as much as 80% of all fraud activity." In Connecticut, which has embarked on a robust biometric identification program for welfare recipients known as the Digital Imaging System (DIS), the state's Department of Social Services (DSS) "conservatively estimates that in the first year of operation [1996], savings in the range of $5,512,994 to $9,406,396 have been achieved" [42].

In these tight budgetary times when welfare programs are being curtailed and resources are overextended, anyone who is illegally receiving an entitlement payment is, at the bottom line, depriving an honest, needy person of her entitlement because there is simply less money to go around.

To the extent critics have concerns about function creep, two points need to be made: First, as explained above, the critical and key function creep issue is controlling information systems, not controlling a nine digit number or an x-byte numerical template used as a biometric identifier. Secondly, issues specifically related to biometrics can be best addressed within our present legal and policy framework. We do not need a new "Law of Biometrics" paradigm; the old bottles of the law will hold the new wine of biometrics quite well. In this regard, legislative proposals, particularly at the federal level, should be considered and studied, particularly if the threat of function creep or the emergence of an undisciplined secondary market is real. With respect to private sector use of biometrics, viable options exist for our nation's policymakers. For a more detailed analysis of this biometric blueprint proposal, refer to [41,43].

Cultural, Religious, Philosophical Objections

Cultural: Stigma & Dignity

Simon Davies of Privacy International notes that it is no accident that biometric systems are being tried out most aggressively with welfare recipients; he contends that they are in no position to resist the State-mandated intrusion [44]. Interestingly, in the 1995 GAO Report on the use of biometrics to deter fraud in the nationwide Electronics Benefit Transfer (EBT) program, the U.S. Department of the Treasury expressed concern over how finger imaging "would impact on the dignity of the recipients" and called for more "testing and study [67]."

While stigma and dignity arguments tied to the less fortunate elements of society have a strong emotional appeal, the available empirical data suggest that the majority of entitlement recipients actually support the use of biometrics. For example, a survey of 2,378 entitlements recipients in San Antonio, Texas, who participated in a biometric pilot program found that "90% think finger imaging is a good idea and 88% think finger imaging will help make people more honest when applying for benefits" [45]. Survey data in Connecticut and other states suggests similar results [46].

Religious Objections

Several religious groups criticize biometrics on the ground that individuals are forced to sacrifice a part of themselves to a Godless monolith in the form of the State. For example, observing that "the Bible says the time is going to come when you cannot buy or sell except when a mark is placed on your head or forehead," fundamentalist Christian Pat Robertson expresses doubts about biometrics and notes how the technology is proceeding according to Scripture [47]. And at least one religious group has complained that the hand geometry devices used by California were making "the mark of the beast" on enrollees' hands.

Recently, in one of the first legal challenges to government use of biometrics, New York courts upheld a decision of the New York State Department of Social Services to discontinue public assistance payments where a recipient refused to provide her biometric on religous grounds [68]. Similar objections have also been made in the context of the Government's mandated provision of social security numbers. In *Bowen v. Ray*, a leading Supreme Court case dealing with this issue, a Native American objected to the provision of a SSN for his minor daughter's application for welfare assistance as a violation of the family's Native American religious beliefs. The Court refused to sustain this challenge [48].

As these cases demonstrate, the courts are experienced in dealing with similar objections involving the State's mandatory provision of identifiers. The judiciary has an adequate framework to deal with biometrics-related religious concerns if they should arise in this context.

Philosophical: Biometric-Based Branding

Biometrics merits criticism on the grounds that a biometric identifier is nothing more than biometric-based branding or high-tech tattooing. There is an understandably odious stigma associated with the forced branding and tattooing of human beings, particularly since branding was used as a recognition system to denote property rights in human slaves in the eighteenth and nineteenth centuries and tattooing was used by the Nazis to identify concentration camp victims in this century. More than just the physical pain of the brand or tattoo accounts for society's revulsion. Analogizing from these experiences, biometric identifiers are merely a physically painless equivalent of a brand or tattoo that the State will impose on its citizens. While biometrics may lack the performance of a microchip monitor which could be implanted in humans, the biometric identifier will similarly serve the interests of the State [49]. Biometrics are another example of the State taking technology to reduce individuality.

Comparisons of biometrics to brands and tattoos again appeal to the emotions. Essentially these arguments are the ultimate form of the Big Brother concerns outlined above. Slave owners and Nazis forced branding and tattooing on victims who had absolutely no choice. In the private sector realm, citizens are making voluntary choices to use or not to use biometrics. When biometrics is used in the public sector, the use will be for legitimate purposes and will be overseen by democratic institutions.

Actual Physical Harm; Physical Invasiveness

To the author's knowledge, there are no actual documented cases of biometrics causing physical harm to a user. Anecdotally, some users of biometrics have complained that hand geometry systems dry their hands while military aviators participating in an experimental program voiced concern that retinal scanning would damage their 20/20 vision with extended use over time.

Any liability resulting from any proven actual physical harm caused by biometric systems would be addressed by the individual states' tort liability regimes. Eventually, the judiciary will have the opportunity to decide the admissibility of biometric identification as scientific evidence using prevailing legal standards [50].

Biometrics as Privacy's Friend: Support for Biometrics

Biometrics Protects Privacy by Safeguarding Identity and Integrity
While critics of biometrics contend that this new technology is privacy's foe, the opposite is, in fact, true. Biometrics is a friend of privacy whether used in the private or public sectors. Biometrics proves itself as privacy's friend when it is deployed as a security safeguard to prevent fraud.

To consider a specific example drawn from the financial services industry but applicable to almost any fraud prevention scenario, criminals eagerly exploit weaknesses with the present access systems which tend to be based on passwords and PINs by clandestinely obtaining these codes. They then surreptitiously access a legitimate customer's account or ATM. The honest citizen effectively loses control over her personal account information. Her financial integrity is compromised and her finances are gone because a criminal has gained unauthorized access to the information. In effect, she has suffered an invasion of her privacy related to her financial integrity. With biometric-based systems, identity theft, while never completely defeated, becomes more difficult for the criminal element to perpetuate. Biometrics means less consumer fraud which means greater protection of consumers' financial integrity.

Biometrics Used to Limit Access to Information
Biometrics becomes a staunch friend of privacy when the technology is used for access control purposes, thereby restricting unauthorized personnel from gaining access to sensitive personal information. For example, biometrics can be effectively used to limit access to a patient's medical information stored on a computer database. Instead of relying on easily compromised passwords and PINs, a biometric identifier is required at the computer workstation to determine database access. The same biometric systems can be used for almost any information database (including databases containing biometric identifiers) to restrict or compartment information based on the "Need to Know" principle.

Biometrics also protects information privacy to the extent that it can be used, through the use of a biometric log-on explained above, to keep a precise record of who accesses what personal information within a computer network. For example, individual tax records would be much better protected if an Internal Revenue Service official had to use her biometric identifier to access them, knowing that an audit trail was kept detailing who accessed which records. Far less snooping by curious bureaucrats would result.

Biometrics as Privacy Enhancing Technology
Beyond protecting privacy, biometrics can be seen as enhancing privacy. There are several newly-developed biometric technologies which use the individual's physical characteristic to construct a digital code for the individual without storing the actual physical characteristics in a database [51,22,24].

The applications of this type of anonymous verification system are extensive. Most notably, such a biometric-based system would seem to provide a ready commercial encryption capability. Moreover, rather than technological advances eroding privacy expectations as we saw, for example, with the EPA's use of a special aerial camera in *Dow*, biometrics, as used to create an anonymous encryption system, would provide for privacy enhancement.

Many of the criticisms of biometrics discussed above are either off the mark in that they should really be aimed at contemporary information systems which are the result of economic, political and technological change or the criticisms fail to acknowledge why knowing an individual's identity is necessary. As the next section explains, the use of biometrics might provide for even further individual privacy protections through a phenomenon known as biometric balkanization.

5. Biometric Centralization vs. Biometric Balkanization: Which Protects Privacy Better?

It is important to address whether a specific biometric technology will come to dominate biometric scanning systems. In other words, will the biometric future feature biometric centralization whereby one biometric would dominate multiple applications, or will we see biometric balkanization where multiple biometrics are used for multiple applications? At present, finger imaging has an early lead in terms of industry presence and received an important seal of governmental approval when it was endorsed by the GAO. The popularity of finger imaging is explained primarily by its consistency and uniqueness, the fingerprint's long acceptance by the public, and extensive competition in the finger imaging market leading to rapidly decreasing user costs, among other factors.

For example, with regard to public acceptance of finger imaging, a survey of 1,000 adults revealed that 75 percent of those polled would be comfortable having a finger image of themselves made available to the government or the private sector for identification purposes. This high acceptance is arguably underscored by over half of those surveyed saying they had been fingerprinted at some point in their lives. Only twenty percent thought that fingerprinting stigmatizes a person as a criminal [52].

Despite this early lead, however, it is not clear that finger imaging will emerge as the biometric of choice. It is tempting to predict that finger imaging will dominate or that another biometrics will come to monopolize the market because of its perceived advantages. However, this view overlooks one of the great strengths of the current biometric market: It offers many robust technologies which allow maximum choice for users. A more likely outcome is that "biometric balkanization" will result: Multiple biometrics will be deployed not only by various public and private sector actors but multiple biometrics will be deployed by the same actor depending on the specific mission.

Arguably, biometric balkanization, like its Eastern European namesake, can take on a sinister spin. Individuals will be forced to give up various identifying "pieces" of themselves to countless governmental and corporate bureaucracies. In an Orwellian twist, the retina, the iris, the fingerprints, the voice, the signature, the hand, the vein, the tongue and presumably even the body odor will all be extracted by the State and stored in databases.

Yet, biometric balkanization offers at least two key advantages for the protection of privacy. First, biometric balkanization offers maximum flexibility to the private or public actor that will use the technology. The actor can tailor a specific biometric program to meets its own unique mission within its resource constraints. Depending on the situation and the degree of accuracy in identification required, the optimal biometric for that use can be selected. For example, the best biometric used to verify access to a government entitlements program might differ from the best biometric used by a university to ferret out undergraduate examination fraud, which in turn might differ from the best biometric needed in a prison environment where hostile users will go to extreme lengths to foil identification efforts. Similarly, voice verification might be ideal for determining account access over the telephone while signature dynamics might be better suited for the tax authorities monitoring returns.

Secondly, biometric balkanization might actually mean a synergy of the actors' interest and the individual's concerns. Consider, for example, the public sector use of biometrics: Government agencies basically want dependable, workable biometrics to achieve their primary purpose -- verifying or identifying an individual. The individual essentially wants the same thing, plus protection of private information. If different technologies are used for different situations, citizens will not face the necessity of reporting to the government's "biometric central" for enrollment. By allowing the agencies maximum choice of biometric technologies, the individual gains greater protection for private information.

Biometric balkanization could also lead to the safeguard of biometric compartmentation which would be achieved through the use of different biometric identifiers. For example, an iris pattern used for ATM access would be of little use to the Connecticut Department of Social Services which uses finger imaging just as a finger geometry pattern captured at Disney World would be of little value to tax authorities investigating phony signatures on fraudulent tax returns from the Sunshine State.

From the privacy enhancement perspective, biometric balkanization is the equivalent of being issued multiple identification numbers or PINs or passwords with the important difference that biometric-based systems provide better security and greater convenience.

On balance, however, the greater threat to privacy will likely not arise from the use of advanced technology to monitor but rather from sloppiness in database management. The potential for a breach in database security increases greatly as shortcuts are taken, budgets are slashed, trained personnel are few and leaders do not draft and implement a biometric blueprint, or plan to safeguard biometric identification information for which they are responsible. Accordingly, limited government regulation should be viewed as biometric technology-promoting and not biometric technology-opposing.

6. Conclusions

Biometrics is a new technology which is being deployed in a variety of creative public and private sector applications. As biometrics gains in popularity and grows in uses, the law, or at least a modern-day equivalent of Judge Learned Hand, will likely take notice. As this paper has suggested, while biometrics is a new technology, it does not require a striking new legal vision to regulate it. Rather the situation is more akin to new wine in old bottles in that existing legal doctrines can deal with the challenges biometrics present. The situation is compounded in that the American approach to privacy matters has tended to be ad hoc and piecemeal. While the question of whether America needs a comprehensive approach to privacy concerns is beyond the scope of this paper, the legal and policy challenges posed by biometrics are not so novel and extraordinary that they cannot be dealt with under existing processes.

Before succumbing to the criticisms of biometrics as privacy's foe, the countercase needs to be made: Biometrics is privacy's friend. Critics of biometrics are too quick to kill the biometric identifier when it is really the "information society" and the technical underpinning of computer matching that should be the focus of their aim. To the extent biometrics raises important legal and policy issues, the existing institutional framework can address these concerns.

Biometrics protects information integrity in both the private and public sector context. By restricting access to personal information, biometrics provides effective privacy protection. Biometric balkanization further safeguards privacy by allowing maximum choice for the organization using biometrics which also makes biometric compartmentation viable.

We are eyeball to eyeball with a new technology reality that promises greater security and efficiency for both its public and private sector users. Biometrics can be used in worthwhile ways and, at the same time, safeguard legitimate privacy concerns. Now is not the time to blink.

Acknowledgments

The author gratefully acknowledges the invaluable assistance he has received from Arthur S. DiDio, M.D., J.D., Ivan Fong, Esq., Professor Steve Goldberg, Jonathan Massey, Esq., Professor Julie R. O'Sullivan, and Shirley Cassin Woodward, Esq. who contributed comments to earlier versions of this chapter.

References

[1] President William J. Clinton, Commencement Address at Morgan State University, May 18, 1997.
[2] R. Chandrasekaran, "Brave New Whorl: ID Systems Using the Human Body Are Here, But Privacy Issues Persist," *Washington Post*, March 30, 1997.
[3] A. Davis, "The Body as Password," *Wired*, July 1997.
[4] J. D. Woodward, "Biometric Scanning, Law & Policy: Identifying the Concerns; Drafting the Biometric Blueprint," *University of Pittsburgh Law Review*, Fall 1997.
[5] S. Goldberg, *Culture Clash: Law & Science in America*, New York University Press, NY, 1994.
[6] *The T. J. Hooper*, 60 F.2d 737 (2d Cir.) *cert. denied*, 287 U.S. 662 (1932)(Hand, J.).
[7] B. Miller, "Everything You Need to Know About Automated Biometric Identification," *Security Technology & Design*, April 1997.
[8] P. T. Higgins, biometric consultant, in Washington, D.C. (Jan. 13, 1998).
[9] C. Tilton, "Put a Finger on Your Security," *Security Advisor*, Premiere, 1998.
[10] K. McManus, "At Banks of Future, An Eye for an ID," *Washington Post*, May 6, 1996.

[11] G. Roethenbaugh, "Biometrics: A Global Perspective," *in BiometriCon '97 Conference Proceedings,* March 12-14, Arlington, VA, 1997.

[12] C. Fried, *An Anatomy of Values,* Harvard University Press, Cambridge, MA, 1970.

[13] R. B. Parker, "A Definition of Privacy," *Rutgers University Law Review,* Vol. 27, pp. 275, 1974.

[14] T. Gerety, "Redefining Privacy," *Harvard Civil Rights-Civil Liberties Law Review,* Vol. 27, pp. 233, 1977.

[15] *United States v. Westinghouse Elec. Corp.,* 638 F. 2d 570 (3rd Cir. 1980) (holding that medical records of a private sector employee, while within the ambit of constitutional privacy protection, could nonetheless be disclosed to a government agency upon a proper showing of governmental interest).

[16] *Griswold v. Connecticut,* 381 U.S.479, 1965.

[17] *Smith v. Maryland,* 442 U.S. 735, 1979.

[18] P. Mell, "Seeking Shade in a Land of Perpetual Sunlight: Privacy as Property in the Electronic Wilderness," *Berkeley Technology Law Journal,* (footnote omitted), 1997.

[19] M. Rotenberg and E. Cividanes, *The Law of Information Privacy: Cases & Commentary,* 1997.

[20] *Whalen v. Roe,* 429 U.S. 589, 1977.

[21] *Smith v. Maryland,* 442 U.S. 735, 1979.

[22] A. Cavoukian, Assistant Privacy Commissioner of Ontario, "Go Beyond Security -- Build in Privacy: One Does Not Equal the Other," (May 1996) *available at* http://www.microstar-usa.com/tech_support/faq/privacy.html.

[23] *United States v. Miller,* 425 U.S. 435, 1976.

[24] "Privacy and Data Security Targets of Mytec's Commercialization Strategy," *PR Newswire,* June 20, 1997.

[25] E. Alderman and C. Kennedy, *The Right to Privacy,* (1995).

[26] S. G. Davies, "Touching Big Brother: How Biometric Technology Will Fuse Flesh and Machine," *Information Technology & People,* 1994.

[27] M. Skoler, "Finger and Palm Prints: A Window on Your Health," *Glamour,* Apr. 1984.

[28] H. Chen, *Medical Genetics Handbook,* W. H. Green, St. Louis, MO, pp. 221-226, 1988.

[29] "Gastroenterology: Fingerprinting GI Disease," *Johns Hopkins Physician Update,* pp. 5, April 1996.

[30] S. LeVay, *Queer Science: The Use and Abuse of Research into Homosexuality,* MIT Press, Cambridge, MA, 1996.

[31] B. Bates, *A Guide to Physical Examination and History Taking,* 5th edition, 1991.

[32] R. Clarke, "Human Identification in Information Systems: Management Challenges and Public Policy Issues," *Information Technology & People,* December 1994.

[33] J. Hall, "For New ATM, the Eyes Have It," *Trenton Times,* September 19, 1995.

[34] "FutureBanking," *American Banker,* October 21, 1996.

[35] U.S. Department of Health, Education and Welfare, *Records, Computers and the Rights of Citizens: Report of the Secretary's Advisory Committee on Automated Personal Data Systems,* MIT Press, Cambridge, 1973.

[36] J. Toland, *Infamy: Pearl Harbor and Its Aftermath,* Anchor, NY, 1992.

[37] *Korematsu v. United States,* 323 U.S. 214 (1944).

[38] *Olmstead v. United States,* 277 U.S. 439, 479 (1927) (Brandeis, J. dissenting).

[39] *Greidinger v. Davis,* 988 F.2d 1344 (4th Cir. 1993).

[40] *Dow Chemical Co. v. United States,* 476 U.S. 227 (1985).

[41] J. D. Woodward, Testimony on "Biometrics and the Future of Money" before the Subcommittee on Domestic and International Monetary Policy, Committee on Banking and Financial Services, U.S. House of Representatives, May 20, 1998.

[42] D. Mintie, "Report from Connecticut," *Biometrics in Human Services User Group Newsletter,* Vol 3, No. 1, March 1997. http://www.dss.state.ct.us/faq/bhsug031.htm.

[43] J. D. Woodward, "Private Sector Use of Biometrics: The Need to Safeguard Privacy Concerns -- The Need for a Biometric Blueprint," in *CTST '98 Proceedings*, Washington DC, 1998.

[44] "Foolproof Identification Methods Create Privacy Worries," *(National Public Radio Broadcast*, Segment number 2360, October 8, 1996.

[45] C. Edwards, "Reports from the States: The Texas Lone Star Imaging System," *Biometrics in Human Services User Group*, May 1997. http://www.dss.state.ct.us/faq/bhsug04.htm.

[46] D. Mintie, "The Connecticut DSS Biometric Project and EBT Card: Implementation Issues," in *CTST '96 Government Conference Proceedings*, Arlington, VA, 1996.

[47] "Biometrics: Chipping Away Your Rights?," *The 700 Club Fact Sheet*, VA, October 9 1995.

[48] *Bowen v. Ray*, 476 U.S. 693 (1986) (holding that the Free Exercise Clause of the First Amendment does not compel the Government to accommodate a religiously-based objection to the provision of a Social Security Number for Little Bird in the Snow, a minor welfare recipient).

[49] R. E. Smith, "The True Terror is in the Card," *New York Times Magazine*, September 8, 1996.

[50] *Daubert v. Merrell Dow Pharmaceuticals*, 509 U.S. 579 (1993).

[51] "Test Center Comparison," *Infoworld*, June 16, 1997.

[52] "People Patterns: Fingerprints? No Problem," *Wall Street Journal*, January 31, 1997.

[53] R. Nanavati, Presentation on "Top Ten Trends in Biometrics" at CardTech/SecurTech Conference, in Washington, D.C, April 27, 1998.

[54] F. James, "Body Scans Could Make ID Process Truly Personal," *Chicago Tribune*, June 4, 1997.

[55] D. R. Richards, "Rules of Thumb for Biometric Systems," *Security Management*, October 1, 1995.

[56] G. Roethenbaugh, Biometrics Explained, 1998. Available at: http://www.ncsa.com/services/consortia/cbdc/explained.html.

[57] *United States v. Miller*, 425 U.S. 435 (1976).

[58] S. D. Warren & L. D. Brandeis, "The Right to Privacy," *Harvard Law Review*, 1890.

[59] A. F. Westin, *Privacy and Freedom*, 1967.

[60] R. E. Cytowic, "All in the Genes," Washington Post, September 1, 1996, (book review).

[61] N. Hawkes, "Fingerprint Clue to Health," *Times of London*, February 26, 1996.

[62] "Briefs," *Biometric Technology Today*, April 1998.

[63] Dr. F.P. Nasrallah, Assistant Professor of Ophthalmology at George Washington University, and Dr. Arthur S. DiDio, M.D. in Washington, D.C., personal communication, Apr. 4, 1996.

[64] M. Barthel, "Banks Eyeball Sci-Fi Style Identification for ATMs," *American Banker*, Sept. 22, 1995.

[65] *United States v. Knotts*, 460 U.S. 276 (1983) (holding that governmental surveillance by beeper placed in a container with the consent of the owner of the container did not violate the reasonable expectations of privacy of the defendant who placed the container in his car and drove over public highways).

[66] L. Tribe, "The constitution in cyberspace: Law and Liberty beyond the electronic frontier," Keynote address at The first conference on Computers, Freedom, & Privacy, 1991. Available at http://www.eff.org/pub/Legal/cyber_constitution.paper (viewed Nov 28, 1997).

[67] United States General Accounting Office, *Electronic benefits Transfer: Use of biometrics to deter fraud in the nationwide EBT program*, GAO/OSI-95-20, pp. 6-7, Sept. 1995.

[68] Buchanan v. Wing, __N.Y.S.2d__ (N. Y. App. Div. 1997).

Index